GLOBAL 2000 REVISITED

Mankind's Impact on Spaceship Earth

GLOBAL 2000 REVISITED

Mankind's Impact on Spaceship Earth

Edited by
Hugh W. Ellsaesser

An ICUS Book

PARAGON HOUSE
New York

Published in the United States by the
International Conference on the Unity of the Sciences
4 West 43 Street
New York, New York 10036

Distributed by Paragon House Publishers
90 Fifth Avenue
New York, New York 10011

An ICUS Book

The International Conference on the Unity of the Sciences (ICUS) convenes international, distinguished scientists and scholars from every field of study to pursue academic discussion of theoretical and practical concerns. ICUS seeks an integrated world-view based on absolute values generated through multi-disciplinary, academic dialogue.

Library of Congress Cataloging-in-Publication Data

Global 2000 revisited: mankind's impact on spaceship earth / edited by Hugh W. Ellsaesser.
p. 438
"An ICUS book."
Includes bibliographical references and index.
ISBN 0-89226-101-3 (pbk.)
ISBN 0-89226-100-5.
1. Environmental policy. 2. Environmental health. 3. Pollution.
4. Pollutants.
I. Ellsaesser, Hugh W.
RC87.5.R45 1992
617.1'22–dc2090-23411

CIP

TABLE OF CONTENTS

PREFACE

Hugh W. Ellsaesser

The International Conference on the Unity of the Sciences (ICUS) is sponsored by The International Cultural Foundation, which was founded by the Reverend Sun Myung Moon in 1968 and is now headquartered in New York City, with branches throughout the world. The ICUS Conferences were launched in 1972 with the meeting in New York City of twenty participants representing eight nations. Largely due to the vision and faith of its Founder, the Reverend Sun Myung Moon, ICUS was established to produce an integrated world view founded on absolute values generated through interdisciplinary academic dialogue. Through the two themes of ICUS, Unity of Knowledge and Science and Values, the Founder is challenging the academic community to contribute constructively to the formation of a new world culture—one nearer to his vision of an ideal world—the goal he is pursuing in what he has termed the *new cultural revolution.*

The ICUS Conferences grew each year from the twenty participants from eight nations of ICUS I in 1972 to the 808 participants from 100 nations of ICUS X in Seoul, Korea, in 1981. Since then, the number of participants has been reduced to around 250. The proceedings of each of the first eleven ICUS Conferences were published in a single volume. Beginning with ICUS XII in 1983, the format was converted to a Commemorative Volume containing the highlights of the conference such as the program, participants, plenary addresses and a selected paper and summary from each of the six or so committees of the conference, with each committee chairperson left to decide whether the papers of his or her committee warranted publication as a proceedings-type volume just for their particular committee.

The present volume is essentially a Proceedings Volume for Committee VII of ICUS XV, held at the J.W. Marriott Hotel in Washington, D.C., 27–30 November 1986. The subject or charge of Committee VII was the title of this book: *Global 2000 Revisited; or Re-Assessing Man's Impact on Spaceship Earth.* Included are the papers of the twelve Paper Writers and two of the Discussants of Committee VII. Again, in departure from previous practice, discussants at ICUS XV were not asked to formally write up their remarks. However, both because of the stature of the discussants and because of the very controversial nature of the papers on the hormetic effects of radioactivity and the disposal of radioactive wastes, special efforts were made to incorporate the discussants' comments on these papers. These have been supplemented by an Editorial Introduction and an Epilogue. The former provides much of the thinking that went into the selection of the charge, the paper writers, and the subjects which they were invited to address, and the latter attempts to include highlights of the discussions which followed each paper and to synthesize an answer to the question raised in the committee charge.

I wish to express my heartfelt gratitude to Kenneth Mellanby, ICUS XV Chairman, for the opportunity of a lifetime which he provided by inviting me to organize and chair Committee VII of ICUS XV. I am also grateful for the assistance, advice, and support provided throughout this endeavor, not only by Chairman Kenneth Mellanby, but also by Vice-Chairman Alvin M. Weinberg, Executive Director James A. Baughman, and Deputy Director Gregory Breland and their staffs. And, of course, very special thanks go to Committee VII Honorary Chairman S. Fred Singer, to all of the paper writers, discussants, and other participants in Committee VII to whose talents and cooperation are due any success that can be attributed to this endeavor.

Livermore, California
24 September 1987

INTRODUCTION

Hugh W. Ellsaesser

When first asked by Dr. Kenneth Mellanby to consider undertaking the task of organizing Committee VII of ICUS XV on "something to do with the environment," I saw it as the opportunity of a lifetime to continue and enlarge upon my studies of environmental questions extending over the past 20 years. Since most such questions have been introduced as man-induced degradations of the natural environment threatening man's continued existence and even the suitability of the planet for life itself (man's fouling of his nest), this was adopted as the original theme, unifying Committee VII without placing undue restrictions on the subjects which it might eventually consider.

In view of the wide discussion of environmental issues, one might legitimately wonder, why another? My interest in such questions was sparked early by the large disparity between the claims of those portraying themselves as environmentalists and what I had come to accept from my own training and observations (I have the somewhat unique experience of having resided in southwestern Kansas during the dust bowl era of the 1930s, in Los Angeles in 1946–47 during the early discovery of "smog," and in London, England, during the so-called Air Pollution Episode of December 1952). These experiences, plus innate characteristics (of which I am frequently reminded by colleagues), apparently combined to create a uniquely skeptical mind with regard to air pollution, in particular. In any event, I have been spurred to delve rather deeply into the available scientific literature in an effort to evaluate for myself the seriousness of the hazards regularly trumpeted in both the popular and the scientific press.

The consequence has been a growing conviction that all environmental issues are characterized by a sharp dichotomy between the "consensus" view and what can be logically and scientifically defended, based on the available information. To render this plausible, one need merely objectively appraise the consistency with which environmental issues are presented as moral issues, portraying the sins of mankind and three of his major creations: civilization, technology and the free market-free enterprise capitalistic economic system. The ecotacticians of the 1970s in a thousand variations, "peddled the proposition that 'only man is vile'" (Richard Newhaus, 1971).

The organization of Committee VII of ICUS XV thus was taken as an opportunity to further explore and expose this dichotomy of views by examining new problem areas and by providing a platform for individuals known to disagree with the "consensus" view in their area of expertise and to have spent considerable time and effort in developing and substantiating their own views. The problem was first formulated as a comparative evaluation of all the threats that have been cited to our continued existence on planet Earth in an effort to decide which is the more threatening or more imminent and thus should be prioritatively addressed. (In this sense, this volume is a sequel to Singer, 1989, 1990 & 1991; the proceedings volumes for the environmental committees of ICUS's XII, XIII, and XIV.) The simultaneous comparative consideration of these diverse threats was considered a key aspect of the Committee VII deliberations, designed to evoke objectivity and scientifically defensible appraisals, as well as to reassure each participant that he was not alone in feeling alienated from the mainstream of his discipline.

However, this seemed to create two serious problems for the project. How could all of the threats to our survival which have been raised, be covered with only twelve papers? And how does one gain credibility for viewpoints which depart from the "consensus" view? Here, "consensus" is defined as the view reflected in the popular and scientific press which by now generally has been "legitimatized" in at least one report from a scientific survey committee and, in turn, "refereed" by the National Academy of Sciences. Bluntly, how does one claim to be more scientific than the National Academy of Sciences without turning off his audience? If you, the reader, are still with me, I will endeavor to address this question seriously. As a quick thrust, let me note that "The Emperor's New Clothes" was not a fairy tale—it was an allegory intended to unveil human nature, just as actual historical events earlier had revealed it to its author.

A modern and well-known example of the dichotomy I am describing is provided by the *Global 2000 Report to the President* (Council on Environmental Quality and the Department of State, 1980) and the response which it *provoked* entitled *The Resourceful Earth* (Simon and Kahn, 1984). I use the word "provoked" since in his preface Julian Simon expresses the gratitude of himself and Herman Kahn (who sadly died during the project) "to those of our authors who, out of passion for truth and outrage at bad science, have given their time and effort to prepare chapters." The sharp contrast between these two reports cannot be missed. Simon and Kahn's introduction states categorically:

> —the original Global 2000 is totally wrong in its specific assertions and its general conclusion. It is replete with major factual errors, not just minor blemishes. Its language is vague at key points and features many loaded terms. Many of its arguments are illogical and misleading. It paints an overall picture of global trends that is fundamentally wrong, partly because it relies on non-facts and partly because it misinterprets the facts it does present.

This harsh condemnation is somewhat mitigated by the following parenthetical statement which is all too typical of committee reports on environmental issues:

> (In partial defense of the writers who prepared the *Global 2000* work, the summary Volume I—which was the main basis for the news stories—egregiously mis-stated, for reasons which we can only surmise, many analyses and conclusions in the working-paper Volume II, thereby turning optimistic projections into pessimistic ones.)

For an outstanding earlier example of this deft ploy of rewriting a report by adding a summary, introduction or letter of transmittal, I invite you to compare the report of the Donora Air

Pollution Episode (Schrenk *et al.*, 1949) with the news release incorporated in it as an introduction and summary.

Aside from its introduction, *The Resourceful Earth* is not primarily an evaluation or criticism of *Global 2000.* For the latter, see "Entering the Twenty-First Century—The Global 2000 Report to the President" (Clawson, 1981), "Half Truths About the Future" (Dubos, 1981), "Globaloney 2000" (Kahn and Schneider, 1981), or "Global Confusion, 1980: A Hard Look at the Global 2000 Report" (Simon, 1981), in which the titles tend to tell the story.

I am indebted to Dr. Alvin Weinberg for suggesting that the Committee VII program be linked with the already known contrast provided by these reports. By adopting the title *Global 2000 Revisited* and selecting a keynote speaker to set the stage for our deliberations by reviewing this contrast, we both improved our claim to legitimacy for diverse views and were relieved of any obligation to discuss *all* threats to man's survival which have been raised. *The Resourceful Earth* is by no means the only publication provoked "out of passion for truth and outrage at bad science." My first disillusionment with the "consensus" view came in the field of air pollution (Ellsaesser, 1971, 1973, 1975, 1979, 1981). This led me to collect statements such as the following from specialists of the highest credentials in their field at the time.

> It is extremely tempting to try to motivate people to act in support of air conservation by appealing to them on grounds of health. This is an area in which the public is sensitive. . . . But caution must be used. First, scare tactics are unwarranted, and they seldom produce the desired result. Second, it is extremely difficult to demonstrate a clear health effect of air pollution. In the most acute cases [the Air Pollution Episodes], such an effect is obvious. However, unequivocal proof of chronic effects is still wanting and may not become available for many years. (Air Conservation Commission of the AAAS Committee on Science in the Promotion of Human Welfare—see Dixon and Lodge, 1965.)

> It was only recently that the Surgeon General of the U.S., in his testimony before Congress, admitted that substantive evidence of the direct effect of common day-to-day air pollution on health is lacking (Sterling *et al.*, 1967).

> The obvious discrepancy between the alleged disastrous effect of air pollution on health and the inconspicuous concentration of SO_2 measured in the air have taxed the imagination of toxicologists for the past twenty years (Battigelli, 1968).
>
> The fact that millions of Los Angeleans have been exposed for years to above-threshold limits of oxidant, without any serious health effects, raises doubt that we are correctly assessing the toxicity of this pollutant in humans as it exists in smog (Tabershaw *et al.*, 1968).
>
> Are we asking the right questions? I happen to think that we are not. I see a lot of interest in how to meet this standard or how to live with that standard, and I think the proper question is: Should we? In other words, are these standards of sufficient scientific merit (Goldwater, 1976)?
>
> Also, in my opinion, the hydrocarbon standard is patently too stringent and is unnecessarily costing the American public hundreds of millions of dollars because of its influence on the automobile exhaust emission standard for hydrocarbons (Stern, 1976).
>
> At the present time, it has not been possible to definitely relate an increase in mortality to any specific air pollutant (Tabershaw, 1976).

The view of these experts obviously did not prevail. The ink was hardly dry on the National Air Quality Standards adopted on the last day of the legislatively mandated deadline by the five-month-old Environmental Protection Agency when there appeared in the prestigious journal *Science* an article bearing the title "Sanity in Research and Evaluation of Environmental Health" (Stokinger, 1971). Among many similar exhortations in it were the two following:

> When the pollution-oriented health administrators and the public alike begin to focus clearly on the enormity of the bill that would be required to reduce pollution to meet unnecessarily severe standards (*versus* commandment 3) precipitously prepared from undigested, dubiously-related facts (*versus* commandments 1 and 4), on which the public has been ill-advised or misled, (*versus* commandment 2), then will come the day of reckoning and rude awakening to the folly of past anti-pollution actions.

> The ruinous concept of "zero tolerance" for pollutants must go! Man has never, before he was a man or ever after, survived in an unpolluted void.

Considering the timing, the title, the content, and the prestige of H.E. Stokinger at the time, I do not see how this article can be considered to be other that a public appeal for a *return to sanity*. Yet, so far as I am aware, it has never been so characterized by anyone other than myself.

Still, my disenchantment with the role of the National Academy of Sciences in the early 1970s has been echoed by others. At the 3rd Interuniversity Consortium for Environmental Studies Invitational Symposium, Chapel Hill, N.C., 24–25 February 1975, Dr. I.T. Tucker (1976) had this to say:

> Dr. Stern, you refer to Volume II of the National Research Council document and to Volume III which Mr. Carpenter referred to also. I think these are typical of the sort of overall views that are presented by groups with the advice and counsel of advisory committees. I have looked at several of these criteria documents. When the statements, the summaries, the conclusions and the data are analyzed, they are frequently inconsistent. . . . I found that recently in a lead document *Health Effects of Lead.* The data were very fuzzy and yet there was a conclusion reached and this was widely publicized. When one looked back of it, there really wasn't any substance to support that particular point of view.

In addition, after publicly casting aspersions such as the above myself, I was informed privately by two scientists that they had resigned from National Academy of Science Committees because they did not agree with and saw no hope of influencing what was being done in their committees. Even worse, they did not record the true reasons for their resignations. One of them went so far as to say that this last action was taken to preclude the possibility of any reprisal against his institution! (Because of his unfortunate demise, I can reveal that the latter statement was made by the late Professor Helmut E. Landsberg, immediately recognizable by many as an eminently respected and honored member of his discipline and certainly not an unreasonable dissident or crank.)

Beyond the specific field of air pollution, numerous concordant voices have sounded on this issue. I will cite a few of the better-documented ones that have come to my attention. John Maddox, a long-time editor of *Nature*, was an early entrant in this field with *The Doomsday Syndrome* (Maddox, 1972), which Ron Arnold (1982, p. 35) described as "a politely scathing denunciation of environmentalist Cassandras who seemed to enjoy perverting the truth for fun and profit." Maddox forthrightly stated his was "not a scholarly work, but a complaint"—a complaint that since the appearance of Rachel Carson's (1962) *Silent Spring*, the western world had

> been assailed by prophecies of calamity. . . . Although these prophecies are founded in science, they are at best pseudo-science. Their most common error is to suppose that the worst will always happen. . . . Scientists have not played an expecially creditable part in the environmental movement.

With his usual prescience, he dubbed the writing of this book an "unpopular pursuit." Just consider the number of times you have heard mention of *The Doomsday Syndrome* compared to *The Limits to Growth* (Club of Rome, 1972), which he demolished in his postscript.

In a similar vein, *The Disaster Lobby* (Grayson and Shephard, 1973, p.1) in the following words labeled the decade following the appearance of *Silent Spring* "The Age of Unreason":

> In such a way was the book *Silent Spring*, by Rachel Carson, introduced to a segment of New York's window-shopping public. And in such a moment was born a period in American history that might be described as the Age of Unreason. For at no other time—not even during the aberrant decades following the Civil War and World War I—have so many Americans succumbed so thoroughly to the influences of illogic, of hysteria and of sophistic nonsense as they did in the 1960s and early 1970s. . . . In essence, the Carson opus was an impassioned, one-sided and largely unscientific denunciation of the growing use of chemical pesticides, particularly DDT. . . . It was an attack on the business establishment, . . . It was an attack on scientific and technological progress, . . . It was an attack on the United States, . . . And most of all, it was an attack on man himself. . . . To the left-wing academic brigade, . . . *Silent Spring* was reveille and the cavalry charge rolled into one. And thousands and thousands of

> professors and graduate students mounted for the assault. . . . The news media—the television and radio networks, the magazines and the big metropolitan-area newspapers—did their part with talent and zeal. Since the 1940s, these organs of information had been tilting further and further to the left as their conservative owners shifted their focus from editorial content to business survival. Caught in the crunch of fiscal matters, they had stopped supervising the material that went into the newscasts and the talk shows and the feature stories, and into the vacuum they left flew a gaggle of ultra-liberal editors who soon began to shape all journalistic output in their own slanted image. The war against DDT was a natural for them, and they waged it with single-minded ruthlessness.

In *The Liberal Crack-Up,* R. (Bob) Emmett Tyrrell, Jr. (1984) describes the following decade as even worse:

> American foreign policy became a ribald joke before the world. American domestic policy wheezed and groaned as more and more flapdoodle was piled onto it. Galoots and wretches manned our government. Our culture flaunted the worst of fourth-rate vulgarity. The Great Republic seemed to be taking on the appearances of a banana republic. Europeans long given to patronizing us suddenly fell silent and began to worry. . . . Happily, some areas of American life have remained tolerably invulnerable; but other areas—for instance, academe—have almost wholly capitulated and so gone to seed.

In his review of this work, George Melloan (1984) adds to the diagnosis:

> And surely any truly sympathetic shrink would indulge all those others of us who have been waiting so long to hear someone with Mr. Tyrrell's skills tell them that they were not insane during the 1970s. Rather, a large part of the public policy debate of that era was insane.

Cy Adler's (1973) *Ecological Fantasies: Death from Falling Watermelons* continues in the same vein with a strong dose of satire. Ben Wattenberg (1974), in *The Real America,* attempts a diagnosis and a setting straight of the record. According to him, the environmentalists, adopting Rachel Carson's now proven technique of calculated exaggeration, have used lurid and extravagant rhetoric to exploit the doubts of average Americans.

These career reformers have organized as never before. They launched the "Failure and Guilt Complex." And—most effective and disturbing of all—they captured not just the pulpits, the small-town editors, and the League of Women Voters, "but they have arrogated 'the media,' the reputed public conscience of America, a weapon that never even existed in the 1920s."

A landmark in this field of literature is the meticulously documented tome by Edith Efron (1984), *The Apocalyptics—Cancer and the Big Lie.* She states in her preface:

> This book is an intellectual detective story, I am the detective, and I shall not give away the "plot." I shall simply say here that I discovered a cultural crime which should not be possible in a free society: a complex corruption of science and a prolonged deception of the public. The crime emerged from the sciences of environmental cancer and cancer prevention, and it has all the superficial characteristics of "The Purloined Letter." It has been committed under our very eyes; its details are publicly recorded in documents which are within hand's reach, and yet it remains invisible to most of the people of this country who are its victims. . . . The cultural crime I discovered *was* perpetuated by ideology in a white smock. . . . There is one aspect of this corruption of science which can and should be identified in advance. It is a manifestation of a much wider cultural problem—an insidious assault on reason, science and the value of objectivity, which has taken place in every field of scholarship since the 1960s. In all branches of the humanities and the social sciences, there has been a rejection of the criteria and disciplines of the field and a slide into a subjectivity which serves as a vehicle, overt or covert, for ideological values. . . . Most humanists, however, have not realized that the same trends have infiltrated science. I did not know it. In my reading in the environmental sciences, and later environmental cancer, I bumped into evidence of such hostility to the objective disciplines of science, evidence of so aggressive a rejection of facts and logic, that I could scarcely credit my senses.

Even Philip Abelson (1987), former editor and now deputy editor of *Science*, now admits:

> For more than ten years, the public has been subjected to a media barrage leading to widespread, misinformed fear of chemicals.

In her epilogue, in speaking of the press (in which she started her career), Edith Efron (1984, p. 478) says:

> *They "know" primarily what is false or unknown—and they rarely know how to identify the actual threats to their lives* [italics hers]. . . . The ostensible protector of the community actually disarmed the community.

In speaking of the self-cleansing mechanisms by which superb science is never threatened by mediocre science, since one authentic discovery has the power to wipe out decades of intellectual trash, she says:

> That is entirely true—for a closed scientific community. It is not true, however, for a nation. Countries do not have automatic self-cleansing mechanisms. The intellectual trash invades the law and the press. The intellectual trash piles up. The intellectual trash *becomes* the culture (Efron, 1984, p. 475).

Among citations in support of her indictment, Edith Efron references a speech by then-President of the National Academy of Sciences, Philip Handler (1980), warning against the "anti-scientific" and "anti-rationalistic" trends which were infiltrating both the humanist and scientific cultures, in which he called upon scientists to avow their ignorance, to make no claims to knowledge they did not possess—and to confront and identify the "charlatans" in their midst. He said it was time for them to resist the "intimidation" to which they had been subjected for so many years; it was time to return to

> the ethics and norms of science.
>
> Scientists best serve public policy by living within the ethics of science, not those of politics. If the scientific community will not unfrock the charlatans, the public will not discern the differences—science and the nation will suffer.

And he said these problems were particularly acute in the environmental sciences.

This is perhaps a good place to note that *The Resourceful Earth* contains a one-page Statement of Dissent by Bernard L. Cohen, the essence of which follows:

> As a scientist, I see no barriers to a bright future for America and for mankind. . . . However, in the past ten years, science has come under irrational attack from the forces of ignorance, and is losing public support. . . . Our government's science and technology policy

is now guided by uninformed and emotion-driven public opinion rather than by sound scientific advice. Unfortunately, this public opinion is controlled by the media, a group of scientific illiterates drunk with power, heavily influenced by irrelevant political ideologies, and so misguided as to believe that they are more capable than the scientific community of making scientific decisions. . . . Unless solutions can be found to this problem, I believe that the United States will enter the twenty-first century declining in wealth, power and influence, and within the next century will become an impoverished nation. I, therefore, find it difficult to share in the optimism that characterizes this report. That does not mean that I sympathize with *Global 2000*; indeed, some of those who were most influential in its preparation have been among the leading perpetrators of the policies that are ruining us. The coming debacle I foresee is not due to the problems they describe, but to the policies they advocate (Simon and Kahn, 1984, p. 566).

Another scholarly work in this area is Elizabeth Whelan's *Toxic Terror.* She begins by saying:

I am incensed by the possibility that lies, exaggeration, and innuendo about the relationship of our health and environment might change the whole fabric of the lives of Americans. As a consumer who appreciates the current bounty and great potential this country has, as a taxpayer who is no longer willing to tolerate expenditures for problems which don't exist, and as a scientist who recognizes that the truth is the only antidote for the poisoned pens, I feel an obligation to set the record straight (Whelan, 1984, p. 17).

According to the foreword by Norman E. Borlaug, she then:

unmasks the pseudo-scientific efforts that resulted in the banning of EDB [ethylene dibromide] and reveals the frightening gap which exists between the rhetoric of those consumer activists who try to hold the country in a state of "toxic terror" and the scientific realities about such topics as EDB, DDT, dioxin, 2,4,5T, formaldehyde, and acid rain. In this book you will get the facts behind the headlines about Love Canal, Three-Mile Island, and the use of urea formaldehyde foam insulation. This book may make you angry—indeed it should outrage you—because you have been the victim of a hoax. You have been told that as a result of your country's high state of technology and economic development, your environment is making you sick, when the exact opposite would be closer to the truth (Whelan, 1984, p. xvi).

After this she poses the same three questions, asked by so many of us:

> 1. Why are the media . . . so eager to give us bad news? . . . Why do they demonstrate such gullibility—or willingness—to accept the views of that small cadre of scientists who sing the refrain of gloom and doom for a living? Why hasn't the investigative-reporting instinct of the media tuned its skills to an expose of the doomsayers, calling their bluff and allowing us a welcome gust of airwaves filled with good news?
>
> 2. Why do the same scientists, whether the focus is Love Canal or Three-Mile Island, show up in reports to confirm the bad news? . . . Why aren't American scientists outraged that science is being distorted, and their whole profession discredited and abused?
>
> 3. Who *are* these self-styled environmentalists, those bearers of all this bad news? . . . And since they so frequently dismiss those who disagree with their anti-technology views as "hired guns of industry," it seems appropriate to ask, Who funds *them*? What do they *really* want for social and economic change (Whelan, 1984, p. 278)?

In attempting to answer these questions, Elizabeth Whelan cited other workers. Regarding the role of the media, she cites among others, Ben Wattenberg (1984) who in three pages in *Readers Digest* and in more than 400 pages in his book cites examples to prove his thesis that *The Good News Is that the Bad News Is Wrong.* He also demonstrates that, insofar as the media is concerned, "good news is no news" unless it can be recast so that "good news is bad news." Regarding the role of the missing scientists, she cites Claus and Bolander (1977) and Philip Handler (1980a,b). Her most promising solution:

> First, scientists must come to appreciate how irresponsible the mass media are in disseminating information, or misinformation, about the relationship of health and the environment. . . . Physicians and scientists should not deny or fight the power of the media; they should join it. If a charlatan is in town, a qualified professional should be demanding a debate. To stand back is to permit facts to be distorted—a dangerous error of omission (Whelan, 1984, pp. 193–4).

In trying to unravel the controversy surrounding James Watt, Ron Arnold (1982, p. xix) concluded that setting the record straight is not enough; "it is important to explore the motives and methods of those who are doing the distorting."

Regarding who are the environmentalists, what are their goals, and who funds them, Whelan (1984) cites Herbert Meyer (1979) and Ron Arnold (1982). From these and other sources she concludes that the environmental and public interest leaders are *not* worried about our future, are *not* representing the interests of the vast majority of Americans, are *not* mainstream in their political orientation and are *not* selfless as they would like us to believe. In *The War Against Progress,* Meyer (1979) writes, this war

> is being fought by people who themselves enjoy a high standard of living and who believe that to protect their own comfort they must prevent others from becoming equally comfortable. . . .

Arnold (1982, p. 37) concluded that some of those behind the movement were not simply anti-intellectual, but actually *anti-intelligence,* "deploring the biological fact that reason had enabled us to become the dominant species on earth."

Elizabeth Whelan (1984, p. 298) had considerable difficulty trying to trace funding sources because these do not have to be revealed. She did find that the funding is substantial, that most environmental groups receive substantial contributions from a small group of foundations, including the Rockefeller Brothers Fund, the Charles Stewart Mott and Ruth Mott Foundations, the George Fund Foundation, and, surprisingly, some tobacco family foundations, including the Mary Reynolds Babcock and Arca Foundations. Arnold (1982, p. 43) cited contributions to environmental groups by the Atlantic-Richfield, Ford, Mellon, and Rockefeller Family Foundation interests alone of over twenty million dollars in 1979 and over eighteen million dollars in 1980.

While apparently missed by Whelan (1984), Arnold (1982, p. 45) appears to have plumbed most deeply the question:

> Why did the science of ecology become so confused with political motivations? Why can't environmentalists see through their own "socio-psychological needs"? Why do some of these people "delight in the collapse of things"? We may see the "hows" ever so plainly, but the "whys" elude us. And that's what really matters—why?

He found the answers in the works of Ronald Inglehart (1977), Herman Kahn (1976), and Abraham Maslow (1977). Inglehart (1977) found from massive surveys that the real changes of the last two decades boiled down to two things: values and political styles.

> The values of western publics have been shifting from an overwhelming emphasis on material well-being and physical security toward greater emphasis on the quality of life, . . .

And, as a result, an increasingly large proportion of the public is coming to have sufficient leisure to take interest in national and international politics and to want to participate in decision-making at this level. He found six major causes for these shifts: *technical innovation* was the driving force, spurring unprecedented productivity in the late 1940s; this, in turn, spurred *economic growth* shifting the *occupational structure* to 30 percent producers and 70 percent service workers—creating a need for *expanded education* to supply more technical skills; the new technology made expanded education both necessary and affordable and at the same time created *mass communications.* Superimposed on top of this, the post-World War II "baby boom" bulge in the population grew up in this affluent, highly-educated, communications-rich and indulgent society, with unprecedented physical security creating the *distinctive generational experiences* of the "generation gap."

Economic growth was a force for change with unexpected results. We didn't realize that higher affluence would reduce our concern with material values and that we would evolve into "post-materialists," leading to an increased demand for participation in decision-making, that the now possible differences in lifestyle would generate furious disputes over everything from protection of the environment and the role of women to the redefinition of morality, drug use and broader claims to participation in the political process. The *changed occupational structure* unexpectedly created serious divergences of outlook between the new technical, professional, and scientific majority and the old managers who were guided by profits, economic growth and the welfare of their individual firms or bureaucracies. Herman Kahn (1976) discovered that despite the service majority's higher education, its vision had narrowed to see only its own goals and problems; they acquired "educated incapacity"—an "acquired or learned inability to understand or see a problem, much less a solution." Increasingly, Kahn said, the more expert, or at least the more educated a person is, the more likely he is to be affected by educated incapacity. This led to a new kind of class warfare, not between the haves and have-nots, but between the knowledge class and the producer class. The expansion of education did not simply give additional information to the students. It changed values as well—making them more liberal, less dogmatic, less authoritarian, more interested in political matters, and taught them to think independently. [In the latter, unfortunately, emphasis was placed on mere novelty, with little attention to creativity and the imparting of the skills needed for testing against reality through logic, consistency, and the scientific method.] The *development of mass communications* was the real sleeper. For one thing, it is nationwide, which means local traditions are constantly being invaded by alien viewpoints. This clearly makes it tougher for parents to pass their values on to their children in unaltered form. News coverage is a particular force for change by presenting dissatisfaction, alternative lifestyles and dissonant signals—and particularly so when passed through media filters. Studies such as that of Lichter and Stanley (1981) showed that journalists and broadcasters almost all had college degrees; two-fifths came from Eastern upper-middle-class families; more than 80 percent voted liberal in the previous four presidential elections and more than half agreed that "the U.S. exploits the Third World and causes poverty" and that its use of natural resources is "immoral." As a group, media people were overwhelmingly for abortion, affirmative action, women's rights, sexual freedom, and environmental protection. The *distinctive generational experiences* had more influence on our changing values than we realized. To the new generation of the 1960s, war was something that happened to other countries. They had less concern about economics and put more emphasis on lifestyle and the "quality of life"—and ideology. They were willing to endure, and certainly to impose, economic penalties in order to obtain environmental values.

Having revealed the cause-and-effect patterns of shifting values and political skills, Inglehart (1977) looked for the mechanisms that would explain *why*. He found it in the landmark study *Motivation and Personality* in which psychologist Abraham Maslow (1970), after years of clinical research, laid out his theory of the "needs hierarchy." Not surprisingly, he discovered that top priority goes to the physiological needs—food, clothing, and shelter, as long as these are in short supply, and that physical security comes next in importance. On further study, he discovered something very peculiar: the basic material and safety needs, once satisfied, no longer motivate as powerfully; discontent sets in, and new, higher-level needs arise. The first of these new needs is for love and affection, followed by the need for a sense of belonging, then the need for self-esteem, and then for what Maslow called "self-actualization"—being "all that one can be." Again to his surprise, Maslow found that when the higher-level needs themselves are gratified, they, too, cease to motivate as strongly; a new discontent sets in, and final, highest-level needs arise: the need for knowledge (or curiosity and intellectual needs), and lastly the aesthetic needs (the need for beauty, for beautiful surroundings, or to create beautiful things). These last two also encompass spiritual or philosophic needs.

However, Arnold (1982) went even further than Inglehart. A few pages beyond the aesthetic needs, Maslow (1970) reported some serious consequences of rising up the needs hierarchy. Once we have gotten one or two stages into the non-material needs, we begin to feel an "independence of and a certain disdain for the old satisfiers and goal objects." The lower-material levels begin to look boring and then even repulsive. As Maslow warned:

> In a word, we tend to take for granted the blessings we already have, especially if we don't have to work or struggle for them. The food, the security, the love, the admiration, the freedom that have always been there, that have never been lacking or yearned for tend not only to be unnoticed but also even to be devalued or mocked or destroyed. This phenomenon of failing to count one's blessings is, of course, not realistic and can therefore be considered to be a form of pathology.

Maslow called it "postgratification forgetting and devaluation," and predicted it to be "of very great potential importance and power." From this, Arnold (1982, p. 50) concluded that Inglehart (1977) was wrong:

> The needs hierarchy alone does not explain environmentalism. These two treacherous pitfalls of forgetful devaluation and learned inability to see economic problems have twisted environmentalism from a sane and insightful protection of the whole environment, civilization included, into a mad scramble for political power and economic control by self-appointed saviors who don't care what it costs.

Johnson (1987) provides strong support for the above analysis. He begins his article entitled "The Heartless Lovers of Humankind" by saying:

> In the past two hundred years, the influence of intellectuals [read theoreticians] has grown steadily. It has always been there, of course, for in their earlier incarnations as poets, scribes and soothsayers, intellectuals have laid claim to guide society from the very beginning. From the time of Voltaire and Rousseau, the secular intellectual has filled the position left by the decline of the cleric, and is proving more arrogant, permanent, and above all, more dangerous than his clerical version. . . .
>
> I believe the reflective portion of mankind is divided into two groups: those who are interested in people and care about them; and those who are interested in ideas. The first group forms the pragmatists and tends to make the best statesmen. The second is the intellectuals [theoreticians]; and if their attachment to ideas is passionate, and not only passionate but programmatic, they are almost certain to abuse whatever power they acquire. For, instead of allowing their ideas of government to emerge from people, shaped by observation of how people actually behave and what they really desire, intellectuals reverse the process, deducing their ideas first from principle and then seeking to impose them on living men and women.

As examples of intellectuals passionately attached to ideas and insensitive to the needs and views of others, he names Shelley, Marx, Lenin, Stalin, and Freud and cites incidents from the lives of each, revealing "the same dichotomy: an unbridgeable gap between theory and practice, between ideas and people"—each "another case of a man who never allows his ideas to penetrate his private relationship or improve his dealings with people." He also named Ayn Rand in this category but claimed:

> There are good reasons why most intellectuals [theoreticians] share common ground with socialists.
>
> It is not the formulation of ideas, however misguided, but the desire to impose them on others that is the deadly sin of the intellectual. That is why they so incline by temperament to the left. For capitalism merely occurs if no one does anything to stop it. It is socialism that has to be constructed, and as a rule, forcibly imposed, thus providing a far bigger role for the intellectuals in its genesis. . . .
>
> In the twentieth century, building upon nineteenth century foundations, the appetite for violence in the pursuit and realization of ideas has become the original sin of the intellectual. . . .
>
> While the armchair men of violence in the West applauded and condoned, intellectuals elsewhere participated and often directed the great slaughters of modern times. Many helped to create the Cheka, the progenitor of the present KGB. Intellectuals were prominent at all stages in the events leading up to the Nazi holocaust. The events in Cambodia in the 1970s, in which between one-fifth and one-third of the nation was starved to death or murdered, were entirely the work of a group of intellectuals, who were for the most part pupils and admirers of Jean-Paul Sartre—"Sartre's children," as I call them.
>
> So one of the lessons of our century is: beware the intellectuals. Not merely should they be kept well away from the levers of power; they also should be objects of peculiar suspicion when they seek to offer collective advice. Beware committees, conferences, leagues of intellectuals! For intellectuals, far from being highly individualistic and non-conformist people, are in fact ultra-conformist within the circles formed by those whose approval they seek and value. This is what makes them, en masse, so dangerous, because it enables them to create cultural climates, which themselves often generate irrational, violent and tragic courses of action.

In this last thought we find a recasting of the words of Robert M. Maciver (1952):

> The greatest evils inflicted by man on man over the face of the earth are wrought not by the self-seekers, the pleasure-lovers, or the merely amoral, but by the fervent devotees of ethical principles, those who are bound body and soul to some larger purpose, the nation, the "race," the "masses," the "brethren" whoever they may be. . . . In the heat of devotion to that larger but exclusive purpose, there is bred the fanaticism that corrodes and finally destroys all that links man to the common humanity. In the name of the cause, they will torture and starve and trample underfoot millions upon millions of their fellowmen. In its name they will cultivate the blackest treachery. And if their methods fail, as fail in the end they must, they will be ready, as was Hitler, to destroy their own cause or their own people, the chosen ones, rather than accept the reality their blinded purpose denied.

In a parallel study, Peter Metzger (1979) and Isaac and Isaac (1983) arrived at a similar diagnosis. After citing a number of signs of the diverse conflicts within our contemporary society, they state that these reflect the activity of:

> the coercive utopians, whose programs if implemented, will accelerate the decline in our standard of living, erode our democratic system, and may well result in the loss of genuine national independence. . . . Most of the diverse groups we will describe are utopian because they assume that man is perfectible and the evils that exist are the product of a corrupt social system. They believe that an ideal social order can be created in which man's potentialities can flower freely. They are "coercive" because in their zeal for attaining an ideal order they seek to impose their blueprints in ways that go beyond legitimate persuasion.
>
> If they believe that society is perfectible, the utopians also believe that the society in which they live is deeply flawed, indeed hateful. And if one has to identify the single aspect of American life that they find most repugnant, it is our economic system. . . . The source of their strength has been their invocation of purposes that Americans, to their credit, want to achieve—social justice, peace, a pollution-free and safe environment, equality between the races and sexes, the reduction of risk, greater control of the individual over the decisions that affect his life (Isaac and Isaac, 1983, pp. 2–4).

Who are the coercive utopians? According to Isaac and Isaac (1983, pp. 4–9), many are intellectuals—people who wield the power of the spoken and written word and do so in the absence of any direct responsibility for public affairs. They dominate the leadership and professional staff of the mainline Protestant denominations and their related organizations. They are the leaders of almost all of the peace groups. They are the intellectuals in a number of institutes and "think-tanks" that have flourished in the soil of so-called "revisionist history." They people the community action organizations. They are in government bureaucracies. They are prominent in the legally independent, but wholly government-funded Legal Services Corporation and in the similarly constituted Corporation for Public Broadcasting. They play an important role in environmental organiza- tions. They lead the consumer organizations established by Ralph Nader. They are in the colleges, and are particularly prominent in the law and social science faculties of elite universities. Their major think-tank is the Institute for Policy Studies. Perhaps most important, they have had an enormous impact on the way we think. They write many of the articles we read in opinion-setting magazines, on the Op-Ed pages of the New York Times and Washington Post. We watch them on the panels that interpret current events for us on television. We hear them from the pulpit and our children listen to them in college classrooms. During the Carter administration, they were invited into the government in droves.

Insofar as they fit the traditional political framework, the coercive utopians lean toward anarchism. But anarchism is not the absence of government. It is a form of government in which decisions are the product of unanimous agreement among the governed. That is, far from dispensing with coercion, anarchism has to avail itself of a coercive and oppressive process to achieve unanimity. While drawn to anarchism, the utopians also are attracted to the one teaching which provides a comprehensive secular explanation for all social issues—namely, Marxism. Church bureaucracies and peace movement leaders have been especially prominent in providing support for Third World Marxist liberation movements, whose victory, they believe, is essential for the creation of a global, just, peaceful, social order. Insofar as the utopians have real-life models, they are places like Cuba, Vietnam, Nicaragua, and China. Rural communal life, rhetoric of egalitarianism, the block level of civic control, all are seen as expressions of neighborhood solidarity and self-governance. The very absence of consumer goods makes these societies immensely appealing. That their chief distinguishing characteristic is the desire of vast numbers of their citizens to leave at almost any price does not daunt the utopians.

While anarchism can claim to be a native American radicalism, there are important differences between the new anarchism and the old. One is that the utopians are not open with us. Their goals are put forward in such a way as to evoke a broad favorable consensus. Because much of the agenda of the new utopians is known only to the initiated, the public is in effect coerced into courses of action that a majority would reject if it were aware of their full implications. The anarchist model, with no central organization, with all decisions made in face-to-face local meetings, with technology scaled down to "appropriate" and "human scale," would scarcely seem the most suitable with which to counter the Soviet war machine. But by defining the Soviet Union's actions as essentially defensive, the utopians can push for unilateral disarmament, emasculation or outright abolition of our intelligence agencies, the elimination of whole energy technologies and the severe restriction of others. Thus, they employ a double standard. The goal, as is quite clear in the literature of these groups, is to bring down the system through the system.

> Although the public has not been aware of the role of the utopians, the Republican victory was in part a repudiation of their achievements. The Reagan administration was committed to an anti-utopian thrust. Nonetheless, if the utopians were downcast at the Republican victory, it made them redouble their efforts, and they have been successful in undercutting key administration initiatives in the area of environmental regulation, defense and foreign policy. The utopians remain in key government bureaucracies. They have a strong, even dominant, voice in the media. The utopians will remain a powerful force in our midst—it is time we knew more about them. (Isaac and Isaac, 1983, pp. 13–14)

This review has not been exhaustive. Hopefully, it will have demonstrated that the traditional role of scientists in our society as the developers and archivists of basic knowledge, who make continuing efforts not only to extend the boundaries of that knowledge but also to continually update the distinction between that which we know and that which we merely believe has been seriously eroded. This has come about because of widespread and sustained efforts to replace science with what has been termed "social relevance" in the decision-making process. Thus, when one hears differences of viewpoints among scientists, it is well to keep in mind that one side may be presenting what it believes to be factual information, while the other side may be more concerned with influencing decisions in a predetermined direction. Moveover, scientific viewpoints are not mere ends in themselves. For the media, in general, "good news [the absence of a problem] is no news." Such news also does little to spur legislative bodies to appropriate funds for further research in such areas. It is also relevant that the coercive utopians have had a virtual monopoly over our communications media for at least two decades. These points are not lost on scientists. To continue to hold out against such pressures requires strength of character, as well as a relinquishing of the expectation of rewards in the form of fame and fortune.

There is one further criterion that is generally reliable in evaluating the merits of conflicting views. It was offered in the House of Representatives in May 1824 by John Randolph of Roanoke, Virginia:

> The old proverb says, God works good, and always by degrees. The devil, on the other hand, is bent on mischief, and always in a hurry. He cannot stay; his object is mischief, which can best be effected suddenly, and he must be gone elsewhere.

I ask you to keep these points in mind as you peruse the following papers. Max Singer, in the keynote paper, again addresses this issue of dichotomy of viewpoints from his own inimitable point of view, as does Margaret Maxey in the final paper on "Managing Technological Risks: A Bioethical Perspective." Regardless of how one interprets this review in its effect on science and the role of science in the advancement of the social culture of the species, there is another reason to be apprehensive in this regard. As recently pointed out by

Westheimer (1987), while the role of science in our culture has been expanding rapidly in such areas as nuclear physics, information processing, geology, astronomy, properties of matter, antibiotics, genetic engineering, etc., the amount of time devoted to scientific subjects required by non-science majors in our major educational institutions has been declining. And because of the vertical structure of science, one subject (say thermodynamics) built on top of another (say calculus), this does not bode well for the non-scientific portion of our population being able to communicate with the scientific portion to decide issues such as those raised in this volume.

References

Abelson, Philip H. 1987. "Cancer Phobia." *Science* 237:473.

Arnold, Ron. 1982. *At the Eye of the Storm: James Watt and the Environmentalists.* Chicago: Regnery Gateway.

Battegelli, M.C. 1968. "Sulfur Dioxide and Acute Effects of Air Pollution." *Journal of Occupational Medicine* 10:500.

Carson, Rachel. 1962. *Silent Spring.* New York: Fawcett Crest.

Claus, George and Karen Bolander. 1977. *Ecological Sanity.* New York: David McKay.

Clawson, Marion. 1981. "Entering the Twenty-First Century—The Global 2000 Report to the President." *Resources* 66 (Spring):19.

Club of Rome. 1972. *The Limits to Growth: A Report for the Club of Rome's Project on the Predicament of Mankind.* New York: Universe Books.

Council on Environmental Quality and Department of State. 1980. *Global 2000 Report to the President.* Washington D.C.: Government Printing Office.

Dixon, J.P. and J.P. Lodge. 1965. "Air Conservation Report Reflects National Concern." *Science* 148:1060.

Dubos, Rene. 1981. "Half Truths About the Future." *Wall Street Journal.* 8 May editorial page.

Efron, Edith. 1984. *The Apocalyptics—Cancer and the Big Lie.* New York: Simon and Schuster.

Ellsaesser, H.W. 1971. "Air Pollution: Our Ecological Alarm and Blessing in Disguise." *EOS, Transactions, American Geophysics Union* 52(3):92.

_________. 1973. "Air Pollution Control: Does Our Program Fit the Problem?" *Bulletin of the American Meteorological Society* 54(2):117.

_________. 1975a. "The Upward Trend in Airborne Particulates That Isn't." S.F. Singer (ed.). *The Changing Global Environment.* Dordrecht, Holland: D. Reidel, 235–269.

_________. 1975b. "Where We Are Now in Air Pollution." *Proceedings Twelfth Space Congress* 2–5. Cocoa Beach: Canaveral Council of Technical Societies.

_________. 1979. "Air Pollution: A Different View." *Water, Air and Soil Pollution* 11:115.

_________. 1981. "Statement Prepared for Presentation at the Hearings on Consolidation Coal Company's Petition to Amend the Primary Ambient Air Quality Standards for Sulfur Dioxide." *The Science of the Total Environment* 19:13.

Goldwater, L.J. 1976. "Strengthening Environmental Standards: Introduction." *Journal of Occupational Medicine* 18(5):291.

Grayson, Melvin J. and Thomas R. Shepard, Jr. 1973. *The Disaster Lobby—Prophets of Ecological Doom and Other Absurdities.* Chicago: Follett.

Handler, Philip. 1980. "Science and the American Future." Speech at Duke University, 6 March 1980, partially published as "Public Doubts About Science." *Science* 208:1093.

Inglehart, Ronald. 1977. *The Silent Revolution: Changing Values and Political Styles Among Western Publics.* Princeton: Princeton University Press.

Isaac, Rael Jean and Erich Isaac. 1983. *The Coercive Utopians—Social Deception by America's Power Players.* Chicago: Regnery Gateway.

Johnson, Paul. 1987. "The Heartless Lovers of Humankind." *The Wall Street Journal.* January 5, 1987. (Reprinted 1987. *The St. Croix Review.* 20(2):27.)

Kahn, Herman. 1976. *The Next 200 Years: A Scenario for America and for the World.* New York: William Morrow.

_________ and Ernest Schneider. 1981. "Globaloney 2000." *Policy Review* Spring:129.

Lichter, S. Robert and Stanley Rothman. 1981. "Media and Business Elites." *Public Opinion 1981*(Oct/Nov).

Maciver, Robert M. 1952. "The Deep Beauty of the Golden Rule." *Moral Principles of Action.* Ruth Nanda Anshen (ed.). New York: Harper and Brothers.

Maddox, John. 1972. *The Doomsday Syndrome.* New York: McGraw-Hill.

Maslow, Abraham H. 1970. *Motivation and Personality.* 2nd Ed. New York: Harper & Row.

Melloan, George. 1984. "The Crisis Continues. Review of *The Liberal Crack-Up,* by R. Emmett Tyrrell, Jr. (New York: Simon and Schuster, $16.95)." *Policy Review* (Fall 30):77.

Metzger, H. Peter. 1979. "The Coercive Utopians: Their Hidden Agenda." Denver: Public Service Company of Colorado. (First presented at the 47th Meeting of the American College of Nuclear Medicine, Colorado Springs, Colorado, 28 April 1978.)

Meyer, Herbert E. 1979. *The War Against Progress.* New York: Storm King.

Neuhaus, Richard. 1971. *In Defense of People: Ecology and the Seduction of Radicalism.* New York: Macmillan.

Schrenk, H.H., H. Heimann, G.D. Clayton, W.M. Gafafer, and H. Wexler. 1949. "Air Pollution in Donora, Pa.; Epidemiology of the Unusual Smog Episode of October 1948." *Public Health Bulletin No. 306.* Washington D.C.: Federal Security Agency.

Simon, Julian L. 1981. "Global Confusion, 1980: A Hard Look at the Global 2000 Report." *The Public Interest* 62 (Winter):3.

__________ and Herman Kahn. 1984. *The Resourceful Earth.* Oxford and New York: Basil Blackwell.

Singer, S. Fred, ed. 1989. *Global Climate Change: Human and Natural Influences.* New York: Paragon House (An ICUS Book).

__________, ed. 1990. *The Ocean in Human Affairs.* New York: Paragon House (An ICUS Book).

__________, ed. 1991. *The Universe and Its Origin: From Ancient Myth to Present Reality and Future Fantasy.* New York: Paragon House (An ICUS Book).

Sterling, T.D., S.V. Pollack, and J.J. Phair. 1967. "Urban Hospital Morbidity Air Pollution." *Archives of Environmental Health* 15:362.

Stokinger, H.E. 1971. "Sanity in Research and Evaluation of Environmental Health." *Science* 174:662.

Tabershaw, I.R. 1976. "Oxides of Sulfur." *Journal of Occupational Medicine* 18(5):360.

__________, F. Ottoboni, and W.C. Cooper. 1968. "Oxidants: Air Quality Criteria Based on Health Effects." *Journal of Occupational Medicine* 10:464.

Tucker, I.T. 1976. "Discussion of 'Air Pollution Standards: An Overview.'" *Journal of Occupational Medicine* 18(5):302.

Tyrrell, R. Emmett, Jr. 1984. *The Liberal Crack-Up.* New York: Simon and Schuster.

Wattenberg. Ben J. 1974. *The Real America.* New York: Simon and Schuster.

__________. 1984. *The Good News Is that the Bad News Is Wrong.* New York: Simon and Schuster. (A condensation under the same title appeared in *Reader's Digest* the same year.)

Westheimer, F.H. 1987. "Are Our Universities Rotten at the 'Core'?" *Science* 236:1165.

Whelan, Elizabeth M. 1984. *Toxic Terror.* Ottawa, Ill.: Jameson Books.

ONE

ALTERNATIVE PERSPECTIVES ON THE EARTH'S PROSPECTS

THE GLOBAL 2000 REPORT (GTR) VS. *THE RESOURCEFUL EARTH* (TRE)

Max Singer

Introduction

There are two alternative perspectives about the Earth's physical prospects. The first perspective is very familiar—I call it the "scarcity perspective" or the "doomsday perspective." It is most authoritatively represented by the official document of the U.S. government, *The Global 2000 Report* (Council on Environmental Quality and the Department of State). The current round of great public attention to this perspective began with the report *The Limits to Growth* (Club of Rome). And there is now a group of organizations called The Global Tomorrow Coalition engaged in putting forward this perspective.

The best way to evaluate this perspective is to ask what is an opposite view of the world? This paper will show why the opposite view is correct—and why, therefore, the doomsday perspective should be completely rejected. (Which, of course, is not to deny that doomsayers sometimes present true facts and real tasks requiring attention.)

The major part of the paper will be a presentation of such an opposite perspective, which I call the "Passage to a Human World Perspective"—a perspective which is supported by books such as *The Resourceful Earth* (Simon and Kahn), *World Economic Development* (Kahn), *The Ultimate Resource* (Simon), and *The Next 200 Years* (Kahn *et al.*), written or edited by Herman Kahn and Julian Simon.

The Scarcity Perspective

The outlines in Tables 1-1 and 1-2 are an efficient way to summarize the scarcity perspective.

The general sense of the situation expressed in Table 1-1 is the opposite of the truth. And all of the specific propositions in Table 1-2 are incorrect. Correct statements are shown in the appendix. But to get to the heart of the doomsday perspective, we must go beyond refuting some specific propositions, because the scarcity perspective grows out of a real human concern. I want to address that underlying concern, not just expressions that it takes. So I use the phrase "scarcity perspective" to cover all the ideas and arguments that grow out of that concern, which is felt in one way or another by a great many people. It is the fear or belief that modern society is riding for a fall—that our way of life uses so much material and creates so much waste, and our population is growing so fast, that sooner or later we have to run into dangerous shortages, or hopelessly pollute our environment. After all, the Earth is finite.

Another form of the same concern is the common idea that somehow the question of whether the rapidly increasing billions of people in the Third World can catch up depends on whether there are enough raw materials left for them at reasonable prices.

Another form of the scarcity perspective is the feeling that it is somehow unfair that the U.S. should be using a quarter of the

Table 1-1: The Doomsday Perspective (Tone)

The world will be:
- More crowded
- More polluted
- Less stable
- More vulnerable

People will be poorer
Life will be more precarious
Threats to the future of mankind
Population approaching carrying capacity of the earth
Better projections would show more intensifying stress

Table 1-2: The Doomsday View (Propositions)

1. The overall *physical condition of the world is getting worse* as more and more people use it more and more intensely.
   - Raw material scarcity
   - Pollution
   - Dangers to health (chemicals, nuclear wastes, etc.)
2. As raw materials are consumed at ever-increasing rates, the world is eventually *doomed to have great raw material scarcity* problems (perhaps combined with worsening pollution).
3. Current world population trends *threaten to exceed the earth's carrying capacity* in the next few centuries. If they continue, there would be 10 times the current population in only a few centuries and there would be a shortage of living space.
4. Already, the amount of *arable land is shrinking* and losing quality as farmland is converted to urban use and to roads, and as topsoil is eroded.
5. Therefore, the basic long-term problem for mankind will be *how to adjust to increasing scarcity.*

raw materials taken out of the Earth, that we are taking more than our share. This charge of "greediness" can relate either to poor countries or to future generations. But, obviously, it only has meaning if the result of our taking is that there isn't enough left for others.

How do you get at something as broad as the scarcity perspective? People rarely say things like, "I think there will be a big scarcity problem in one hundred years because. . . ." They don't think specifically about long-run scarcity because it is too difficult and it doesn't seem relevant to any practical issues. They just *assume* that the scarcity perspective is correct. Few people even realize that the scarcity perspective is a particular idea, that there is an alternative to it, and that, therefore, it might be wrong.

The scarcity perspective is as natural as breathing, so people would say that they have never thought about long-run raw material supply problems. But, in effect, they have. Anyone who has any of the concerns I quoted above is acting as if he thought about the question of whether there will be raw material scarcity problems in the future and reached the conclusion, "yes, sooner or later, there will be a problem." (Why else would it be unfair for us to take all the raw materials we want, if the supply remaining for others were not reduced?)

Of course, if people are specific enough to point to a particular problem and a particular date they can be answered specifically, and shown why that particular scarcity fear is unrealistic. This has happened repeatedly, and many specific predictions have come due and been proven wrong by events. The problem is that many people observing, for example, a debate about whether GTR is correct, have the following kind of reaction: (from a *Washington Post* report on meetings where the GTR was strongly criticized) "a consultant to Earthwatch acknowledged that the original GTR projections may turn out to be wrong but said, 'that is not the point. The idea was to call attention to the fact that we are destroying our habitat. That is still true.'" (Sometimes people get angry at one for criticizing studies like GTR. They say, in effect, "you are quibbling over details and trying to prevent people from facing up to one of the world's greatest problems. By arguing about the specific numbers in these studies you are giving people an excuse for continuing their selfish and dangerously wasteful practices. It is infamous to treat such studies like academic exercises; if people don't recognize and do something quickly about the danger of exploding world population and the ecological limits of Spaceship Earth, we may pass a point of no return.")

That is the perspective I am trying to define and that I will show is wrong. It is not only that the particular argument in GTR or *The Limits to Growth* or *The Population Bomb* (Paul

Ehrlich) is wrong. (Although the particular arguments and conclusions in these books are wrong; they are as wrong as the book *Famine—1975*, which was published in 1967 and sold tens of thousands of copies.)

The important thing is that the whole perspective is wrong. There isn't anything there. We are not destroying our habitat. (We might, but we don't have to.) And we don't have to sacrifice our freedom or our standard of living to avoid destroying our habitat. We don't even have to make drastic changes in the way we make decisions. Of course there are practical problems of how to grow more food, enable poor countries to get rich enough so that their people can live decent lives, and many other urgent tasks. But these problems have essentially no connection with long-run scarcity or the finiteness of the Earth.

To really deal with the human concern that underlies the scarcity perspective, I need to show that there are enough raw materials—and enough places to dispose of waste without polluting the environment—forever, not just for a few centuries or a few thousand years. The best way to really get at the heart of the scarcity perspective—not just one expression of it—is to show that there is no reason to worry that mankind's safety or standard of living need ever be threatened by resource shortages or pollution problems. (Of course, it is not necessary to say that there are no dangers and problems to reject the scarcity perspective. The fact that mankind faces some dangers and problems doesn't invalidate the statement that there is essentially no truth to the scarcity perspective.)

Such an extreme claim—that we can be confident that there are enough resources forever—is necessary because if I just show that we have enough for a thousand years people will say, "How can you be sure? Maybe your calculations aren't exactly right and there is only enough for five hundred years, or even a hundred. And, anyway, if we are going to get into trouble in a thousand years maybe we have a responsibility to begin to move at least a little way in the direction that will need to be taken eventually." It seems somehow shabby and second-class to be going against history, against the long-run imperative facing mankind. It seems selfish and shortsighted to try to get away with what we can take, and leave the ultimate clean-up job to our remote descendants, just because they aren't here to complain or to vote.

That is why I am going to attempt the seemingly silly task of answering the question, "Do we have enough forever?" (Of course, I don't really take care of "forever." For example, I don't deal with the question of geological changes, or the sun becoming a red giant. These are matters of millions or billions of years. I am only concerned with human history, which so far has been measured in tens of thousands of years.)

But before discussing the "technical problems" of providing resources and protecting ourselves from pollution, we should first take a look at what is happening in history—because it is a dramatic story that provides the essential context.

The Passage to a Human World Perspective

Until recently, asking the basic questions about human life on Earth always produced the same answers; there was surprisingly little news. Life expectancy scarcely changed until the last century, when it was still less than thirty (having risen only a few years). This single grim statistic tells much about the fundamental nature of people's life experience. It says that few children grew up with two parents; that most adolescents had more brothers and sisters dead than alive; that the person who got to know his grandchildren was rare indeed. So human life was always full of the death of loved ones, and the likelihood of dying young.

Throughout human history, most people produced what they consumed by monotonous, arduous, outdoor work, and lived in cramped and dirty surroundings. Most of them spent their entire lives not far from where they were born and spoke to no more than a few hundred people before they died—even the rare ones who lived a long life. They were at the mercy of powerful forces which they did not understand. Violence was common in their lives. Most of the time their thoughts were about the unpleasant side of nature: dirt, disease and death.

By 1800, people had learned to create large complex societies; a small minority had learned to read and write (an accomplishment that affected many others); and many people had come to believe in God. But when confronted with a realistic description of the nasty details of the way all the ordinary people actually lived as recently as the last century, most of us would exclaim, "they lived like animals."

This is too extreme. People always have been significantly different from animals. And during our long history we have produced great human creations and achievements which should still give us pride. But, compared to the human life we experience here today, the human life of the past—or of parts of the poorest countries today, indeed of almost everyone who has ever been born—seems so different that one can easily think of it as more like the lives of animals than like the human life we know.

But now, one-quarter of the people live modern wealthy lives, living nearly their full lifespan. In the wealthy countries, few people experience much death in their family, and most people know both their grandparents and their grandchildren.

In the modern part of the world, people are doing completely different things than they used to do. Most of their thoughts no longer are about the unpleasant aspects of nature. They are about humankind's creations: science, art, politics, commerce, bureaucracy, entertainment, and, sadly, crime and war. Nature is primarily something to protect and enjoy. Sometimes it is a disturbance, but only rarely is it a serious threat.

Ordinary people in the modern parts of the world know something of their own and the world's history. From the media they know something of the other people who share the Earth with them. Sometime during their life they can go to see almost any place they want. They have a reasonable degree of knowledge of the basic facts about nature. And their lives are filled with choices: Where should they live? What kind of work should they do? The character and quality of most people's lives depend on the choices they make. In brief, most of these people live the kinds of lives we are familiar with.

No other word as aptly describes the character of the new world into which we are passing as the word "human." The new features of this world are being created by people. In this new world the most basic fact of life—how much of it each person has—will be determined by the genes of our human species, not by external natural forces that cut human life short. Humans will be plentiful, and their lives will be the kind of lives that we think of as normal human lives—concerned mostly with human creations, not with a constant struggle against nature. Formerly, nature dominated man—but not completely; after we finish our current brief passage, man will dominate nature—but not completely.

We must still use the future tense to speak of the "human world," because it is not yet true that almost all people in the world live in what we call "human conditions." But we are well on our way to a human world and, since many people have already arrived, there are samples of this world that we can see now. We can get a feel for what a human world might be like by looking at the lives of the people in countries which already have crossed the threshold from poverty to wealth.

Despite the inevitable uncertainties about the future, already we can see that in the "wink of an eye" (i.e., a few centuries), the world is being changed more drastically than it had been changed in all human history. We are living in the middle of that wink. We can see examples of what life will be like in the future, and we can still see examples of what all human life was like in the past. We live today just about at the middle of a period of only a few hundred years, during which mankind is changing the world as it has never been changed before, into the way it will probably be from now on.

This perspective can help us to balance the normal static, snapshot view of the world with a broader understanding. For example, we are often told that the world is divided into rich countries and poor countries. This is a little like describing a family as being divided into little people who have no money and have to go to bed at nine o'clock and big people who are allowed to drive the car and stay out late. The big people were once little, and the little ones will become big.

In a family, the current arrangements don't mean nearly as much as a sense of the process that the family group is going through. And since adults are so familiar with the pattern of development and family life, most of us don't usually see families as snapshots, we see them in relation to the pattern of the generations. (Of course the 10-year-old, lacking the perspective that comes from experience, sees things differently. She is much more aware of the "injustice" of the moment, and the moment seems much longer.)

If we look at the world with the kind of dynamic, historic, perspective with which we look at a family, we would give a more realistic description by saying, "The world is passing from poor to rich. All countries were poor, some countries are already rich, others are nearly there, and others still have farther to go." Because we are now in the middle of the development process that is responsible for our passage, some countries are now widely separated from others. And since we are barely conscious of the process, the current distance between countries seems permanent, and, therefore, highly unfair and immoral. But to have a reasonable perspective on the moral question, the world, like a family, also should be seen as a dynamic process in which today is only a long moment.

The confident statement that we are creating a human world sounds optimistic and arrogant. To see why it isn't, two points need to be emphasized. First, although we can be confident that the world will be human, we cannot know whether it will be better. We humans have amply demonstrated our power to do evil and create unhappiness. Human domination of the world is a challenge, not a seal of success. Change is not necessarily progress. The prediction that we are making a human world is a realistic description, not an optimistic hope. (And some people will feel that it is a profoundly pessimistic prediction.)

The second point to emphasize is that the word "human" is not used here in the sense of being the opposite of God. God created people as much as He created the rest of nature. While man is special, man is also part of that nature. There is no reason not to think of man's creations as part of the work of God. There is no reason one has to think that people, by changing the world from nature-dominated to human-dominated, are denying or challenging God—rather than carrying out God's will.

How do we know that the rest of the world also will become modern and wealthy? Because we can recognize the powerful process that has been shaping events, and see that that process is still working, and that there is no reason for it to stop now. So it is not very bold to predict that the process will be completed.

It is as if a thrown ball were in mid-flight—no one would hesitate to predict where the ball is going. This paper describes a process now dominating history which is almost as clear as the path of a thrown ball.

Of course, there can be no guarantees of the future. Just as a ball can be knocked off course, so the process now bearing us toward a human world could come to an abrupt end.

This approach of predicting the future by trying to understand the processes now at work is very different from a crude extrapolation of trends. If I were merely extrapolating trends, I would have to say that the ball that is thrown upward would continue upward forever. But if I understand the forces working on the ball as it flies through the air, I can use its path in the past to predict its path in the future, even though the direction will change.

The process that is working now to build a human world is the spread of understanding of how to create wealth.

The term "wealth" refers to economic resources, which can be physical, like machines or houses; or they can be financial, like stocks and bonds; and some are even less tangible, like education or leisure. The idea of economic resources (or wealth) is so broad that it is hard to realize how much is covered by such a simple phrase, everything from a gigantic steam shovel to a computer program, from an hour's time to a house at the beach. "Poverty" means "the lack of things that could be bought with money."

The Spread of Wealth

About two centuries ago, some societies began to learn how to increase the productivity of human work and, thus, to become wealthier and wealthier. Since then, the first societies to learn have continued to apply what they learned, and continued to learn more, so that their productivity continued to grow and their wealth accumulated. Also, one after the other, other nations have begun to learn how to steadily increase productivity. Some are doing it fast and catching up; others are going more slowly. Some haven't started yet and others have started and then fallen back. But all nations are going along essentially the same path, starting from poverty, learning gradually to make the changes necessary to be more productive, accumulating wealth, and adjusting painfully to the changes.

Until about one hundred years ago, there were no wealthy societies. Today one-quarter of the people of the world live in wealthy societies. (A wealthy society is one where most people have enough resources to live decent human lives—say, at least $3,000 per capita.) Because one can understand the process by which this tremendous amount of wealth was produced, one can have great confidence that the much greater amount of wealth required in the future also will be produced.

In about a century—or perhaps two—the great majority of people will live in countries that have crossed the threshold from traditional poverty to wealth that is sufficient to provide human lives for their people. (Statements here about the whole world do not include the non-wealthy communist countries—80 percent of the population of which is Chinese—because I do not wish to make any predictions about their future.) No one knows how much further the wealth accumulation process will go after that, or what patterns it will follow.

Tables 1-3 and 1-4, at the top of the next page, give a quantitative description of the passage to a human world.

Ideas vs. Things. Two different perceptions about the nature of productive wealth are possible. One perception is the old-fashioned and obvious one: Wealth is things like fertile land, rich oil deposits, and large stocks of gold. The alternative perception, which is the modern view of productive wealth, is that wealth is mostly intangibles; for example, the characteristics of a people, such as their culture and their education, that enable them to create wealth gradually from whatever things they have. In a word, the modern view is that most productive wealth is not "rocks" but "ideas."

Look at which countries have been most economically successful since World War II. The outstanding performers have included Japan, South Korea, and Singapore—a large, a medium, and a small country, all relatively poor in natural resources, all with large numbers of people compared to their arable land. None of these countries have done well because of their natural resources. Neither has Holland, Denmark, Switzerland, and many other countries.

Table 1-3: Some Catchy Numbers to Describe the Passage to a Human World

	Before the passage (1780)	*Now (1980)*	*After the passage (2180?)*
Gross World Product (billions of 1980 $)	$200	$10,000	$250,000
World Population (millions)	800	4,000	10,000
Average Income (GWP/person, 1980 $)	$250	$2,500	$25,000
Life Expectancy (worldwide)	30 years	60 years	80 years

Table 1-4: Comparison of the Amounts of Change

	1780 to 1980	*1980 to 2180 (?)*
Gross World Product grows: that is, overall it:	2% per year multiplies by 50	1.6% per year (?) multiplies by 25
World Population grows: that is, overall it:	.8% per year multiplies by 5	.5% per year (?) multiplies by 2.5
Average Income grows: that is, overall it:	1.2% per year multiplies by 10	1.2% per year (?) multiplies by 10
Life Expectancy grows: that is, overall it:	.35% per year doubles	.14% per year (?) multiplies by .3

On the other hand, other countries with great natural wealth, such as the Sudan, Zaire, or Argentina, have not grown at all well and are relatively poor.

Much of the "edifice of error" which blocks our view of the real world is built on the unexamined, old-fashioned, perception of wealth as "things." This out-dated perception has many important implications. If wealth is "things," it is natural to wonder whether more can be created. Where can they come from except from other things? Any thing can only belong to one country at a time. If wealth comes from things, it is natural to think that the poor can only get wealthy if they get some things from rich countries. And the suspicion is very easy that rich countries' wealth came from taking things that somehow belonged to, or should have gone to, poor countries. (After all, the countries of the world used to be much more equal.) All in all, the old-fashioned perception of the nature of wealth leads easily to a paranoid view of the world, full of fears and pessimism. For people in rich countries the old-fashioned perception also can lead easily to feelings of guilt because our things seem to be what others need.

Of course, seeing wealth as primarily coming from things also leads to the fear that resource scarcity will prevent wealth from spreading to the rest of the world. The next section of the paper will show why this fear is unrealistic.

The modern perception of wealth, which sees it as mostly coming from ideas, suggests that wealth is unlimited, that it can be created everywhere at the same time, that our wealth doesn't conflict with others' wealth but is more likely to add to it, and other very nice, and mostly correct, ideas.

The most significant implication of seeing that wealth comes from ideas is that a nation's path to wealth is primarily a "national learning experience." (Of course, words like "ideas" and "learning" are intended as metaphors standing for a very broad range of activities.)

"National learning" is not primarily a matter of technical information. Social and cultural change is a much more critical part of the national learning experience than acquiring information and technology. Economic growth comes from people learning to work better, but many people are not really free to work better without gradual and often subtle changes in their society. The culture that admires truth and effectiveness, at least at work, will produce more than the culture that prefers observance of traditional forms and cares little about truth and effectiveness.

Most of the ideas that produce higher productivity and, thus, more wealth are not brilliant or dramatic. Most are barely noticeable. Maybe the typical example to keep in mind is the shipping clerk who figures out how to stack the boxes in the storeroom in a different way so that he can handle more orders. The result is more output with the same input. That is what economic growth is all about. That is a good example to keep in mind also because it illustrates that motivation is likely to be a big influence on productivity. In fact, the rate at which new technology increases productivity is much less determined by the amount and quality of new technology developed in the laboratories than it is by the attitudes and relationships in the organizations that can use the new technology, and on the degree of competitive pressure that they face.

Social and cultural change are an inevitable result of increasing wealth, and they are also necessary in order to be able to become more productive and, thus, wealthier. Such change is the most painful part of the process.

Let's think a minute about why the United States is so wealthy. When Abraham Lincoln was a young man, 150 years ago, this was a poor country. Of course, the land had more gold and coal and iron and oil and top-soil and trees than it does now, and there were fewer people to share this great wealth of resources. But the people were poor. They could not afford to live what we think of as decent human lives.

You might say, "the country was poor because it did not have all the roads and dams and factories and power-lines that now make our country rich." But why hadn't the Americans of that time built that equipment if they needed it to be wealthy? There are two reasons, and these are the driving forces behind the spread of wealth. One is that they did not know how. The other is that accumulating capital takes time. Each year, you have to produce more than you consume, year after year.

Since, as I was told by a taxi driver in Mexico City, poverty is the result of people not knowing how to do anything, the way to get rich is to learn. Basically, a nation gets wealthy by learning how to be more productive and then by accumulating capital.

"Productivity" is the key concept; it is the amount of value that an hour's work produces. We are wealthy because an hour of work today produces more than ten times as much as an hour of work in Lincoln's time. Normally, a country can only be wealthy if its people have high productivity.

Obviously, the reason we are more productive than Lincoln's friends is not that we are born stronger or smarter than they were. We are more productive because our society has learned things that theirs didn't know. Some of what we have learned is built into equipment like power plants and computers, but most of what we know that makes us so productive is expressed in the overall system that keeps producing new techniques and solving problems.

What is the difference between the U.S. and India that causes one to be so much richer than the other? Let's do a thought experiment. Suppose that there were a miracle and when everybody woke up one morning, all Americans were in India and all Indians in the U.S., and there were notices on the bulletin boards that the transfer was permanent and everybody had better get used to it. They would have all of our machines and roads and we would have theirs. (And we had also exchanged population size.) What would happen?

Such a thought experiment should not be taken literally. (Nor should anyone think that it reflects smug national or racial superiority.) The point is simply that the Indians would find that they did not know—as a society and, in most cases, as individuals—how to do what is necessary to make our system work. As a result, the income of this country would drop sharply.

Even though we may feel that our system doesn't work very well, it is producing at a very high level. It would be many years before the Indians would be able to make it produce as much as it does today. The proof that much of the learning is in our society, rather than in individuals, is that if an individual Indian comes to the United States, she can quickly learn how to do well, and how to have high productivity in our system. But when she goes back to India, she will not have such high productivity. The skills learned in the U.S. may not be appropriate at home, and she can't be so effective without the rest of our system.

It is hard to think of our country as fabulously efficient and productive. Many of us are working in jobs that don't seem to produce anything at all. But we are doing something right. People buy what we produce, much of it in world markets where nobody does us any favors.

At the bottom is continuously improving technology and methods of production. Generally, this requires dividing work in new ways or into smaller pieces, and using special tools or specialized labor. (These statements use the language of physical production because that is easier to visualize, but essentially similar factors work in the service sector and in the production of intangible products.) This, in turn, creates a need for better ways to coordinate the specialized work.

The key to understanding what's going on is to realize that the complexity we need to be efficient requires a tremendous amount of coordination and delicate decision-making. By analogy, this could be called "thinking." We have to buy the efficiencies of specialization by spending a lot on some kind of "thinking" which is necessary to make these efficiencies possible.

In effect, the system that we are all part of is the substitute for the "thinking" we need. "The system" is what does the job of coordinating all of our specialized efforts. It does it automatically, in the sense that we didn't design it and don't pay attention to it (but like our "automatic" system of digestion, it sometimes produces a lot of rumbling, gas, and pain). And the increases in efficiency from specialization are so great that we can afford a very expensive system to provide the necessary coordination.

The system has the side effect that lots of people are doing things that don't seem to be very useful. A better system would have just as good decision-making and less decadent silliness and time-wasting. But nobody has ever seen a system that was better. Nature pays to be able to avoid thinking by wasteful production; we pay for our avoidance of thinking by having a social system that involves a great deal of waste.

The point of the thought experiment about our changing places with the Indians is that the primary reason the U.S. is richer than India is simply that we already have learned to do a lot of things that India has not yet learned how to do. Our society is well adapted to high productivity. The main thing we have learned that adapts us to high productivity is a social system that lets us work together in constantly changing ways.

Of course, when India learns how to be more productive, it won't learn to do things exactly as we or others do them. They will learn their own way, which will be about as difficult for them to learn as our way was for us to learn.

The word "learn" is misleading because it is a single word standing for many processes. Also, it suggests that there are clear right answers, which is not always true. Here we are using "learning" as a metaphor to sum up many changes in the thinking and feeling and behavior of individuals and communities.

But make no mistake about it, the process now shaping our world—economic development—is best understood as a learning experience—using that phrase in the broad metaphorical way I have described. If you want to understand the spread of wealth around the world, or to predict whether economic development will happen in a country, or how fast, or to understand the obstacles, or what to do to make it go faster, you need to think about how to predict or influence a *learning process.*

Compound Interest and the Accumulation of Capital. Two links are necessary to connect learning to wealth: the accumulation of capital and compound interest. While wealth is not limited by any shortage, the speed with which a nation can become wealthy is limited. Questions about the future of wealth are not "how much is possible?" but "how fast?"; "what rate of income growth is possible, for each nation at any time?"

The accumulation of capital includes both things and ideas. "Capital" means both money and non-monetary resources that are useful for producing. Experience, learning, and machinery are all capital. Mostly, they embody work done in the past and adapted to be able to do work in the future. Capital is created when a country doesn't consume all it produces; also, when the society learns from the environment without work.

The awesome power of compound interest is not generally appreciated. Throughout almost all of human history there was no broad, sustained growth. So a sustained (that is, average) rate of growth of two percent over a long period of time is rapid growth—and a tremendous force for change. Two percent per year means doubling every 35 years and multiplying nearly eight times in only a century. Therefore, although growing at two percent per year sounds almost like standing still, a two percent growth rate has revolutionary effects on people's lives.

Scarcity Doesn't Jeopardize the Passage to Human World

To see why we shouldn't worry about a catastrophical running out of raw materials, we should start by looking at the raw materials we use now. Table 1-5 gives per capita raw material use in the U.S. in 1980. Briefly, it shows that out of total expenditures per capita of about $12,000, $955 was spent to buy fuels, $500 for food, and $170 for all other raw materials. The possibility of dangerous raw material scarcity should be analyzed in these three categories: food, energy and everything else.

Food. The world's food supply now is about two billion tons of grain equivalent. If we are to feed a rich world of ten billion people with American eating habits, we will need to grow five times as much grain equivalent a century or two from now. (Most of the food we take from the ground is grain—and for Americans, most of the grain goes to feed animals. "Grain equivalent" counts actual grain, plus the amount of grain that would provide as many calories as come from other crops.)

The problems that farmers and governments are having now with falling grain prices and growing surpluses demonstrate that famine and hunger today don't come from a lack of food but from a lack of money, or from political inhibitions on distribution.

Table 1-5: Use of Raw Materials in the United States (1980)

Item	*Amount used (per capita)*	*Cost per lb.*	*Total cost (per person)*
Food	2,000 pounds	$ .25	$ 500.
Fuels	16,000 pounds	.06	950.
Drinking Water	2,000 pounds	.0001	.20
Metals	1,000 pounds	.06	60.
Wood	2,000 pounds	.02	40.
Misc. Minerals	2,000 pounds	.02	40.
Stone, sand, & gravel	20,000 pounds	.001	30.
	45,000 pounds	0.036	$1,620.20

The current food supply is grown on 2.8 billion acres of land with an average yield of .7 metric tons per acre. With conventional agriculture, more food must come from either more acres or higher yield per acre.

Number of Acres Harvested for Food. The most careful study of potential additions to the supply of arable land was made by a group of Dutch soil and agriculture experts (Buringh *et al.*). They evaluated some 10,000 soil units in the 222 regions in which the latest soil maps divide the world. Their conclusion was that after taking into account all the basic constraints, there are an additional five billion acres that could be used to grow food crops. These additional acres would not be substantially less suitable than the land now used, and could be made arable at an average cost of some $1,200/acre (1981 dollars), less than the price of much farmland in the U.S. today. There is no substantial technical challenge to the general conclusions of the Buringh group.

The Buringh study also indicated that there are four billion acres that could produce at least one additional crop per year by sequential cropping (double cropping). In effect, sequential cropping could add as much grain as an additional four billion single-cropped acres. (This paper considers sequential cropping with potential sources of increased acreage, rather than the more common practice of defining it as a potential source of increased yield.)

To summarize, we can say that the number of acres of food crop harvests can be multiplied by three.

Yield/acre: We will limit it to what can be done with current agricultural science, that is, with currently known techniques, plus the continued application of current agricultural research techniques. (This excludes scientific breakthroughs, or exotic developments like gene-splicing, etc.)

Since benefits from recent and future scientific advance have not been counted in these estimates, if there are any such benefits, they would be available to make up for unforseen problems. This is a powerful reserve (a very conservative form of calculation).

Effective yield increases can come from the following:

- increased inputs—e.g., fertilizer, herbicide, irrigation, etc.

- improved inputs and techniques (compared to present average practice)
- higher yielding varieties of seeds
- reduction of post-harvest losses (e.g., from rats, spoilage, etc.)
- reduction of the grain requirements of animals used for food (currently four to eight calories of grain are used for each calorie of meat consumed)

Taking all these sources together, a factor of five increase in average effective yield is possible over the next few centuries (Chou *et al.*).

Worldwide average yields grew nearly two percent per year during the 1970s (Sanderson). (The reported growth of yields does not take into account changes in post-harvest losses, nor the efficiency of conversion to meat.) If average yields grow at 1.5 percent per year, it would take 110 years to achieve a factor of five increase in yield, which is unlikely to be needed because some of the increased demand for food is likely to be met by increasing the number of harvested acres or in other ways.

"Artificial" farming, or non-traditional foods (e.g., algae—perhaps for animals only) would be an additional source of supply if they can compete.

Although it is not demonstrated (and not necessary for the issue considered here), if one has to bet whether grain prices in constant dollars in the year 2000 or 2100 are higher or lower than today, I believe that one would do as well to bet "lower" as to bet "higher." Certainly, there is no basis for thinking that they may be many times higher.

In brief, since we can point to the sources of a fifteenfold increase in the renewable output of conventional agriculture, there is little reason to doubt that food production can keep up with demand, on a renewable basis, at least for a world population several times as large as that experts expect after the current surge of population growth ends in the next century—about ten billion (World Development Report). World population growth rates increased over several centuries from near zero to a peak near two percent in the early 1960s. They are now declining and are expected to be near zero again by about the middle of the next century. It is generally expected that the growth rate will have been over one percent for about one hundred years. (The special September 1974 issue of *Scientific American* on population described the "demographic transition" to a high population with low birth and death rates. This has been supported by subsequent data and represents the expert consensus.)

Energy. Energy is overwhelmingly the main supply issue. The cost and sources of energy for the long run are very uncertain. Here we address only the limited question: "Can we be reasonably confident that our huge long-run needs for energy can be supplied at an acceptable cost?" (I assume that fossil fuels will not be sufficient and that nuclear energy might be excluded for political or other reasons.)

Currently, the world spends ten to fifteen percent of GWP on raw energy. In the wealthy worlds of the long-term future (demand would be much lower if the world were not wealthy), we could afford to spend at least thirty percent of GWP for raw energy—probably fifty percent. If energy prices rise, we will use less energy per dollar of GWP. Therefore, we can probably afford unit energy costs four to six times as high as today (assuming the rise is spread over a century or more). (Table 1-6 on the next page shows today's world energy budget and some alternative possibilities for the year 2200.)

Approximately one-tenth of one percent of the sunlight striking the Earth is sufficient to meet all of our energy needs. (Which is not to say that we are limited to the solar energy normally striking Earth.) Solar energy can be converted to any form of energy we need. The problem about solar energy is cost. The pattern of energy-use in an all-solar energy economy would be so different that it is not very meaningful to compare unit costs with those today. But it is the best we can do.

One answer to our question is that we don't have to worry about being able to meet our energy needs affordably if the average unit cost of solar energy two hundred years from now is no more than four to six times current average energy costs (about $4/MMBTU [Million British Thermal Units]). Nobody has squarely attempted such a specious cost-estimate. But there are a number of programs today to try to make several forms of solar energy competitive with current sources of energy in the

Table 1-6: World Energy Budgets

	GWP	*Energy effic.*	*Energy amount*	*Price of energy*	*Total cost of energy*	*Energy's share of total GWP*
	(1)	(2)	(3)	(4)	(5)	(6)
			col. 1 / col. 2		col 3 X col 4	col. 5 / col. 1
	(tril.* dol.)	(GWP per Q)	Q**	(tril $ per Q)	(trillion dollars)	%
Actual Budget						
1980	$10	30	.3Q	$4	$1.2	12%
Alternative Hypothetical Budgets for Year 2000						
"W"	$500	120	4 Q	$16	$64	13%
"X"	$500	60	8 Q	$60	$128	26%
"Y" (basecase)	$500	60	8 Q	$8	$64	13%
"Z"	$500	30	16 Q	$8	$128	26%

* A "trillion" is one million times one million (it is also one thousand times one billion).
** A "Q" is a very large amount of energy. It is equivalent to the energy in 180 *billion* barrels of oil, or 10 to the 18th power BTU; that is, the output of about 30,000 big electric power plants used full-time all year long.

next few decades. One must be skeptical about whether these programs will succeed, but it is reasonably conservative to believe that in two hundred years large-scale solar energy costs will not be more than four to six times as high as average energy costs today (that is, not in excess of $20/MMBTU in constant dollars).

Saying that solar energy can provide the energy we need does not imply that we will or should use much solar energy, but solar energy can do the job if there is no better solution.

Therefore, there is not much basis for thinking that we eventually have to be in trouble because of inadequate energy supply.

Everything Else. The $170 per year per capita spent on other raw materials includes $100 for all minerals (Mineral Commodity Summaries), plus $40 for wood and $30 for stone, sand and gravel. The only individual minerals on which more than $6 per year is spent are iron and copper. (See Table 1-7 at top of next page.)

All metals together cost about $60, including iron and copper, which comprise about half of the total cost and 90 percent of the quantity of metal. Most of the other half of the cost of metals is for the fourteen metals (including precious metals), each of which takes between almost one percent and ten percent of the total. This leaves only about $1 per person for all other metals.

Other non-fuel minerals require $40, most of the cost of which is divided more or less equally among the following six minerals: lime, phosphate, potash, cement, gypsum, and sulfur.

For at least one hundred years, the cost of minerals as a group, and most individual minerals, has been declining (Barnett and Morse; Smith).

Iron, aluminum, and other minerals capable of doing the jobs done by almost all of the minerals we use are so widely present that it not conceivable that we would run out of them. (Iron is five percent, and aluminum eight percent of the Earth's crust.) Competing products put a ceiling on the cost of copper—probably below $5 per pound.

Thus, the key point is that the U.S. is now spending at most $10–20 per year per capita for all the minerals that we might conceivably "run out of" and for which there are no good substitutes. It is very hard to imagine circumstances in which costs for a major mineral—or for doing without it—would rise by a factor of ten. It has never happened. But we would not be seriously hurt even if the cost of all such minerals, or the cost of doing without all of them, rose by as much as one hundred times.

In summary, because we now get the minerals we need with such a small share of our efforts (e.g., $100 per capita for Americans), and since many of them are common materials,

Table 1-7: Approximate Per Capita Use of Minerals in the U.S. (typical year in early 1980s*)

	Quantity	*Price*	*Cost*
Iron	600 lbs	$ .02/lb	$ 12
Copper	20 "	.75 "	15
Other high-volume metals bauxite, lead, titanium, zinc, and manganese	80 "	.15 "	12
Expensive metals nickel, tin, cobalt, moly tungsten, and magnesium	3 "	3.00 "	10
Precious metals gold, silver, platinum	.6 oz	20.00/oz	12
Misc. non-metals (primarily: lime, cement, sulfur, potash, gypsum, and phosphate)	2,000 lbs	.02/lb	40
Total (wtd. avg. price)	3,000 lbs.	$ .03/lb	$100

(columns do not add because of rounding)
Excludes stone, sand, and gravel (20,000 lbs, $30).

* Many of these prices and quantities fluctuate substantially from year to year. Since the only purpose of the table is to give a general feel as to the relative importance of different items, and a sense of scale, this table smoothes the fluctuations and calls itself a "typical year."

the outer limit of the effect of scarcity of minerals, taking the worst case possible, is equivalent to an annual tax of less than five percent of GWP (and most of the probability is an order of magnitude less).

Conclusion. The conclusion is that raw material scarcity is not a serious danger to society and, therefore, need not concern the citizen. Raw material supply is a matter for businessmen and experts to work out, and for them it presents a number of fascinating questions and great challenges—which are not covered here, and which must be analyzed quite differently.

Pollution Doesn't Jeopardize the Passage to a Human World

Thousands of scientists are studying questions related to pollution problems. And the problem is getting bigger and more complicated every year as the number of people and the amount of production increases and the number of new chemicals multiplies. No one can hope to understand all of these complexities. But let's see whether we can get a reasonable handle on the human implications of pollution without going very far into these technical issues.

Before we get started on a realistic discussion of the pollution problem, we can do an exercise to gain perspective:

How clean is your house? If you spent more time and money on keeping your house clean, wouldn't it be cleaner?

How dangerous is your house? How many smoke detectors and burglar alarms do you have? Have you checked to see whether there are internal air contaminants? Is all your electric wiring properly grounded?

Even though the world's problems are not so simple as keeping a house clean and safe, we can learn something about the general nature of the world's problems by thinking about people's houses. Basically, people's homes are as clean and safe as they choose to make them. They could be made safer by spending either time or money (or effort to learn how). So we choose our level of safety by deciding when to stop spending time and money on making our houses safer.

Although the world environment is infinitely more complicated than our homes, our environment, like our home, will be as clean and safe as we choose to make it. Almost certainly the thing that will keep it from being cleaner and safer than it is will not be some inherent limit of nature, but human decisions not to exert additional effort to make it cleaner and safer.

But there are three "catches" that ought to concern us about the analogy between the world's pollution problem and our home environment.

Catch number one: Sometimes we fail to avoid a danger because we don't know about it. Who knew, for instance, that installing asbestos insulation might be dangerous? I call this problem "sneaky pollution," and it is discussed below. There is no guarantee against sneaky pollution, and no clearly safer strategy for minimizing the risk.

Catch number two: We know that the cost of keeping our house clean is reasonable, but how can we know that the cost of keeping the environment clean is reasonable, especially in the distant future? Good question. See below for a pretty good answer.

Catch number three: I can decide to spend the necessary effort and money to keep my own house clean (at least if I live alone), but our individual decisions aren't enough to keep the environment clean. Protecting the environment requires cooperation among many people, sometimes people in a number of different countries. This is the "common property" problem, and it makes things much harder. But it is not by any means the only or worst "common property" problem. We are dependent on decisions of others for many other things that are as important to us as the quality of the environment.

What Does Experience Suggest about Pollution in the Future? There are no obviously decisive statistics. Some places are getting better at the same time other places are getting worse. And since pollution includes many different things, any one place may be getting better in some ways and worse in others. Since so many facts can be shown pointing in opposite directions, the only way to think about what the trends are is to try to understand the overall patterns.

One common pattern is old forms of pollution disappearing and new forms arising. Horse manure in city streets and soot from residential use of soft coal were once big problems, but are no longer. Dysentery germs in urban water supply were once an even bigger problem, and now have essentially disappeared from advanced countries. But now sulfur dioxide and carbon monoxide from cars and modern industrial plants are problems.

Another common pattern results from economic development. When a country moves from primitive, very poor conditions to the very first stage of economic development, pollution usually is reduced, as plumbing and sewer systems and other basic measures for cleanliness become understood and affordable. The next stage of development often brings industrialization and increases in pollution. Later, the industrialization often produces enough wealth so that the country wants to, and can afford to, take measures that reduce pollution.

President Carter unintentionally illustrated this pattern with a story he once told about Japan. He said that he had heard that Tokyo's massive program against air pollution was initiated after the Emperor had noted sadly that there were no longer butterflies in the Imperial Gardens. And now, as a result of those programs, the butterflies are back in the Emperor's garden. President Carter's story illustrates a basic pattern that is likely to be repeated in many countries: first butterflies and poverty, then development and no butterflies, then wealth and butterflies.

How Should We Understand the Basic Idea of "Pollution"?

The idea of pollution includes two different kinds of things: dirtiness and danger. We spend money to reduce pollution because we want to be cleaner and safer. And we can include wilderness preservation, the protection of scenery, and other concerns for the amenity values of the environment, as forms of cleanliness.

While safety is often talked about as an absolute requirement, modern science can detect such infinitesimally small "dangers," that the decision about how much "health-threatening" pollution to allow also becomes a matter of preference or efficiency, not an absolute.

Taking all these considerations into account, it turns out that there are two quite different kinds of pollution problems that have to be dealt with. I call these "mass pollution" and "sneaky pollution." With "mass pollution," you know what is the pollutant and that you don't like it; the problem is spending the resources to get rid of it. With "sneaky pollution," you don't realize how bad it is until it is in the environment; the problem is knowing in time.

As a practical matter, of course, it is not so simple. Usually some people know the facts while others are ignorant or disagree. And sometimes only part of the harm from a pollutant is understood. Sometimes there is genuine technical uncertainty; sometimes there is controversy because people have different interests and values. And people can even disagree about whether something is harmful, harmless, or useful—like fluoride in the drinking water. But even though in some cases a mass pollutant also will be somewhat sneaky, it is useful to think separately about: (a) the problem of how to get something that is recognized to be "garbage" out of our environment; and (b) the problem of how to recognize what really is "garbage" (because if we let too much of it get into our environment, it will cause harm. Where "harm" means either dirtiness, ugliness, unhealthiness, or danger).

Mass Pollution. Most pollution is "mass pollution"; that is, we know what is causing our problem—either dirtiness, danger or both. But it is expensive to get rid of or prevent the pollution.

The basic fact about mass pollution is that in the long run we will have only as much as we "decide" to have. It is our choice. Cleanliness is like other products: you can buy as much as you want and have money to pay for. Of course, the avoidance of pollution, which is community cleanliness, is a "common property," like defense or a stable currency. Individuals can't buy such common properties, only groups can.

There are several problems that come from the fact that community cleanliness is a "common property." The community may buy more or less than any individual would prefer. You can't please everyone. Also, the purchase of cleanliness is more complicated than buying potatoes. It is often hard to know what is the efficient way to buy it, and the best way may be one hundred times cheaper than the obvious way.

Each country or region (and for a few pollutants the world) has to decide how much cleanliness and safety it wants, and what actions are the best way to buy it. In practice, obviously, it is much more complicated to make this set of decisions than I make it sound, and often they are not made nearly as well as they might be.

Decisions about how to produce a common property like environmental cleanliness are made by a political process. And political processes are not usually good at finding the efficient answer to technical problems because they are designed to preserve domestic peace and achieve democratic or other political values—not to do good research. Therefore, we may have a lot more pollution than many people want, and/or waste a lot of money on foolish anti-pollution measures. But there is no reason to believe that political incompetence will produce disaster in pollution control before it produces disaster in war avoidance or tyranny avoidance or depression avoidance.

The Cost of Mass Pollution. In 1975, Resources for the Future (RFF) arranged for Ronald Ridker and William Watson to do a four-year study of the resource and environmental consequences of different growth paths that the U.S. might follow during the next fifty years. The results of this study were published with the title *To Choose a Future* (Ridker and Watson).

The Ridker & Watson team (R&W) estimated the amounts of materials and energy that would be needed for many possible combinations of population and economic growth in the U.S. during the fifty years from 1975 to 2025. They then estimated the amount of the major forms of mass pollution that might be produced by all of the necessary mining, manufacturing, power generation, etc. They also considered the effect of many different pollution control laws.

Of course, the mass pollution problem imposes two kinds of costs on us. First, there are the "pollution control costs" spent to prevent the pollutants we produce from getting into, or staying in, the environment. Pollution control is like a production task. It involves designing and procuring pollution absorbers, pollution sinks, pollution-avoiding production technology, etc.

The second kind of cost of the pollution problem is "pollution effects costs," including things like cleaning costs, damage to property as a result of pollution, health-care costs, and the value of reductions in health caused by pollution.

Most of both kinds of pollution costs are very difficult to estimate, even when you know all the facts. But even though no confidence can be placed on any one estimate (for example, the cost of removing sulfur dioxide produced by three hundred big coal-burning electric generators in 2025), the sum total of the many thousand estimates is reasonably useful, even if each separate estimate is almost worthless.

Many possible mistakes and unpredictable factors produce underestimates, while others produce overestimates. If the estimators are not trying to push the results one way, the errors in each direction tend to balance each other.

What Ridker and Watson did to pull everything together was to make about seven different scenarios for the next fifty years. For each possible combination of production scenario and pollution control law, Ridker and Watson estimated (a) the total cost of the pollution control programs, and (b) the expenses caused by the pollution that was allowed to remain in the environment. The sum of these is the total cost of pollution, as best as can be calculated.

For the United States in 1975, for example, where R&W had real costs to observe, pollution control costs were about $18 billion and pollution damage costs were about $48 billion. In other words, R&W estimated that in 1975 pollution cost the U.S. a total of $66 billion, which was like a tax of seven percent on everything we bought.

R&W calculated that the "pollution tax" will be between 4.4 percent and 6.4 percent in the year 2025. Their detailed analyses show total pollution costs going down regardless of which growth path our economy follows, and whether our pollution control laws are at the relaxed or strict end of the range of possibilities that seemed reasonable to them.

The R&W conclusions were: ". . . environmental damages from the mass pollutants covered in this chapter are likely to remain the same or fall over time despite the growth in the economy and greater number of people at risk. . . . In all cases, after 1985 they fall over time on a per capita basis and as a percentage of consumption."

In other words, no matter how much the U.S. grows, we can have less pollution damage than now by paying the costs required to keep clean—and the costs won't be very high.

Ridker and Watson stands today as by far the single most thorough and substantial effort to evaluate long-term pollution costs. It is the best evidence we have, and it says quite clearly that for the next fifty years in the U.S., regardless of how fast we grow, mass pollution will cost us less than it does today.

If R&W are correct that pollution costs are going down over the next fifty years in the richest large area of Earth, it is a powerful piece of evidence that further economic growth in the world in the long term does not have to increase the pollution tax that people need to pay to maintain the quality of the environment.

The fact that it is difficult to make good decisions about buying cleanliness doesn't change the basic situation; we the people, not nature, determine how much mass pollution there is. Nature will allow us to have all the raw materials we want at reasonable costs with about as little pollution as we choose and without spending a noticeable share of the economy for pollution control. While many people *seem* to contradict this basic point, in fact they have not examined this question and do not challenge this conclusion.

There is no evidence or sensible theory that we will have to pay very much to have an environment that is in better condition than the one we have today. This conclusion holds for tens of billions of people and very high living standards, with very large consumption of energy and all else that goes with it.

"Sneaky Pollution." With mass pollution the problem is deciding how much cleanliness to buy, and buying it efficiently; with sneaky pollution, the problem is knowing what's happening.

Here is where all those who insist that there must be something to worry about can have their day. "Sneaky pollution" is defined as a category to give room for most of the "what if . . . ?" questions that can't be answered.

We can't know for sure all the effects of what we do. I define "sneaky pollutants" as any phenomena that turn out to cause more trouble than was originally thought. So the idea of "sneaky pollution" brings together all the harmful effects that we are unable to predict. The concept is broader than just chemicals. Many germs or bacteria as well are good examples of sneaky pollutants. Until we find out about them, they cause many deaths.

One dramatic example of a sneaky pollutant was methyl mercury. At one time, people believed that it wasn't absorbed by the human body and, therefore, that industrial wastes containing methyl mercury were not dangerous. The error was discovered only after hundreds of Japanese and others received deadly or disastrously harmful doses. Once the cause of "Minamata disease" was discovered, the discharge of methyl mercury was virtually stopped almost overnight.

Various forms of natural pollution that produce malaria or dysentery or other diseases have been quickly limited once the danger from them was recognized, that is, as soon as they were recognized as pollution. Such natural sneaky pollutants have caused millions and millions of deaths.

We are especially interested in man-made sneaky pollutants. We want to know whether we will make things or do things that are surprisingly dangerous, perhaps cumulatively over a long time, or to later generations. Will we do things that, if we knew what the results would turn out to be, we wouldn't do them? To be blunt, will we make mistakes that kill many people? The answer is "yes." Our mistakes have caused millions of deaths in the past, and probably will in the future.

By definition we cannot stop sneaky pollution—because we don't know what it is. As soon as we recognize it, it becomes mass pollution. So how can we estimate how big the problem of sneaky pollution will be and what can be done about it, since by definition, we don't know what the problem is? More important for our concerns, does the problem of sneaky pollution get worse if the world population gets very large and very rich?

The short answer is that we can't know how bad sneaky pollution might be. It is entirely possible that some form of sneaky pollutant will kill everybody and end the human species. And it is quite likely that sneaky pollutants will in the future, as they have in the past, kill millions of people.

Therefore, I want to be very clear that I do not mean to say that we don't have to worry about sneaky pollutants. There is no assurance that they will not cause immense harm. But, on the other hand, there is no reason for thinking that sneaky pollution is anything like the most dangerous problem we face. (The phrase "no reason for thinking that" means exactly what it says; it does not mean "it is definitely wrong to think that.") And there is no way of knowing whether more population and economic growth will increase or decrease the total harm from sneaky pollution.

Three things determine how much harm we will suffer from sneaky pollution. First is the number and nastiness of sneaky pollutants. Second is the speed with which technical people expose the sneaky pollution, that is, how quickly they learn that a mistake was made. Third is how quickly the society acts to correct the mistake once it is discovered.

It is reasonable to think that the more the world economy grows and advances technically, the more sneaky pollutants of every degree of nastiness will be produced. But that doesn't mean that sneaky pollution will cause increasing harm. The decisive question is whether our ability to respond to the sneaky pollutants grows faster than the sneaky pollutants do.

Presumably, our ability to find the sneaky pollutants will increase as the level and quantity of our science and technology rises. But after scientists found that feeding citrus fruits to sailors would prevent scurvy, it was thirty years before the British Navy began providing limes to their sailors. And it was another forty years before the "new technology" of limes was transferred to the British Merchant Marine, although frequently 25 percent of a ship's crew might die of scurvy if the ship did not provide limes or some other source of Vitamin C.

It is harder to guess whether, when the world economy gets bigger and richer, we will become quicker or slower to act against pollution dangers after they are uncovered by scientists. It depends on what might be called "social wisdom." And anyone who thinks he can predict whether social wisdom will increase or decrease in the future is welcome to the job. But, other things being equal, people are more able to recognize dangers if they can see what needs to be done to deal with them and if they can afford to do it. So increasing wealth means that a little less social wisdom is necessary to respond to newly-recognized pollution mistakes. Also, as people get wealthier they usually decide to spend increasing shares of the income on safety and aesthetics. This is another way in which wealth may speed the response to sneaky pollution.

In brief, sneaky pollution is a problem to which there are no guaranteed answers. It will be costly and it may be disastrous. But we have no good reason for thinking that it is a problem that will get worse as we complete the passage from natural to human worlds. Clearly, we are likely to produce more sneaky pollutants, but equally clearly, our ability to reduce the harm from any particular sneaky pollutant also will grow. It is the balance between these two effects that is unpredictable—although I would bet that our ability will increase more than the challenge.

Conclusion

The necessary concluding comments are contained in Table 1-8 and Table 1-9.

Table 1-8: Integrating the Two Perspectives

Dilemma

1. We are rapidly changing from a natural to a human world.
2. But all these "terrible problems and dangers" exist.

Resolution

1. For each "terrible problem or danger" ask:
   "Does it really threaten the passage to a human world?"
   (It almost certainly won't.)
2. If not:
   Work on the problem,
   But put it in the correct perspective.
   - Physically, we can expect to be okay
   - Work and wisdom can speed the passage and reduce the costs
   - Overreaction and radical change usually make things worse
   - The U.S. is part of the solution not the problem

Table 1-9

Some things I am not saying

1. We're safe. There is no danger.
2. Nothing needs to be done.
3. We know the future.
4. We now know exactly how to solve all our problems.
5. "Progress" is inevitable. (I.e., "the world keeps getting better.")
6. The world will be better in the future.

What I am saying

1. The threats to mankind from scarcity, pollution, or any limits imposed by the physical condition of the earth are small.
2. Our basic expectation should be that almost all countries will be wealthy within a century or possibly two.
3. Mankind's real worries are human—not physical.
   - Decadence (too much wealth, power, freedom, & knowledge)
   - War
   - Tyranny (especially communism)

Appendix

Here are facts that refute the Doomsday Propositions:

The overall physical condition of the world is not getting worse. Obviously, some things are getting worse, but the following indicators seem to be the most significant, and they show the world getting better for people:

- Worldwide life expectancy is increasing (also in almost all countries except the Soviet Union).
- Amount of economic resources available each year per person is increasing (worldwide and most countries).
- Per capita food production is increasing.
- Famine deaths are decreasing.
- Most raw material costs—with the possible exception of fuel and forest products—are not increasing.
- In at least some wealthy countries, overall pollution levels are declining—which suggests that as other countries become wealthy, their pollution levels also will be reduced.

The world is not getting more crowded—except in the meaningless sense that Greenland would be more crowded if its population doubled. There is no measure of "crowding," if the word implies harm or discomfort. Usually, the amount of living space available to a person depends on his or her income and preference. Therefore, as real incomes rise, any increase in crowding will be that preferred by people.

Increasing resources—and knowledge—means reduced vulnerability.

It may be possible that the world will have raw material scarcity problems; but there is no evidence that it is inevitable, and the best evidence is that it will not happen at all (e.g., *Scarcity and Growth,* Barnett & Morse). Of course, some raw materials may become scarce, but not enough to be troubling to human life.

All demographers agree that current population trends are in the direction of something like a leveling-off of world population below twenty billion people. If "carrying capacity" is a meaningful concept, twenty billion is way below it.

Standard U.N. and other figures show that arable land is increasing (net of losses).

Since the share of effort and resources used to get raw materials (and protect against pollution) is declining, and probably will continue to do so, adjusting to scarcity will not be a major problem. The major problems the world will face are the human problems of adjusting to wealth and power. (And the major physical problems will concern economic organization, not raw material supply.)

References

Barnett, H.J. and C. Morse. 1963. *Scarcity and Growth.* Baltimore: Johns Hopkins University Press.

Buringh, P., H.O.J. van Heemst, and G.J. Staring. 1975. *Computation of the Absolute Maximum Food Production of the World.* Wageningen, Netherlands: Agricultural University.

Chou, M., D.P. Harmon, Jr., H. Kahn, and S.H. Wittwer. 1977. *World Food Prospects and Agricultural Potential.* New York: Praeger.

Club of Rome. 1972. *The Limits to Growth: A Report for the Club Of Rome's Project on the Predicament of Mankind.* New York: Universe Books.

Council on Environmental Quality and Department of State. 1980. *Global 2000 Report to the President.* Washington D.C.: Government Printing Office.

Douglas, Mary and Aaron Wildavsky. 1982. *Risk and Culture: An Essay on the Selection of Technological Dangers.* Berkeley: University of California Press.

Efron, Edith. 1984. *The Apocalyptics—Cancer and the Big Lie.* New York: Simon and Schuster.

Ehrlich, Paul R. 1968. *The Population Bomb.* New York: Ballantine Books.

Kahn, Herman. 1980. *World Economic Development, 1979 and Beyond.* Boulder: Westview Press.

__________, W. Brown, and L. Martel. 1976. *The Next 200 Years.* New York: Morrow.

Mesarovic, M. and E. Pestel. 1974. *Mankind at the Turning Point: The Second Report of the Club of Rome.* London: Hutchinson.

Mineral Commodity Summaries. 1985. Washington D.C.: U.S. Department of the Interior Bureau of Mines.

Paddock, William and Paul Paddock. 1967. *Famine—1975! America's Decision: Who Will Survive?* Boston: Little, Brown and Co.

Ridker, Ronald, and William Watson. 1980. *To Choose a Future: Resource and Environmental Consequences of Alternative Growth Paths.* Baltimore: Johns Hopkins University Press (for Resources for the Future).

Sanderson, F.H. 1984. "World Food Prospects to the Year 2000." *Food Policy* 9(4).

Simon, Julian L. 1981. *The Ultimate Resource.* Princeton: University Press.

__________ and Herman Kahn. 1984. *The Resourceful Earth.* Oxford and New York: Basil Blackwell.

Smith, V. Kerry (ed.). 1979. *Scarcity and Growth Reconsidered.* Baltimore: Johns Hopkins University Press (for Resources for the Future).

World Development Report. 1982 (and subsequent years). Oxford: Oxford University Press. (Published for the World Band).

TWO

MAN'S IMPACT ON THE ENVIRONMENT IN HISTORICAL FUJIAN (Southeast China)

Fred Henglun Sun

Introduction

In the past quarter-century, deterioration of the environment due to industrialization has come to be perceived as an important problem throughout the world. There has been much pollution of air, water, soil, and life. To more clearly diagnose the environmental problems and to seek a way to solve them, it is not unprofitable to review the process of historical evolution.

In China, there is a recorded history of five thousand years. The Chinese initiated the idea that "Man is interrelated with the earth" early in the Chunqiu period (about 700 B.C.). At that time, the authorities promulgated a series of laws for protecting biological resources, known as "Prohibitions in Four Seasons" (they concerned prohibitions of hunting, fishing for special species of animals and cutting down of trees in each season), appointed some officers, *"yusi,"* for protecting forests and provided for surveys of mountains, rivers, lands, and forest resources. In part, these resulted in the preservation of a number of historical documents about the environment and made possible the study of environmental changes over the last millennia.

Based on the research results to date of historical geography, archaeology, and Quaternary geology, the author will review the environmental processes since ca. 4000 B.P. in Fujian (Southeast China), discuss the evolution of environmental problems in different periods, and search for ways to protect the environment.

Some Aspects of Environmental Changes

Fujian is located on the southeast coast of China. Opposite is Taiwan, separated by a strip of sea, Taiwan Strait (Figure 2-1). The latitude of the region is between 23–28°N. It is a subtropical monsoonal area and the climate is warm and moist. The topography in the region is rugged and only five percent of the area is plain. Most of the rivers flow from northwest to southeast following the terrain, through the coastal plains and into Taiwan Strait. The shoreline of Fujian is zigzag with considerable extensive islands and is well-known as a typical ria coast. The proportion of forest-covered area is about 39 percent, including evergreen broadleaf forests in the north and monsoon forests in the south.

This physical environment is a consequence of environmental evolution over the historical period, as well as a background of modern environmental problems. With the development of human society in the past millennia, some of the conditions in the environment were improved to meet the needs of human beings, but others were aggravated.

Climatic Trends

On the basis of pollen statistics of several Holocene sediments from the Fuzhou Basin (H. Zheng) and the south Fujian coast (C. Chen), the trend of climatic change over the last 5,000 years was from warm-wet to cool-dry. At 5000–3000 B.P., sporopollen association in the sedimentary cores from Fuzhou

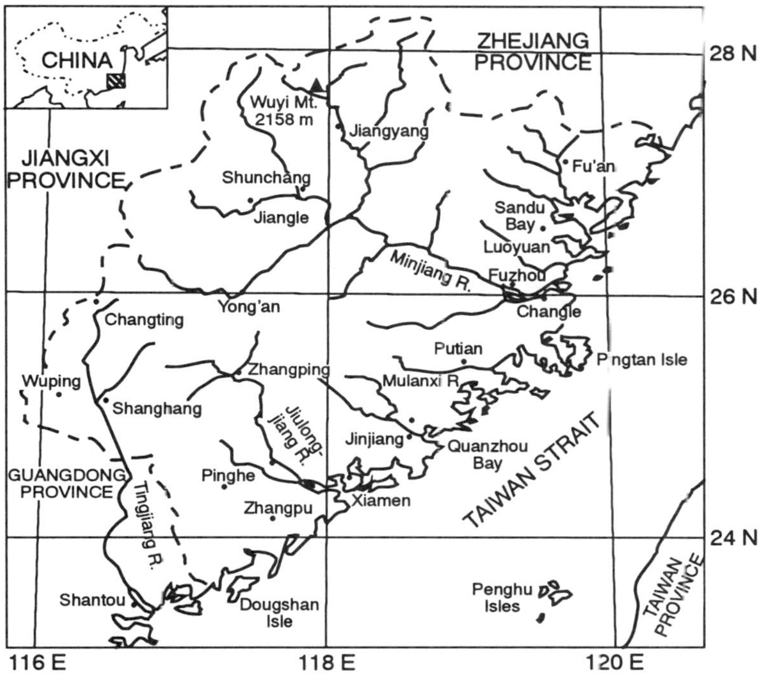

Figure 2-1: A map of Fujian Province, Southeast China.

was characterized by a *Castanopsis-Quercus* zone. The content of arbor pollen is 53.1–76.9 percent, showing a south subtropical type with evergreen monsoon forests. At Xitou neolithic relic site in west Fuzhou (Figure 2-2), some fossils of tropical mammals, like *Rhinoceres*, *Elephas maximus*, and *Rusa unicolar Keer*, were excavated and dated at 4300 B.P. (FPM). It is estimated that at that time the average annual temperature was 1–2°C higher and the annual precipitation was 200–300 mm more than that of today (Zhang and Lin). Similar fossils of tropical mammals were found in Tangshishan neolithic relic near Xitou with a carbon 14 date from shell of 3274–155 B.P. This compares well with the records inscribed on bones or tortoise shells, that there was an elephant hunt in about 1365–1386 B.C.

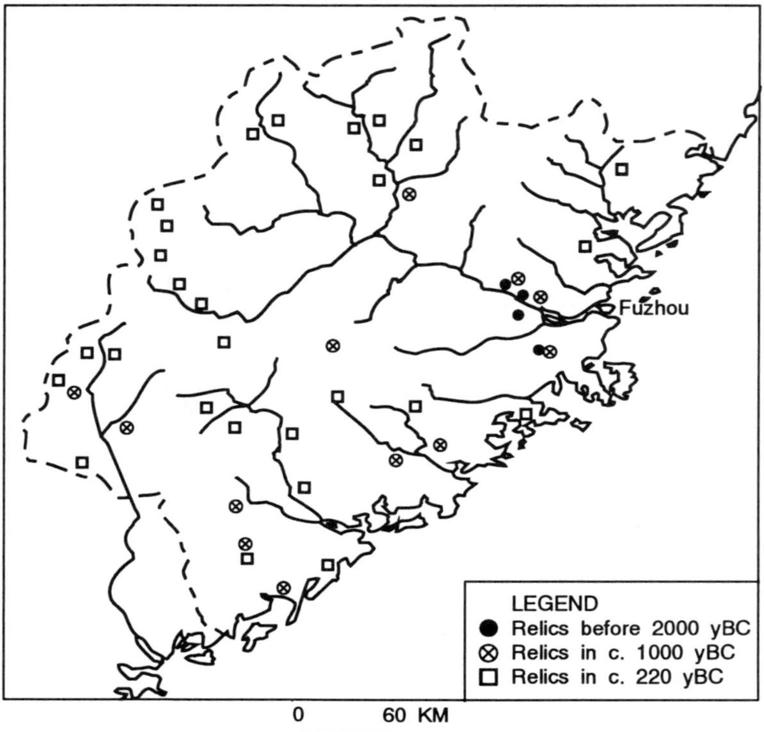

Figure 2-2: A distribution map of ancient relics in Fujian Province. 1 Tangshishan 2 Xitou 3 Hongwei (revised after Fan Zeng)

in Henan of North China plain (K. Zhu *et al.*), indicating that this period might have been hot-wet in climate. Since about 3000 B.P., the climate in Fujian has tended to be dry and cool. In Fuzhou Basin, the pollen association in sedimentary cores was characterized by a *Quercus-Pinus-Gramineae* zone. This shows a sub-tropical landscape of coniferous, broadleaf mixed forest-steppe. A similar discovery was made on the south Fujian coast.

In the course of the cooling, there were some stages when the climate became relatively warm and wet. During the Tang-North Song Empire (about 600–1100 A.D.), there was a remarkable stage. At the time, rivers did not freeze in many winters and orange trees and plum blossoms were widely viewed in Xi'an in north China; from the historical records many wild elephants were active in south China. In east Guantong, which is bounded by south Fujian, it was said that there were so many elephants that local people hunted them for their trunks as delicacies. "In July, 931 A.D., an elephant moved into Quxian county, Zhejian Province, and was tamed and raised by soldiers" (M. Zhu). In Zhangzhou (south Fujian), "there were many elephants habitually (during the Song Dynasty); the groups with ten or more elephants were harmless, but one who met with a single elephant would be caught and trampled to pieces" (W. Zhu). Litchi is a typical fruit in subtropical areas and very sensitive to low temperature and frost. Wang Anshi, a famous politician in the Song Dynasty, recorded in his work "Lin Chuang Ji," "(Litchi) was transplanted to Changan (today's Xi'an) during Han Wu Empire (about 202 B.C.), it was obtained from Jiaozhou (today's Guangxi, south China) during Yongyuan Age of the East Han Dynasty (about 89 A.D.), from Peizhou (today's Sichuan) during Tianbao Age of the Tang Dynasty (about 742 A.D.) and from Fujian during the Song Dynasty" (W. Zhu). That the litchi-growing area moved from north to south, then from south to north, seems to correspond to the process that the climate changed from warm to cool, then from cool to warm. A well-known calligrapher, Caixiang, wrote in "Litchi Manual" (ca. 1059 A.D.), "(litchi) was planted only in four prefectures of Fujian. Fuzhou was the first in output, but Xinghua's (Putian) was the best for taste and Quanzhou and Zhangzhou had been well-known at that time." "(Litchi) was most extensively planted in Fuzhou area and spread over open country" (W. Zhu). The climate of that period may have been relatively warm. In the twelfth century, a cool age occurred when litchi in Fuzhou suffered seriously from freeze injury and Taihu (Tai Lake) froze in winters. In light of records on natural calamity, winter temperatures obviously dropped and the frequency of freezing damage increased in east China after the sixteenth century. Some major cold periods occurred in the early sixteenth century, mid- to late-seventeenth century, and the late nineteenth century (D. Zhang).

A drying tendency was apparent in southeast China, according to the statistical record on flood and drought calamities over the last two thousand years. Particularly in the past one thousand years, dry stages were prolonged and wet stages shortened (S. Zheng *et al.*). These may correspond to the records of desertification and the results of pollen analysis.

The climate in the twentieth century was cool, following the cooling age of the late nineteenth century. The rainfall was less than before. 1928–1954 were decades when the climate became relatively warm and dry. From the 1950s to the 1970s, the climate turned cool and dry again, the average annual temperature decreased by 0.5°C and rainfall decreased by fourteen percent comparing with that of the 1940s (FMO).

Retreat of Forests

The conditions of water, heat, and soil in Fujian are advantageous to the growth of forests and, thus, also to reforestation. Based on pollen analysis from Holocene sediment, evergreen broadleaf-deciduous broadleaf mixed forest, which was dominated by *Quercus* and *Castanopsis,* covered even the Fuzhou Basin and South Fujian coast. The ancient inhabitants of the neolithic age lived on valley terraces, twenty to eighty meters above sea level (Zeng). The animal fossils excavated from Tangshishan relic show that ancient inhabitants living at the foot of the hill beside the stream lived on hunting and fishing, indicating the existence of dense primeval forest which was essential to their subsistence.

The earliest record of deforestation may be recognized from the pollen analysis of Xitou relic. Of 54 pollens discovered in the specimen, 83 percent are pteridophyto spores (60 percent are *Ptoris* and *Pteridium*), and only nine percent are woody pollens (FPM). That fact implies that some valleys were opened up by human beings, forests were replaced by herbs such as pteridophyto and migrated to the adjacent mountains. The charcoal found in Hongwei neolithic nearby (carbon 14 age 3470±220 B.P.) is also evidence of forests being burned out by mankind.

In the light of the distribution of archaeological relics (Figure 2-2), ancient people inhabited the valleys along the lower reaches of the Min River four thousand years ago, spread into the big valleys and basins along the Min, Ting, and Jin rivers about three thousand years ago, and spread over all the valleys and coastal plains by about two thousand years ago. The extension of the areas of habitation implies the process of forest retreat at that time.

It is evident that primitive vegetation was not severely destroyed before the Tang Dynasty (before 618 A.D.). *Han Shu* (A History of the Han Dynasty) reported a campaign in Fujian (about 135 B.C.) which had to march thousands of miles of passageway between dense forests and groves of bamboo where there were various snakes and beasts. Immigrants from north China entering Putian (a coast county, later densely populated), where there were "remote, thickly forested mountains" (ca. 420–589 A.D.) had to "carve trees to mark their routes." "*San Shan Zhi* (A History of Fuzhou)" records that in early Tang Dynasty, "since the population was sparse, cultivated area was limited to within the city walls. There were extensive mountains and valleys covered with dense forests. It was an untraversed region, but a habitat for tigers, leopards and monkeys" (W. Zhu).

In the period of Tang and Song Dynasty, the population in Fujian increased rapidly. It was recorded that the total population in 607 A.D. was about sixty thousand, and in 1300 A.D., more than 6.21 million (J. Chen). Encouraged by the government, reclamation was common. Forests on gentle slopes along the valleys might have withdrawn. It was a rare period when the people of Fujian were more than self-sufficient in grain. Based on the sequence of the establishment of *zhou* and *xian* (two sorts of administrative division, similar to county), the trend of population moving in during the Tang Dynasty was from north Fujian to south and west Fujian, and from inland to coastal areas. This may imply a sequence of destroying primeval forests. The records concerning the activities of elephants in south Fujian during the Song Dynasty suggest that the land was thickly forested at that time and later was placed under cultivation.

Economic activities at that time requiring lumber may reflect the destruction of forests. There was large-scale construction including various city walls and temples that were set up in the period of the Tang-Song Dynasty. Fujian was a ship-building center of China over a long historical period. Quanzhou became the largest harbor in the East during the Song-Yuan Dynasty, handling the main exports of metal, silk fabrics, porcelains, tea, fruits, and crude drugs. These facts, and many related historical records and archaeological relics, imply a rapid deforestation at the time.

Wars and turmoils were said to be periodic catastrophes to forests, but a sustained factor of deforestation was slash-and-burn, a conventional way of reclaiming land in historical Fujian. The Taimu mountainous area (a famous scenic area in northeast Fujian) once was described as follows: "Some *She* people (a minority nationality in Fujian) set on fire and burned the grass on the mountains. The west wind blew swift and violent. Bamboos and trees burned and exploded like thunder. When I traveled through, it was clear that the forests for miles around would soon be ashes" (W. Zhu). The author, himself, saw the scene of burning and reclaiming in the mountainous area of Shunchang county in the early 1970s. Instead of setting an effective fire lane, the farmers just burned up large areas of forests, especially bamboo and coniferous forests. To his surprise, on the bare slope, there were extensive terraced fields which had been built in some ancient time. In the area of Bao Mountain, (with a summit height about 1,300 meters) near the boundary of Shunchang and Jiangle county, these ancient terraced fields covered the slopes of 30–40 degrees up to the elevation of 1,000 meters. It was said that these were tea plantations opened up in the period of the Qing Dynasty (about one hundred and more years ago).

The population in Fujian increased rapidly during the Qing Dynasty. It was 1.45 million in 1661 A.D. and 21.07 million in 1861 A.D. (J. Chen). The net increment was almost 20 million in 200 years. The population moved away from the coastal plain to the inland mountainous area in that period. The following are cited from historical books: "The farmers tilled plain fields before and till mountain land now. What they cultivated were sweet potatoes, tea, pine, bamboo, tung tree and China fir. They farm wherever they can to seek a livelihood. They are not afraid of precipitous cliffs or remote wildnesses." "Tea plantations spread all over the region of Yan Jian (north Fujian), no one can give their number exactly" (Huang). Fuzhou became one of the three largest tea markets in China around the 1860s. Annual export of black tea reached 650,000 *dan* (3,250 tons) at its height in 1882. Fuzhou, in the early twentieth century, also was one of the three largest wood markets in China. Paper made of bamboo was one of their major exports. A lot of open forest might have been lost in that period.

Since the 1950s, the scale of industry and agriculture has grown unprecedentedly. Because of insufficient protective measures, forest denudation was common and severe. The area of forest cover, the standing wood crop, and the fractional coverage by forests declined noticeably. In the Min River valley, the percentage of forest coverage was 56 percent in 1958 and 45 percent in 1978. In Xiqi River valley of Jin River, the crop of standing wood was 6 million cubic meters in 1957, 1.31 million cubic meters in 1973, and 1 million cubic meters in 1985. In pace with the retreat of forests, the areas of soil erosion expanded. In 1958, it was 6.75 million mou (0.45 million hectare) throughout Fujian province, in 1965, it was 11.35 million *mou* (0.76 million hectare), and now is about 20.34 million *mou* (1.36 million hectare) (FAH). In addition, the topsoil thinned, the frequencies of flood and drought increased, and the silt contents in surface run-off rose. These were all the results of forest destruction.

Species Extinction

Fujian is located in a wet subtropical region. While the topography is varied and none was ever covered by Quaternary glacier, it was suitable for occupation by and conservation of species. The fossils of various rare Quaternary mammals were found in the west and the south of Fujian and the Fuzhou area. They were east stegodon, giant panda, China rhinoceros, China tapir, and macaque, etc.

Animal bones excavated from Tangshishan relic included not only those of cultured livestock, such as pigs (*Sus domesticus*) and dogs (*Canis familiaris*), but also those of wild animals, such as *Ursus arctos L., Felistigris. L., Elephas maximus L., Cervus nippon Temminck,* and *Rusa unicolor Keer* (Qi). Other mammals found from Xitou relic were *Rhinoceres, Trachy pithecus sp.,* and

Hystrix sp., etc. (FPM). The existence of these wild animals suggests that there was a great variety of wild mammals and that they spread all over the inlands and coasts of Fujian three thousand years ago. By comparison, *Elephas maximus, Cercus nippon Temminck, Rusa unicolor Keer,* and *Trachy pithecus* are now found only in south Yunnan and Guanxi provinces of south China. They have disappeared from Fujian. *Rhinoceres* has disappeared even throughout China. The rock and bone arrowheads excavated from neolithic relics above-mentioned indicate that human beings became a vital threat to the wild animal as early as several thousand years ago.

Historical records have no lack of accounts concerning the existence of wild animals and their misfortunes. The officers in the early Tang Dynasty imposed levies for native products, such as "beewax, tiger's leather and ape's hide," on the *She* people in Zhangzhou area (W. Zhu). It was recorded in "Tingzhou Zhi (A History of West Fujian)" that "during the Dali Period (766–770 A.D.), there were several hundred monkeys gathering in the fir forests at the Gutian ares." "*Wu Xiang An* (no elephant temple)" was built at the foot of Kuifenling hill during the Song Dynasty (about the eleventh century); the story from "A History of Zhangpu County" is as follows: When there were a great number of wild elephants at the hill at that time, Zhuye, a Buddhist monk, built a temple at the foot of the hill. Many people settled around the temple and opened up the wildness. The elephants were frightened and soon ran away. So the temple was named "*Wu Xiang An.*" Similarly, there is an "Elephant Cave" at Zhangping county, an "Elephant Lake Mountain" at Wuping county, and an "Elephant Hill" at Pinghe county (all in southwest Fujian). During the Min-Qing Dynasty (fourteenth–nineteenth century A.D.), some people of the minority nationality lived on hunting at Yongchun, Changting, Fu'an, and other counties. Tiger, leopard, wild boar, hedgehog, goat, deer, and fox were their common prey. In 1612 A.D., a group of tigers at Luoyuan county did harm to local inhabitants. Some She hunters once killed four tigers with poisoned arrows (W. Zhu). It was recorded that some businessmen from Fujian dealt in deer furs with Luzon (Philippines) during the Ming Dynasty (1368–1644 A.D.). Silver was shipped back in return (Huang).

Many wild animals inhabit the forests, so the retreat of forests may imply the destruction of wild animals. Today, wild animals have almost vanished from the coastal areas and are seldom seen along the valleys. But many precious animals still exist in mountainous areas covered with extensive forests in the north and west Fujian. It was reported that six young leopards, one tiger, ten or more black bears, and countless monkeys and pheasants were caught and killed at Fentiangyang village of Yonggan county in 1984–85 (Jingxing *et al.*). In the mountainous area of Shunchang county, the author himself saw farmers go hunting in the winter ten years ago. Wild boar, goat, bear, tiger, muntjac, and pangolin were commonly bagged.

Wuyishan Nature Reserve, well-known as a rare gene store of species, is located in the mountainous area of northwest Fujian. In this area of 572 square kilometers, primitive vegetation is left substantially intact. More than 1,800 species of vascular bundle plant were counted belonging to 191 families or 798 genuses and including plants from both the boreal region and the paleotropical region. Today, residents here include *Panther trigris, Neofelis nebulosa, Elaphodus cephalophus, Macaca muoatta, Viverra Zibetha,* and 100 other species of mammals. *Lophura nyothemera, Tragopan caboti,* and 400 other species of birds, *Vibrissaphora liu,* salamander, and 100 other species of amphibians and reptiles, as well as some 30 more species of fish and 20,000 more species of insects. Six hundred additional new species of animals were discovered during the past century (Li and Zhao). Wuyishan Nature Reserve accounts for less than 0.5 percent of the area of Fujian, but it is typical of the natural ecology of Fujian. In light of the Nature Reserve, we may believe that animal species must have decreased, their numbers must have declined and their area of distribution must have decreased in historical time in Fujian and southeast China.

Deposition in Estuaries

There is a layer of dark grey silt widely covering the surface of Fuzhou Basin. The depth is 4–6 meters, generally with shells of oyster (S. Lin) and abundant fossils of diatom. Based on the material from core ZH9, of 32 species of diatom discovered from the silt layer, 16 are marine species and the others are brackish and fresh water species. Marine cells account for more than 71 percent of the total diatom cells. An assemblage of *Act. ehrenbergi*(m)-*Coscinodiscus Subtilis*(M)-*Gomphonema spp.*(F) was discovered and other characteristics are two open marine species, *Coscinodiscus ouclusiridis* and *Cos. lineatus*, and the marine indicator *Dictyocha* (of *Chrysophyta*) (Xie *et al.*). The layer of silt at the north of Wuyi Road (Fuzhou) was dated at 4310–150 B.P. These findings verify a hypothesis proposed by David K. Lin in the 1950s that Fuzhou Basin once was a sea bay, named "Fuzhou Bay" (D. Lin). Shells of *Ostrea arca sp., Corbicula sp.*, and *Auricula sp.*, which might live in brackish coastal waters, have been found at Tangshishan shell mound in west Fuzhou Basin (Qi). The date of the oyster shells is 3270±155 B.P. Limited to ancient methods of transport, the ancients would have been unable to carry these delicious sea foods from the distant modern coast. The most reasonable explanation is that at that time the sea reached nearly to Tangshishan, some 70 km inland to the west of the modern Min River mouth.

The stratigraphic correlation (Pang) suggests that at about the same period, sea water extended to the west of Fenzhou, 20 km inland from the current shoreline at Quanzhou area and 15 km inland from the current shoreline in the Putian area. Sea water submerged low land of less than three meters above modern sea-level along Juilong River valley and transgressed adjacent to Fu'an town along Sansha Bay. We can infer that a zigzag shoreline with more islands and narrower coastal plains existed along the coast of Fujian at that time. Especially at the lower reach of the big rivers, several bays extended inland just like the current situation at the northeast Fujian coast (Sun). Fuzhou Basin at that time was under water and "Fuzhou Bay" was like today's Sansha Bay. The narrow coastal plain might have been unsuited for habitation because of its larger tidal range and higher wave, so human beings generally dwelled on the terraces of an elevation of 20 or more meters along the inland valleys.

Some estuaries silted up and part of them became land about two thousand years ago. Ye Town was settled to the northwest of Fuzhou Basin by King Minyue in 202 B.C. *Han Shu* (A History of the Han Dynasty) reported that King Minyue "resided in a land that abounded with rivers and lakes" (W. Zhu). Wang Gong, a poet of the Ming Dynasty, wrote: "When Wuzhu (name of King Minyue) established his empire in an area of ancient *man* nation, a long river overflowed around the city wall." It was recorded that seagoing vessels dropped anchors at Huanzhumen during the Han dynasty and docks were built at Aoqiao (today's Wushi Crossing, 4 km north of Min River at Fuzhou City) in about the 260s A.D. (W. Zhu). It is clear that Fuzhou Basin was a region of rivers and lakes at the beginning of the century. In the Putian plain, a sea bay in the period of the Qing-Han Dynasty (ca. 200 B.C.) extended far inland into today's plain between Jiuhua, Gui, and Hugong Mountains. The area at the foot of Jiuhua Mountain was a harbor for taking shelter from the wind. Sea water withdrew to the east of Putian Town in the sixth century A.D. when a harbor moved to the front of Guanghua Temple (a famous Buddhist temple in Fujian) (Wu). Quanzhou Plain was subject to tides even in the fifth or sixth century A.D., when an activity center of human beings was at Fenzhou, west of Quanzhou.

The deposition along the coast became apparent since the Tang Dynasty (after the seventh century A.D.). In Fuzhou Basin, the city wall expanded to the south several times during the Tang-Song Dynasty as the water receded. In the early eleventh century, most of Taijiang District (the south of Fuzhou City) was under water. Late in the century, a sand bank emerged at the area of Zhongting Street due to the deposition of silt from the Min River (W. Zhu). Extensive plain in the south and east of Fuzhou Basin was under water at that time. Because of the huge tidal flux in Fuzhou Basin, high tide (in 1183 A.D.) might reach Daruo (42 km west of Fuzhou Basin). The sandbanks at the southern branch of the Min River were reclaimed after the Song Dynasty (thirteenth century) (Y. Zhu). Most of the plain at Nantai Isle may have been silted up until the Ming Dynasty (fourteenth century). "*Tengshan Zhi* (A history of Fuzhou)" states that Eunuch Deng received a heavy bribe and then permitted foreigners to build a new harbor at Huanchuanpu (south of the new Min River Bridge in Fuzhou) in about 1500 A.D. But silt and sand gradually deposited and turned it into a sand beach in the Wanli Age (1573–1620 A.D.). Spring tide in the Min River reached to Zhuqi in 1906 A.D., 34.5 km lower along the valley than seven hundred years ago (Y. Zhu). This was a result of sedimentation and lessening tidal flux in the Fuzhou Basin. The upper limit of the tide reached only to 7.5 km again in the last seventy or more years. This reflected the fact that the water area in Fuzhou Basin has been reduced rapidly.

In Quanzhou Plain, the course of the Jin River extended to the south four times from the Five Dynasties to The Song-Yuan Dynasty (907–1368 A.D.) (Wang). Quanzhou was known as the largest harbor in the East in the early Yuan Dynasty, when huge vessels could anchor outside the city wall. Since the late Yuan Dynasty (about fourteenth century), Quanzhou Harbor has deteriorated following a severe silting-up of the water area. When an ancient sunken boat (about seven hundred years old) was excavated from Quanzhou Bay in the 1970s, a thick layer of silt (3.5 m in depth) covered the bottom of the boat.

The process of silt deposits in coastal areas always follows man-made projects that enclose tidelands for cultivation. Such projects were prominent during the Tang-Song Dynasty. The Putian Plain was opened up according to the following records: Wuxing, an officer, enclosed the sea beach for cultivation at Dutang village, Putian County, in the 780s A.D. A long dike was built against the strong tide. Yanshoubei (a reservoir) was built to irrigate 400 *qing* (hectare) of farmland. Hugong Plain bordered on the sea in three directions when Peiciyuan was sent by Fujian authorities as a supervisor in 813 A.D. He built a weir at Huangshi town, then brought 322 hectares of tideland under cultivation (W. Zhu). Similarly, there were hundreds of places along the coastal lowland where reclamation was carried out during the period. Today, we find along the coast many places with names such as "*Dai*," "*Pu*," "*Tang*," "*Wei*," "*Yu*," and "*Yang*." They were generally enclosed from the sea and laid out as farmlands. It was these farmlands that made up the major part of the coastal plain.

The coastal plain is one of the richest areas in Fujian. It is of doubtless value. The formation of plains may have been connected with crustal uplift and sea level movement, but the silt and the sand transported by river are fundamental materials of the plains. It was reported that 7.32 million tons of silt were carried into the Fuzhou Basin by the Min River annually in the 1970s and 2.52 million tons into Quanzhou Bay by the Jin River. In addition to the silt deposits in the coastal plain, extensive fertile farmlands along the inland valley were lost and the bays and the beaches with rich aquatic resources were buried in oblivion. The following statistical data show a shortening of the navigable courses of the rivers due to deposition. In Fujian Province, there once were 29 fully navigable rivers with 5,141 km of navigable river course, but only 14 rivers with 3,856 km of navigable course remained in 1981. The navigable course along the Min River which was 2,989 km in 1960, was shortened one-third by 1980 when the remaining navigable course was only 1,947 km.

Desertification of the Coastal Area

The seasonal northeast wind is strong along the coast of Fujian. With a drying of the climate and increasing sand deposits at the river mouth, desertification appears to have become more serious in the last one thousand years or so.

The number of recorded droughts at the coastal area have increased since the Song Dynasty. In the period of the Tang Dynasty, many pools and lakes were dredged for irrigation of the coastal plains. However, during the Song Dynasty most irrigation relied on the construction of dams and diversion works. Thousands of dikes and dams all over Fujian were counted at that time. The best-known was Mulanbei Dike, which was built in 1085 A.D. It channeled water from the Mulan River to the south Putian Plain, which was turned into a rich agricultural area, along with the building of a complete irrigation system.

Desertification became more marked during the Song Dynasty. "A General History of Fujian" reported, "at Hudong, 30 *li* distance from Changle town, a dike was set against the sea. During the Xining Age (ca. 1070s A.D.), the sand blown by the wind silted up the inlet and poled up as a hill." Pingtan Isle, the largest island off-shore of Fujian, was recorded as a grazing land during the Tang Dynasty. A layer of fossil soil, which was dated at about 1200 carbon 14 B.P., was common in the middle part of the island. Many roots and stems of herbs and partly carbonized tree trunks buried in the fossil soil indicate a forest-prairie landscape at that time. But later on, the soil was buried under a layer of sand. A similar phenomenon occurred at Xiacai, Zhangpu County, where a layer of peat (ca. 700 carbon 14 B.P.) is covered by drifted sand.

The records of sand calamities have increased in the last three hundred years. At Pingtan Isle, "there was a strong gale (1749 A.D.), sea sand was carried in by a turbulent tide, all villages nearshore suffered from a heavy cover. Eighteen hamlets at Luyangpu were buried under the sand in only one night's time." "A History of Dongshan County" reported, "the farmlands were rich and the landscape was beautiful during the Ming-Qing Dynasty." "The wind-drifted sand did not intrude until the late Qing period (eighteenth–nineteenth century)." "It became more intense recently." An ancient naval school drill ground was found covered by sand at Shihu, Jinjiang County. It might have been a verdant nearshore village three hundred years ago based on investigation (H. Lin). Many fragments of pottery and porcelain from the Ming-Qing Dynasty Age can be found at the sandy waste of Zhangban, Changle County. Modern aeolian landforms are well-developed along the coast zone, where sand dunes rise to a height of 30–40 meters (even 80 m) at windward slopes and passes, and sandbars were molded by wind and wave near the shoreline. Tree trunks and roots excavated from the sand are common there.

Desertification may have been aggravated by the wars and man-made destruction of vegetation along the coastal zone since the Ming-Qing Dynasty, as well as by a prolonged drought. In the early period of the Qing Dynasty (1660s–1690s A.D.), the authorities carried out a policy called "moving away from the shoreline." The residents were forced to move back from the coast leaving a buffer zone 30 *li* (15 km) in width. The temples of gods and the houses of people left behind were burned by torch (Huang). This policy was a crushing blow to the natural ecology which the people of the coastal zone once lived on. It encouraged the spread of wind-drifted sand. The disastrous effects remain to this day.

Stages of Environmental Changes

The historical period of Fujian may be divided into several stages, such as the evolution of the physical environment and the enhancement of man's effects on his natural environment.

Neolithic Age (before ca. 1000 B.C.)

The climate was warmer and wetter than at present. Almost all of the regions were covered with primeval forests. There were various wild animals, including many species from tropical regions. The plains at the lower reaches of the rivers were inundated by the sea and the coastal plains were narrower. The sparse population of the ancient inhabitants was scattered among the bigger basins along the rivers. They lived on hunting, fishing, and primitive cultivation.

Agricultural Age (ca. 1000 B.C.–1950 A.D.)

The climate tended to be cool and dry. The primeval forests retreated. The wild animals decreased. The plains at the river mouths were enlarged by silt deposits. Three periods may be differentiated in this age:

Pre-Tang Dynasty Period (ca. 1000 B.C.–ca. 600 A.D.). The climate was not as wet or as warm as before. Cultivated fields were scattered along the terraces at a height of 20–80 m above the valley floors. This was the lower limit of the forested area. The total population throughout the province was less than 60,000 (average 0.5 person/km^2).

Tang-Song Dynasty Period (ca. 600 A.D.–ca. 1300 A.D.). The climate was relatively warm and wet. The farmlands enlarged and the population increased rapidly (6.21 million in 1300 A.D., averaging 50 persons/km^2). The forests retreated from all the valleys and coastal plains. Soil erosion intensified in the mountainous area and deposition increased at the estuaries. This was a rare period in the history of Fujian, when its prosperous economy occupied an important position in the economy of China, partly due perhaps to its fertile plains and valleys and numerous harbors.

Late Yuan Dynasty to Pre-industrial Period (1300–1949 A.D.). The climate evidently became dry and cool. Because of wars, turmoil, increased population, and changing economic structure, the vegetation and the land were periodically damaged. The results were that the primeval vegetation suffered a steady retreat, the soil at mountainous areas became impoverished, floods and droughts increased on the plains, the harbors were filled with silt and the lakes were buried in oblivion. Desertification intensified along the coast, especially since the late Qing Dynasty, because of the "moving away" policy, the increased population (exceeded 20 million in the 1860s), and the unchecked development of some commodity economies based on forests and forest land. The whole environment in Fujian has deteriorated.

Industrial Age (from 1950 A.D.)

The indicator of the age is the construction of railways in the 1950s. Owing to the scale, the intensity and the pollution of industrial production, technical innovations in agriculture and the enlarging scale of traditional agriculture, humankind's effects on the physical environment at present far exceeds that of the past. In the mountainous area, the primeval forests have almost disappeared. Along with the reduction in forest area, there has been a decrease in forest quality. As a result of soil erosion, many river courses and reservoirs have become silted up and the areas that are subject to flood and drought enlarged. The Jiulong River valley has averaged a drought every thirty years from 164 to 1908 A.D. and every six years from 1949 to 1985 A.D. Offshore, the aquatic resources have been exhausted due to the deposition of soil and sand at estuaries and beaches. The contamination of air, water, and soil was very serious.

At the same time, many steps have been adopted to protect and harness the environment, e.g., afforestation, building of water conservation projects, establishment of nature reserves, and monitoring and controlling the contamination, etc. Particularly at the coastal area and on some islands, e.g., Dongshan and Pingtan Islands, large-scale windbreak forests have been planted since the 1950s. They are very effective in controlling the desertification and have helped to rebuild a sound agricultural ecology and to improve the living conditions of the native people.

Conclusion

A tendency toward dry and cool climate, retreat of the forests, extinction of species, filling of estuaries, and desertification along the coast were five aspects of environmental problems in historical Fujian. These problems have increased in severity, along with man's ever-growing ability to influence the environment. The key may be increased understanding of the environment and man's acceptance of responsibility to solve its problems. In the historical view, the destruction of the forests may be the outstanding environmental problem in Fujian, since it influences other aspects of the environment, directly or indirectly. The reason for the destruction was the predatory development in forest areas. The increasing population and its growing demands for farmland was a steady factor in the environmental deterioration in historical Fujian.

The modification of the natural environment is a historical problem. Studying the environmental evolution will contribute to an understanding of the environment and help to find a way to solve current environmental problems. Today, humans have created unprecedented social productive forces, as well as unprecedented ability to modify nature, so they must learn to solve the environmental problems and reconstruct a non-"Silent Spring."

Acknowledgments

The author is grateful to Professor David K. Lin for his most respected guidance in the writing of this paper and to Professor Hugh W. Ellsaesser for his careful revision of this paper in English. Thanks also are due to Associate Professors Wu Yougong and Wang Shaohong for their critical readings of the manuscript.

References

Chen, Chenhui. 1982. "Sporo-pollen Associations of Some Holocene Profiles in the Coastal Region of Southern Fujian." *Taiwan Strait* 1(1):45–53.

Chen, Jingshen. 1984. "Preliminary Study of the Population Development in Historical Fujian." *Population of Fujian* (2):27.

FAH (Fujian Association of Hydrology). 1986. "Some Anxious Data." *Fujian Daily.* June 22, 2.

FMO (Fujian Meteorological Observatory). 1977. "Preliminary Analysis on Climatic Variations in 20th Century in Fujian." *Symposium on Climatic Changes and Prediction of Super-long-term Climate.* Science Press, 29–32.

FPM (Fujian Provincial Museum). 1984. "Second Season Excavation of the Xitou Site at Minhou." *Acta Archaeologia Sinica* (4):459–502.

Huang, Gongmian. 1985. "The Characteristics of Economic Development in Historical Fujian" (manuscript).

Jingxing *et al.* 1986. "A Batch of Beneficial Birds and Beasts Were Set Free at Yongan." *Fujian Daily.* June 21, 2.

Li, Wenhua and Xianying Zhao. 1984. *The Nature Reserves of China.* Commercial Press, 60.

Lin, David K. 1959. "A New Survey of Coastal Changes in Fujian." *Quaternary Research Sinica* 2(1):47–60.

Lin, Huilai. 1982. "A preliminary Investigation of Historical Wind-drift Sands along the Western Coast of Taiwan Strait." *Taiwan Strait* 1(2):74–81.

Lin, Shentian. 1984. "On Problems of Quaternary Geology of Fuzhou Basin" (manuscript).

Pang, Guoxiong. 1984. "Holocene Stratigraphic Division and Sea-level Changes in Coastal Fujian." *Taiwan Strait* 3(2):166–178.

Qi, Guoqing. 1977. "On Beastbones Excavated from Tangshishan Neolithic Relic at Minhou of Fujian." *Paleovertebrate and Paleoanthropology* 15(4):301.

Sun, Henglun. 1982. "Is North Section of Fujian Coast a Submerged Coast?" *Symposia of Graduate Students.* FTU Press, 131–135.

Wang, Tianliang *et al.* 1980. "On Quanzhou Harbor in Historical Time." *Journal of Fudan University* (social science edition, and addition), 74–84.

Wu, Menxiong *et al.* 1982. "On the Elevation and Subsidence of Putian Coast." *Seismology of Fujian* 2(3):20–23.

Xie, Zhaituang *et al.* 1984. "Preliminary Study on Transgressions along Fujian Coast since Late Pleistocene" (manuscript).

Zeng, Fan. 1980. "On Approach to Remains of Pre-historical Culture in Fujian." *Acta Archaeologia Sinica* (3):263–283.

Zhang, Deyi. 1981. "Preliminary Analysis on the Winter Temperatures of South China in the Last 500 Years." *Symposia on Climatic Changes in China.* Science Press, 85.

Zhang, Jiacheng and Lin, Zhiguang. 1985. *Climatology of China.* Shanghai: Technology Press, 512.

Zheng, Hun. 1982. "Holocene Pollen Associations in Fuzhou Plain, Fujian Province" (manuscript).

Zheng, Shizhong *et al.* 1977. "The Changes of Climatic Humid Conditions in the Last 2,000 Years in Southeast China." *Symposia on Climatic Changes and Prediction of Super-long-term Climate.* Science Press, 29–32.

Zhu, Kezhen *et al.* (editors). 1982. *Historical Physical Geography* (Physical Geography of China). Science Press, 9.

Zhu, Mindao. 1977. "Some Scales of Climatic Oscillation in China. *Symposia on Climatic Changes and Prediction of Super-long-term Climate.* Science Press, 37.

Zhu, Weigang. 1984. *A History Sketch of Fujian.* Fujian Education Press.

Zhu, Yongkang. 1985. "Evolving River Course at the Mouth of Min River in Historical Period." *Taiwan Strait* 4(2):161–170.

THREE

THE ROLE OF ALKALINE PARTICULATES ON PH OF RAIN WATER AND IMPLICATIONS FOR CONTROL OF ACID RAIN

L.T. Khemani

Introduction

Acid rain is one of the feedback signals of man's overgrazing of his ecological niche on earth which is capable of spreading an alarm throughout his self-encapsulating cities and suburbs. The acid content of rain in many parts of the world has been steadily rising over the past several years as countries became more industrialized and increased their use of fossil fuel. Vast areas of the Northern Hemisphere, including central Europe, Scandinavia, and northeast United States and Canada, are affected by acid rain. There is great concern that acidity of rain can have significant environmental impacts, in particular on acidification of lakes and on plant soil systems.

The main contributors to rain water acidity are sulfuric and nitric acid. Although no direct quantitative relationship between pollutant emissions and the measured acidity has been demonstrated, naturally and anthropogenically produced SO_x and NO_x are considered to be the major acid rain precursors. In addition to SO_x and NO_x, other pollutants may directly or indirectly affect the acidity of rain water. Ammonia, because of its alkaline nature, is able to neutralize acidic components of precipitation. Less understood are the roles of ozone, carbon dioxide, and particulate matter. Natural dust particulates, which are alkaline, may be capable of neutralizing precipitation acidity.

The role of acidic particulates (SO_4 and NO_3) from anthropogenic sources on the decrease of pH values in rain water samples has been shown by many research workers in the western countries. Recently, the influence of alkaline particulates on the increase of pH values has been demonstrated (Khemani *et al.*, 1985c). The influence of alkaline dust particulates can be seen better in India where the entire area is under the influence of aeolian dust blown up from the arid soils in central Asia and the Indian sub-continent, where the rain water has not yet turned acidic (Mukherjee, 1957, 1964, 1978; Handa, 1969, 1973; Handa *et al.*, 1982; Subramanian and Saxena; Das *et al.*; Khemani *et al.*, 1985c); but the pollution levels are rapidly increasing. Soil in India is, by and large, dusty and rich with alkaline components, like calcium and magnesium, and is the major source for the observed high-levels of aerosols in the atmosphere (Khemani *et al.*, 1985a). The concentrations of total suspended particulates (TSP) reported for big cities in India have varied between 200 and 800 $\mu g\ m^{-3}$ (NEERI).

In order to examine the influence of soil dust on pH of rain, chemical composition of rain water samples and TSP have been undertaken at a few locations in India comprised of urban and non-urban regions.

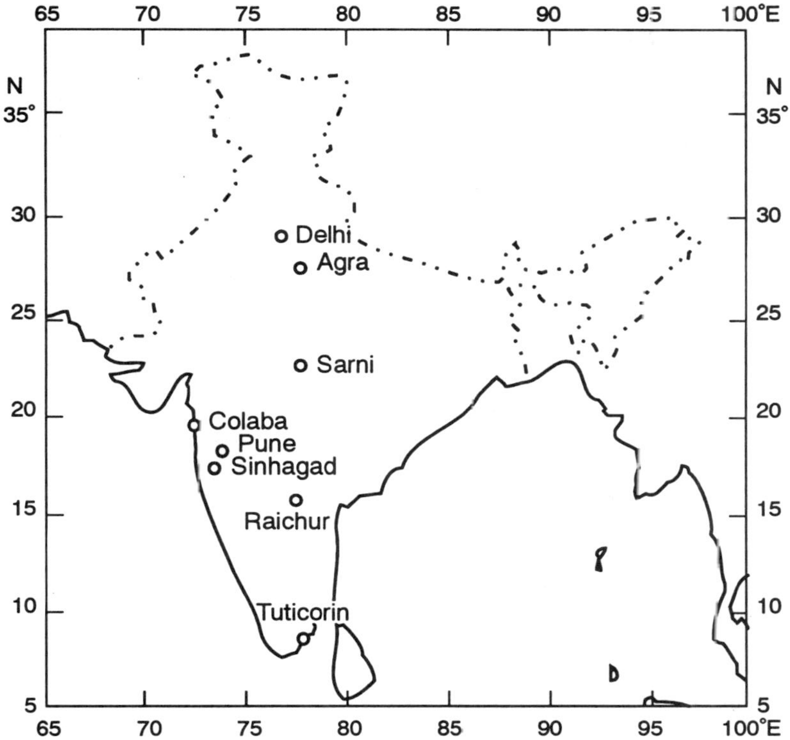

Figure 3-1: Map of India showing the locations of observational sites.

Observational Sites

The locations where observations have been made are shown in Figure 3-1. Environmental conditions of the locations are described below.

Delhi (28° 35' N, 77° 12' E, 218 m asl)

Delhi, an urban station, is situated about 1,100 km away from the nearest coast on the north Arabian sea. The Delhi region is located in the border zone lying between the rich rain-washed Gangetic plains to the east and the semi-arid tracts of Rajasthan to the west and southwest. Being the capital of India, its air pollution problems have attracted much attention from environmental scientists.

Rain water samples have been collected during the monsoon seasons (July–September) of 1965 and 1984. The total suspended particulates (TSP) samples were collected in the afternoons from 1100 to 1500 hours IST, when convective activity was maximum. Altogether, 116 TSP samples were collected between February and October 1980. The TSP and rain water samples were collected from the terrace of the Institute's building (a field station of the institute), situated on a ridge, about 15 m above ground level. The samples obtained from the above location are representative of natural conditions, since the observational site is generally free from obstructions and sources of local pollution.

Agra (27° 10'N, 78° 02' E, 169 m asl)

The world-famous monument Taj Mahal is situated at Agra, which is about 100 km away from Delhi. There are no major industrial complexes located in the vicinity. The Mathura refinery started functioning only in 1980. The rain water samples were collected during the southwest monsoon (July to September) in 1963 and 1984.

Pune (18° 32' N, 73° 51' E, 559 m asl)

Pune is also an urban station and is about 100 km inland from the western coast of India. It is located on the leeside of the Western Ghats. Airflow in the lower troposphere is predominantly westerly during the southwest monsoon (June–September). A large influx of moisture is brought inland from the Arabian sea. The wind in the lower troposphere reverses with withdrawal of the monsoon and easterly flow sets in from October. Continental air masses, rich in nuclei of continental origin, pass over the region during that period.

Rain water samples have been collected during the monsoon season (June–September) for a period of 10 years from 1974 to 1983. The TSP samples were collected from 1100 to 1500 hours for a period of one year from May 1977 to April 1978. The TSP and rain water samples were collected on the terrace of the Institute's building 12 m above ground level. The samples obtained from the above locations are representative of natural conditions since the observational site is free from obstructions and sources of local pollution.

Sinhagad (18° 21' N, 73° 45' E, 1315 m asl)

Sinhagad is a non-urban hill station and is located about 18 km southwest of Pune city. It is in the forest-dense region of Maharashtra state and is sparsely populated. Two samples of TSP, each of four-hour duration, were collected daily in morning (0600–1000) and night (1800–2200) for a period of six days during January 1982.

Sirur (18° 50' N, 74° 23' E, 582 m asl)

Sirur is a non-urban station and is about 130 km inland from the west coast of India and is situated on the leeside of the Western Ghats. This region is known as the rain-shadow area of the Deccan Plateau. West to east orography rises from the coast of Bombay and ends in a plateau at an average height of 0.6 km. The Western Ghats (range of hills) extend from north to south along the west coast. Rainfall during the southwest monsoon is strongly orographic on the windward side of the Western Ghats. Precipitation increases with altitude from the coast of the windward side and falls off sharply on the leeside. Average rainfall during the monsoon period is less than 40 cm at the observational site. The rain water samples were collected during four monsoon seasons from 1979 to 1982.

Raichur (16° 12' N, 77° 21' E, 389.5 m asl)

Raichur is also a non-urban station. It is situated well inland and is free from industrial pollution. Two samples of TSP, each of three-hour duration, have been collected daily from the terrace of the agricultural college building at a height of 10 m above ground level, in the morning and evening hours for a period of 12 days during February 1980. The TSP samples collected from the above site are representative of the natural conditions since the observational site is completely free from local pollution.

Tuticorin (8° 45' N, 78° 11' E, 4 m asl)

Tuticorin is a coastal station and is situated on the east of India in the gulf of Mannar. There is no major industry located in the vicinity of the observational site except petrochemical industry which manufactures fertilizers. The Tuticorin thermal power plant of 620 MW capacity was under construction at the time of measurements. Two samples of TSP were collected daily upwind and downwind of the power plant from the terraces of residential buildings at a height of 10 m above ground level in the morning and evening hours for a period of ten days in February 1978.

Sarni (22° 07' N, 78° 10' E, 436 m asl)

Sarni is a thermal power plant station and is located well inland in the forest area of Madhya Pradesh State of India. A 722.5 MW thermal power plant has been operating at this place without electrostatic precipitators. Three samples of TSP of three-hour duration were collected daily upwind and downwind of the thermal power plant in the morning, afternoon and evening hours for a period of ten days during May 1982.

Sea Areas

Observations over the sea areas around the Indian sub-continent have been made on board the research vessel "Gaveshani" of the National Institute of Oceanography, Panjim, Goa. The track of the ship is shown in Figure 3-2. Three samples of TSP of three-hour duration were taken daily during morning, afternoon and evening hours for a period of 12 days during May 1983.

The sampling equipment was mounted on the upper deck on the upwind side of the ship to avoid ship's exhaust and on-board dust. The sampling level was about 10 m above sea water level and care was taken to avoid any direct sea spray contamination.

Bombay Region

Eighty percent of the industries in the Maharashtra state of India are located in the Bombay region. Most of the industries are situated in the Chembur area, which is the most highly

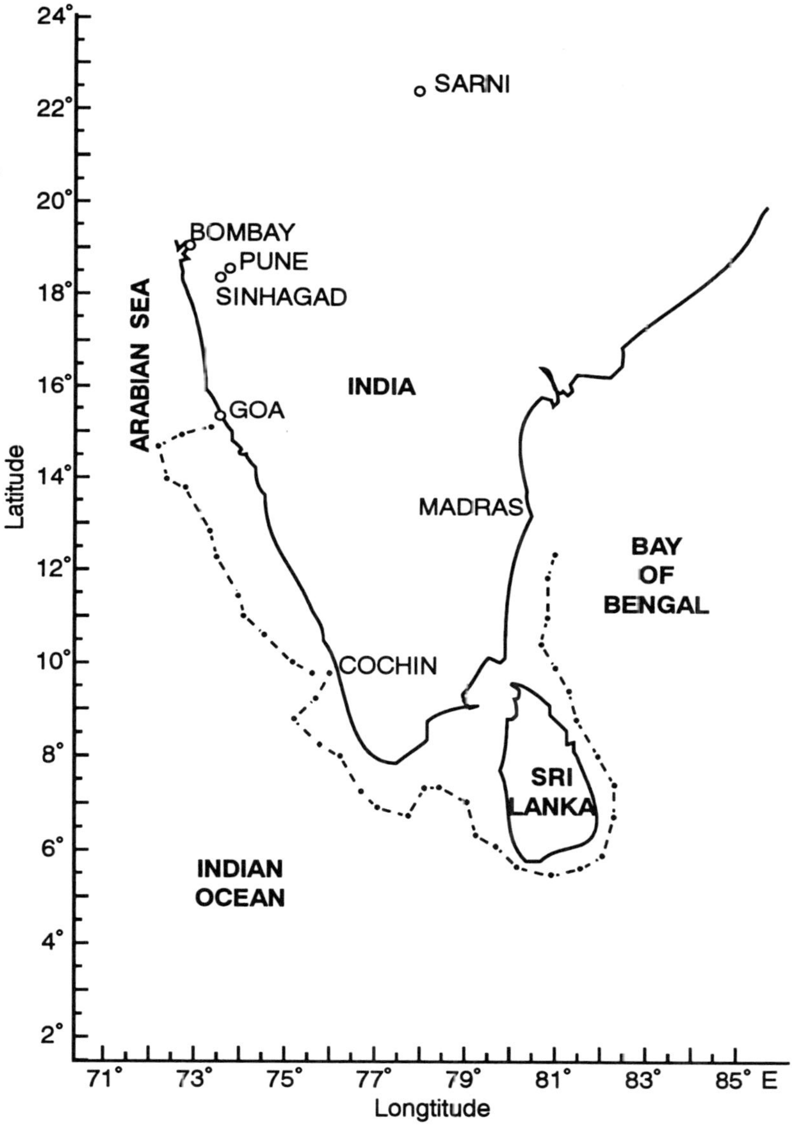

Figure 3-2: Map showing the cruise of research vessel "Gaveshani" in the Arabian Sea, Indian Ocean, and the Bay of Bengal.

industrialized area of India. The locations where rain water samples have been collected are shown by the symbol cross-upon-circle in Figure 3-3. These are Alibag and Colaba, which are situated on the west coast, and Kalyan, which is situated to the east. The areas where major industrial establishments, such as refineries, petrochemical complexes, cotton mills, textile factories, synthetic material plants, and thermal power stations have been situated, are indicated by solid stripes in the figure. Alibag is about 30 km away from Colaba and is free from industrial pollution (Khemani *et al.*, 1980). Colaba, which is located on the southwestern tip of the urban industrial complexes, is, no doubt, on the upwind side of the pollution sources. But, being situated in the immediate vicinity of some of the pollution sources, it may not be free from their influence. Kalyan is surrounded by industrial complexes and is affected by pollution sources of various kinds.

Weekly rain water samples were collected simultaneously at Alibag, Colaba, and Kalyan during the southwest monsoon season of 1974.

Sampling and Analysis

The total suspended particulates (atmospheric aerosols) were collected on Whatman 41 filter paper (ten cm diameter) using a high volume air sampler (Staplex, U.S.A.) at a flow rate of 0.6 m^3 min^{-1}. Whatman 41 filter paper was chosen for its strength to withstand relatively high air-flow rates. Also, this filter paper has adequate capacity to trap small particles with high efficiency and is known for its elemental purity (Dames *et al.*). Each sample was extracted separately by a Soxhlet extractor for chemical analysis of water soluble components of TSP by standard extraction methods. The water soluble extracts have been analyzed for chloride (Cl), sulfate (SO_4), nitrate (NO_3), ammonium (NH_4), sodium (Na), potassium (K), calcium (Ca), and magnesium (Mg).

The rain water samples at a few locations mentioned in the previous section have been collected using stainless steel funnels of 30 cm diameter fitted on to one liter capacity polyethylene bottles previously rinsed with deionized water. The funnels are washed twice daily during morning and evening to

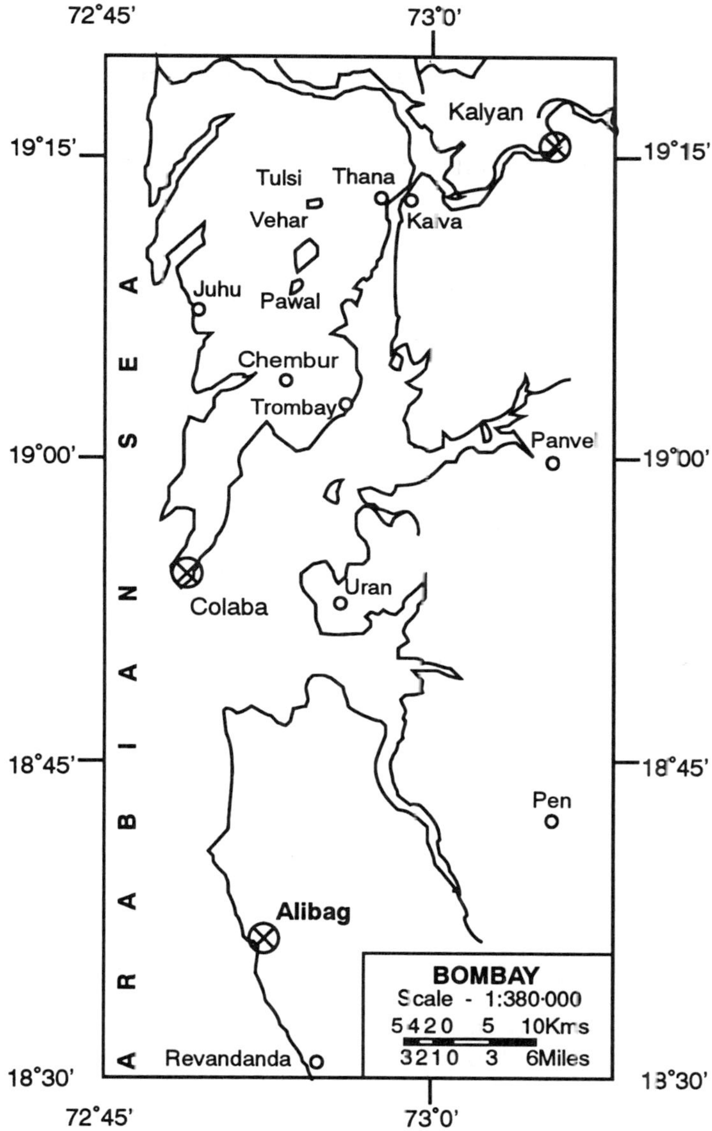

Figure 3-3: Locations of observational sites are shown by the symbol cross-in-circle in the map of the Bombay region.

avoid dry deposition. However, contamination due to dry deposition cannot be completely ruled out. During the southwest monsoon period of 1983, rain water samples were collected on four occasions on board DC-3 aircraft with the help of a stainless steel scoop fabricated for the purpose. Details of the collection from aircraft have been reported (Khemani *et al.*, 1982a). All the rain water samples are filtered through Whatman 41 filter papers and refrigerated at 4°C in the laboratory until all the ionic components are analyzed. The samples have been analyzed for the ionic components chloride, SO_4, NO_3, NH_4, sodium, potassium, calcium, and magnesium. Also, the pH values have been determined in all the samples collected. The chemical composition of rain water at Chembur, a highly industrialized location at Bombay, has been reported (Sequeira, 1976), and these values are referred to in the present study for comparison.

The concentrations of chloride, SO_4, NO_3, and NH_4 were determined using standard colorimetric methods (Khemani *et al.*, 1985c). The concentrations of metallic components were determined by Kipp and Zonen flame photometer up to 1978 and by double beam atomic absorption spectrophotometer (Perkin-Elmer, Model 373) with air-acetylene flame, subsequently. The calibration curves for different chemical constituents were obtained by preparing standard solutions using grade chemicals. These calibrations were periodically repeated to check accuracy. The concentrations of different chemical constituents have been obtained with reference to the blank solutions. The analytical errors are nominal and vary within ± 10 percent.

The pH was measured with a pH meter using reference (KCl) and glass electrodes standardized with pH 4.0 and pH 9.2 reference buffers before and after pH determinations. Hydrogen ion concentration was deduced from the measured pH. Since pH value is influenced by extreme values of rainfall amount, rainfall weighted mean values of pH have been used instead of arithmetic means.

Concept of pH Value

The reference level commonly used to compare acid precipitation to natural precipitation is pH 5.6, the pH that results from the equilibration of atmospheric carbon dioxide with precipitation. This reference level has been chosen because of the ubiquity of carbon dioxide in the global atmosphere and because of the absence of data on other acids or bases. However, Galloway *et al.* believed that there was no single natural pH of precipitation applicable to the whole globe, but rather several natural values, each unique to a region, the size of a continent or an ocean. From the measurements of pH values in rain water collected at five remote areas of the globe, they suggested that the lower limit of natural pH was probably ≥ 5. However, this lower limit of natural pH cannot be applicable in the Indian sub-continent where the pH values are highly alkaline and vary from 6.0 to 7.5 (WMO). These values are much above the carbon dioxide-equilibrated pH. The pH values in Central Europe are around 4 and this value increases towards the oceans and also towards eastern Europe (Georgii). Over the North American continent, high pH values have been observed over the middle and western part of the United States. Only at the East Coast does the acidity of the rain lead to pH 4.5. The pH data in the present study are interpreted in the light of the carbon dioxide-equilibrated value (pH = 5.65) observed in the laboratory, though it may not be a real atmospheric background value.

Southwest Monsoon

An important weather system which has been largely neglected for study of its chemical character is the south Asian summer monsoon. The southwest monsoon sets in over the extreme southwest coast of India in late May and by the first week of July it is established over the entire country. The monsoon current reaching India has had a fairly long journey (6,500 km) over the ocean. Under the influence of strong and persistent surface winds blowing over and from the Arabian sea, vast quantities of airborne sea salt are expected to be transported into the country over a considerable part of the year (Eriksson, 1959, 1960). Giant and large sea salt nuclei carried by the monsoon air flow are known to penetrate deep inland as far as up to Delhi in north India (Kapoor and Ramana Murty). The monsoon weakens by September and by the middle of October withdraws from the entire country.

Results and Discussion

Chemical Composition of Rain Water

The average seasonal concentrations of major ionic components along with pH values in rain water collected simultaneously during 1974 at the five stations, Alibag, Colaba, Kalyan, Chembur, and Pune, are given in Table 3-1. The seasonal variations of ionic components in rain water collected at Pune for a period of a decade from 1974 are shown in Figure 3-4. The average seasonal concentrations of ionic components along with pH values in rain water collected at the non-urban station Sirur for a period of four years from 1979 are given in Table 3-2. Also, the average seasonal concentrations of major ionic components along with pH values, in rain water samples collected at Delhi during 1965 and 1984 and at Agra during 1963 and 1984, are given in Table 3-3.

Chlorine and Sodium. The maximum concentration of chlorine has been observed at Alibag, a non-urban coastal station, and the minimum at Pune (Table 3-1), an inland station on the leeward side of the Western Ghats. The variation of sodium is similar to that of chlorine. It is known that the concentration of sea salt (chlorine and sodium) decreases with an increase of distance from the sea. Chlorine and sodium at Pune have varied from 1.03 and 0.82 mg L^{-1} in 1975 to 2.57 and 1.84 mg L^{-1}, respectively (Figure 3-4). At Sirur, concentration of chlorine has varied from 2.03 mg L^{-1} in 1979 to 3.85 mg L^{-1} in 1981, and the concentration of sodium has varied from 1.02 mg L^{-1} in 1980 to 2.87 mg L^{-1} in 1981 (Table 3-2). The concentrations of chlorine and sodium have not shown much variations during the two monsoon periods at Delhi and Agra (Table 3-3).

The predominant source of chlorine in rain water is sea spray. A small amount of chlorine also is contributed by anthropogenic emissions, i.e., hydrochloric acid. Data on the latter is

Table 3-1: Average Concentrations in mg L[-1] of Major Ionic Components and pH Values in Rain Water Samples during 1974

Location	*Cl*	*SO_4*	*NO_3*	*NH_4*	*Na*	*K*	*Ca*	*Mg*	*pH*
Alibag	5.97	1.11	0.57	0.27	3.84	0.88	2.38	0.60	7.2
Colaba	4.14	2.58	0.40	0.51	3.12	1.13	2.56	0.64	7.1
Kalyan	3.95	5.20	1.92	0.38	2.36	1.02	1.86	0.47	5.7
Pune	1.72	2.20	0.80	0.50	1.39	0.53	1.43	0.36	6.6
*Chembur	5.00	20.20	–	2.10	2.20	1.10	3.10	0.68	4.8

*Sequeira, 1976

Table 3-2: Average Concentrations in mg L[-1] of the Major Ionic Components along with pH Values at Sirur

Year	*Cl*	*SO_4*	*NO_3*	*NH_4*	*Na*	*K*	*Ca*	*Mg*	*pH*
1979	2.03	1.25	1.35	0.06	1.15	0.53	1.46	0.37	6.4
1980	2.29	1.19	2.37	0.05	1.02	0.43	2.27	0.53	6.5
1981	3.85	2.27	5.14	0.14	2.87	1.25	3.87	1.46	7.3
1982	2.39	0.90	1.69	0.04	2.31	1.14	6.51	0.79	7.4

Table 3-3: Average Concentrations in mg L[-1] of Major Ionic Components along with pH Values at Delhi and Agra

Location	*Year*	*Cl*	*SO_4*	*NO_3*	*NH_4*	*Na*	*K*	*Ca*	*Mg*	*pH*
Delhi	1965	2.52	1.23	–	0.23	1.83	0.42	4.83	–	7.0
	1984	2.43	2.73	2.54	0.53	1.75	1.28	2.95	0.62	6.1
Agra	1963	2.24	1.81	0.68	0.20	2.16	2.00	9.36	–	9.1
	1984	2.69	2.48	2.46	0.33	2.32	1.58	4.73	0.82	6.3

sparse, but it is known that the anthropogenic emissions are negligible compared to the natural ones. It is, therefore, assumed that chlorine does not indicate an addition of acid or base in rain water (Granat, 1972a). Sodium in rain water at coastal stations is mainly derived from sea spray, whereas at inland stations soil has been considered to contribute some portion of sodium in rain water. A significant positive correlation ($r = 0.96$) has been observed between chlorine and sodium at all the stations in the Bombay and Pune regions and also at

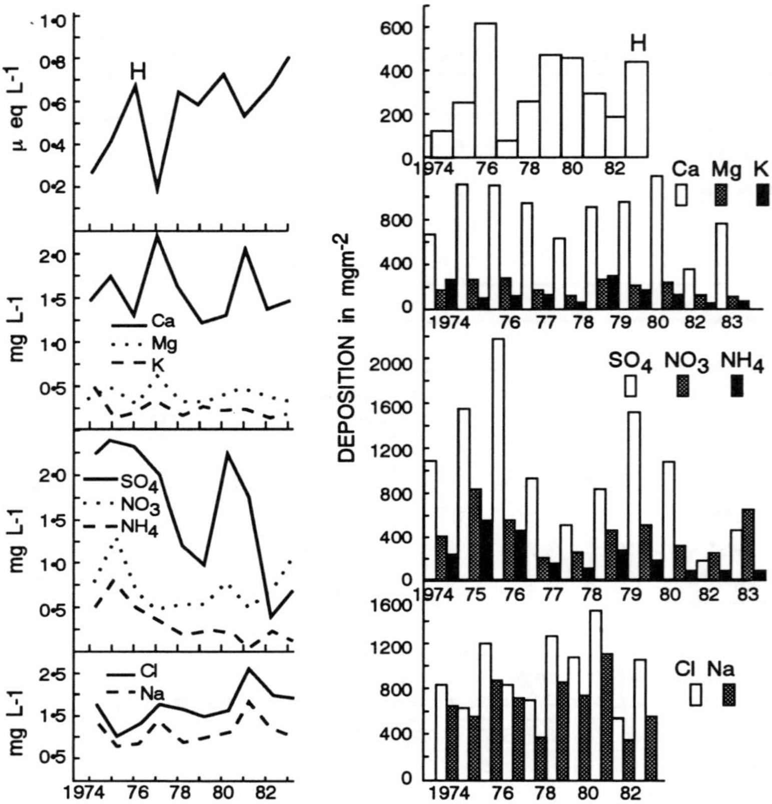

Figure 3-4: **Average concentrations of the major chemical components and their deposition in ten monsoon seasons at Pune.**

Agra and Delhi, suggesting that chlorine and sodium have a common source which is the sea. The sea salt (chlorine and sodium) has been considered neutral salt and the presence of it does not influence the pH in rain water (Granat, 1972a).

Sulfate. The maximum concentration of 20.2 mg L^{-1} has been reported at Chembur, a highly industrialized location in the Bombay region and the minimum (1.11 mg L^{-1}) has been observed at Alibag. Although Colaba is situated on the west coast, the concentration of SO_4 (2.58 mg L^{-1}) at Colaba is twice that at Alibag. This could be due to the proximity of Colaba to the industrial complexes. Low concentration of SO_4 at Alibag indicates that the station is free from industrial pollution. The second maximum concentration has been observed at Kalyan (5.20 mg L^{-1}), which is also situated in the industrial complex area.

The concentration of SO_4 at Pune has varied from 0.42 mg L^{-1} in 1982 to 2.39 mg $^{-1}$ in 1975. Sulfate concentration at the non-urban station Sirur has varied between 0.90 and 2.27 mg L^{-1} during the four monsoon seasons. The concentrations of SO_4 at Delhi and Agra increased by 120 and 37 percent, respectively, after a period of two decades, during which the cities became more industrialized (Table 3-3).

Sulfate in rain water arises mainly from anthropogenic emissions, which is also indicated by the above observations. The sulfate from sea spray is neutral and does not increase the acidity of rain water (Granat, 1972a). However, SO_4 from anthropogenic sources has been considered a major component which increases the hydrogen ion concentration in rain water and decreases its pH (Nagamoto *et al.*). In order to determine the effect of anthropogenic SO_4 at the coastal stations, it is essential to eliminate the contribution by sea spray. Assuming that all sodium is of sea-origin, the contribution by sea to the observed values of SO_4 are calculated. The contribution of SO_4 by sea spray is maximum (86 percent) at Alibag, a non-urban coastal station. The contribution by sea decreases to 30 percent at Colaba, which is located in the proximity of the industrial complexes. However, the contribution by sea in the industrial complex at Chembur and Kalyan is minimum (3 and 11 percent). The contribution by sea at Pune, a station relatively free from industrial pollution, is around 20 percent.

Nitrate. The concentration of NO_3 is minimum at coastal station Colaba (0.40 mg L^{-1}) and Alibag (0.57 mg L^{-1}), which are free from industrial pollution. However, maximum concentration (5.14 mg L^{-1}) has been observed at Sirur, an inland non-urban station. The concentration of NO_3 at Pune varied from 0.45 mg L^{-1} in 1977 to 1.34 mg L^{-1} in 1975. The concentration of NO_3 has been observed, by and large, to be the same at Delhi and Agra (Table 3-3).

Soil seems to be a major source of nitrate in India. Maximum concentration of NO_3 in rain water at Sirur, a pollution-free non-urban station, indicates that soil is the source of NO_3. Also, NO_3 showed significant correlation with soil-oriented components at Sirur (Naik).

Nitrogen oxides are emitted into the atmosphere mainly through the action of bacteria in the ground, but also through oxidation of nitrogen by combustion (Green; Cox; Penkett *et al.*). NO and NO_2 will eventually end up as NO_3 in rain water through various reactions. Thus, the HNO_3 formed in rain water can increase the hydrogen ion concentrations. But it has been observed at Pune that during the monsoon season sea salt (NaCl) reacts with HNO_3 and results in the formation of $NaNO_3$ particles (Khemani *et al.*, 1982b). This reaction tends to neutralize the acidic effects of HNO_3 and helps in reducing the extent of hydrogen ion concentration in rain water. Probably due to this reason, the NO_3 in rain water at Pune has not shown significant correlation with pH (Khemani *et al.*, 1985c). Also, at Delhi and Agra, NO_3 has shown significant correlation with soil-oriented components (potassium and calcium). Hence, NO_3 in rain water in India does not seem to influence the pH values.

Ammonium. Maximum value has been observed at Chembur (2.10 mg L^{-1}) and minimum value at Alibag (0.27 mg L^{-1}) in the Bombay region. At Pune, it has varied from 0.10 mg L^{-1} in 1981 to 0.80 in 1975. Minimum concentration has been observed at the non-urban station, Sirur, where the concentration has varied between 0.04 to 0.14 mg L^{-1}. At Delhi, the concentration was 0.23 mg L^{-1} during 1965; the value became more than twice this in 1984. However, at Agra, the concentration was 0.20 in 1963 and a little higher during 1984 (0.33 mg L^{-1}).

Ammonium in precipitation emanates from ammonia created by the action of bacteria on nitrogen compounds in the ground, perhaps mainly from urine (Healy *et al.*). It is also released from the industrial plants manufacturing fertilizers. The presence of NH_4 in rain water helps in neutralizing the acidic effects caused due to the presence of SO_4 in the form of sulfuric acid. It has been found at Pune that most of the sulfates during monsoon season are in the molecular form of

Table 3-4: Ratio (g/g) of Major Ionic Components to Sodium at Seven Locations and the Corresponding Values for Sea Water

Location	*Cl/Na*	*Mg/Na*	*K/Na*	*Ca/Na*	*SO_4/Na*
Alibag	1.55	0.16	0.23	0.62	0.29
Colaba	1.33	0.21	0.36	0.82	0.83
Kalyan	1.67	0.20	0.43	0.79	2.20
Pune	1.23	0.26	0.38	1.03	1.58
Sirur	1.31	0.34	0.45	1.91	0.82
Delhi	1.41	0.34	0.44	2.17	1.11
Agra	1.05	0.36	0.79	3.14	0.95
Sea Water	1.80	0.13	0.036	0.038	0.25

ammonium sulfate (Khemani *et al.*, 1982b) and the presence of such particles could reduce the extent of hydrogen ion concentration in rain water.

Potassium, Calcium, and Magnesium. The concentration of potassium has varied from 0.88 mg L^{-1} to 1.13 mg L^{-1} in rain water in the Bombay region. At Pune, it has varied from 0.11 mg L^{-1} in 1982 to 0.53 mg L^{-1} in 1974. At Sirur, it has varied from 0.43 mg L^{-1} to 1.25 mg L^{-1} during the four monsoon seasons. At Delhi, it was 0.42mg L^{-1} during 1965 and increased three-fold after a period of about two decades in 1984. At Agra the concentration was 2.00 mg L^{-1} during 1963 and remained, by and large, the same after a period of two decades in 1984.

The concentration of calcium has varied from 1.86 mg L^{-1} to 2.56 mg L^{-1} in the Bombay region. At Pune, it has varied from 1.15 mg L^{-1} in 1979 to 2.17 mg L^{-1} in 1977. At Sirur, it has varied from 1.46 mg L^{-1} in 1979 to 6.51 mg L^{-1} in 1982. At Delhi, it was 4.83 mg L^{-1} in 1965 and dropped nearly to half (2.95 mg L^{-1}) after a period of two decades in 1984. Similarly, at Agra, the concentration was 9.36 mg L^{-1} in 1963 and was reduced to half after a period of two decades in 1984.

The concentration of magnesium has varied from 0.47 mg L^{-1} to 0.6 mg L^{-1} in the Bombay region. At Pune, it has varied from 0.23 mg L^{-1} in 1983 to 0.54 mg L^{-1} in 1977. At Sirur, it has varied from 0.37 mg L^{-1} to 1.46 mg L^{-1}. The concentration at Delhi has been 0.62 mg L^{-1} and at Agra it has been, by and large, similar to that at Delhi.

Compounds containing potassium, calcium, and magnesium are added into the atmosphere by soil erosion, sea spray, and by anthropogenic emissions. It has been considered that these soil-oriented elements are mostly in the form of carbonates (Granat, 1972a). Examples of anthropogenic emissions of potassium, calcium, and magnesium are fly ash from burning of coal or oil and exhausts from processing industries. For urban areas, anthropogenic emissions are of importance, whereas, for the other places soil is the major contributor. These elements are potentially basic in nature and their presence in high concentration in rain water helps in neutralizing the acidic effects of anthropogenic emissions (SO_4 and NO_3) and in maintaining the pH of rain water in the alkaline range.

Ratio of Major Components to Sodium

The ratios of major sea salt components relative to sodium are remarkably constant in the oceans and seas of the world with the exception of coastal and land-locked seas (Horne; Riley and Chester). However, the values in marine aerosol and precipitation have on occasions been observed to be different from their corresponding values in sea water (Junge, 1972; Tsunogai *et al.*). The effects of soil dust, originating from the arid zones of continents, on the chemical composition of atmospheric aerosol particles over the ocean have been observed (Prospero and Carlson, 1972; Prospero and Nees, 1977; Junge, 1979; Prospero *et al.*, 1981; Hirose *et al.*; Khemani *et al.*, 1985b).

Chlorine/Sodium and Magnesium/Sodium. The ratios for all the major chemical components with respect to sodium for the seven stations are given in Table 3-4. Chlorine in rain water is usually thought to be mostly from the sea, like sodium and magnesium and the ratio of chlorine/sodium at the coastal stations (Alibag, Colaba, and Kalyan) as well as at the inland stations (Pune, Sirur, Delhi, and Agra) are found to be less than in sea water. Many comments on the ratio can be found in the literature; citing smaller ratios due to loss of chlorine and/or addition of sodium from continental sources. The ratio of magnesium/sodium, at Alibag (0.16) is closer to that of sea

Table 3-5: Seasonal Deposition of Ionic Components along with Rainfall in the Bombay Region

Location	*Deposition in mg m^{-2}*								*Rainfall*
	Cl	*SO_4*	*NO_3*	*NH_4*	*Na*	*K*	*Ca*	*Mg*	*(mm)*
Alibag	13504	2511	1289	611	8686	1991	5384	1357	2262
Colaba	9609	5988	928	1184	7242	2623	5942	1485	2321
Kalyan	7118	9370	3460	685	4253	1838	3352	847	1802

water (0.13) than at other stations. There is a strong correlation (r = 0.96) between chlorine and sodium and chlorine and magnesium (r = 0.67), suggesting that chlorine, sodium, and magnesium are intimately related and originate predominantly from the sea which is consistent.

Potassium/Sodium and Calcium/Sodium. The ratio of potassium/sodium at all the stations is higher (0.23 to 0.79) than in sea water (0.036). The excess of potassium in rain water is attributable to the local use of potassium-containing fertilizers in agriculture and the widespread use of firewood as domestic fuel.

The ratio of calcium/sodium is significantly higher not only at the inland station Agra (3.14) but also at the coastal stations (0.62–0.82) than in sea water (0.038). Excess calcium in rain water has been reported for coastal as well as for inland stations in India (Ramachandra Murty and Ramana Murty; Khemani, 1974; Sequeira, 1976; Sadasivan; Sequeira and Kelkar, 1978; Khemani *et al.*, 1985c). Excess of calcium, therefore, is a regular feature of monsoonal precipitation. An excess of calcium concentration of 1.1 mg L^{-1} is uniformly distributed over the country (Sequeira and Kelkar, 1978). Due to mixing of maritime and continental air masses in the eastern Arabian Sea (Rao and Desai), it is considered that calcium from soil dust is incorporated into Indian rains. Calcium in rain water samples is present in the form of $CaCO_3$ (Sequeira and Kelkar). Excess calcium has been observed in marine aerosols over the west coast of India, which has been attributed to the transport of soil dust from the Arabian peninsula towards the west coast of India (Khemani *et al.*, 1985b).

Sulfate/Sodium. The ratio of sulfate/sodium is higher than in sea water at all the stations, except at Alibag (0.29) where it is close to the sea water value (0.25). The excess has been found to be smallest (14 percent) at the coastal station Alibag. Excess SO_4 in the marine aerosols over the sea areas around the Indian subcontinent is about 0.21 $\mu g\ m^{-3}$ and its contribution is around 20 percent (Khemani *et al.*, 1985b), whereas the average concentration of excess SO_4 in the marine aerosols over the north Atlantic is about 0.9 $\mu g\ m^{-3}$ and its contribution has varied from 60 to 80 percent (Gravenhorst, 1978). This feature indicates that air around India is quite clean and its sea regions can serve as suitable sites for the study of background pollution in the northern hemisphere.

Wet Deposition of Ionic Components

The wet deposition in mg m^{-2} of the ionic components have been computed by multiplying the total monthly precipitation by the measured weighted mean concentrations of ionic components. The monthly values are added and the seasonal depositions thus obtained for the three stations in the Bombay region are given in Table 3-5. The seasonal depositions of ionic components for Pune have been computed for the 10-year period (1974–1983) and the results are shown in the form of histograms in Figure 3-4.

The sea salt (chlorine and sodium) deposition is maximum at the coastal stations Alibag (13,504 and 8,686 mg m^{-2}) and Colaba (9,609 and 7,242 mg m^{-2}) which is consistent. The deposition from anthropogenic sources (SO_4) is higher at Colaba (5,988 mg m^{-2}) than at Alibag (2,511 mg m^{-2}). The highest deposition of SO_4 (9,370 mg m^{-2}) has been found at Kalyan (Table 3-5). The total deposition at Pune has been maximum in the case of SO_4 (10,148 mg m^{-2}) followed by chlorine (9,686 mg m^{-2}) and calcium (8,923 mg m^{-2}).

Apart from the sea salt, which is neutral and does not, therefore, influence the pH of rain water, the other ionic components which are present in substantial amounts are SO_4 and calcium. The dominance of these components in rain water influence the pH to a greater extent.

pH of Rain Water

The various aspects contributing to the formation of acid precipitation have been discussed by many investigators. Junge and Scheich (1969) found that particles of smallest size range, sampled in rural areas, reacted acid whereas the coarse fraction reacted more alkaline. This tendency also has been found to be valid for marine aerosols present 15 m above the north Atlantic (Gravenhorst, 1975), and in the Arctic (Winkler). The coarse particles, which are also called primary aerosols, are formed by sea bubbling or blown up frequently from basic mineral soil, whereas the smallest particles (secondary aerosols) are formed by gas-to-particle conversions leading to acidic nuclei. Hence, the relative contribution of primary aerosols (calcium, potassium, and magnesium) and secondary aerosols (SO_4 and NO_3) in rain water will decide the pH.

The pH of rain water at the different stations are included, as mentioned earlier, in Tables 3-1, 3-2, and 3-3. The values at Alibag and Colaba have been 7.2 and 7.1 respectively, which are in the highly alkaline range. The value at Kalyan has been 5.7, which is nearer to the neutral value of distilled water. However, at Chembur the pH value has been reported 4.8, which is acidic. At Pune, the pH value has varied from 6.1 in 1983 to 6.8 in 1977, indicating that rain water at Pune is not acidic. At Sirur, the pH values have varied between 6.4 and 7.4, which is in the alkaline range. At Delhi, the pH value was alkaline (7.0) in 1965 and remained alkaline (6.1) about two decades later in 1984. At Agra, the pH value was 9.1 during 1963 which is highly alkaline. About two decades later, the pH has decreased by 2.8 units but still lies in the alkaline range (pH = 6.3).

The correlations (r) worked out between pH and different ionic components from long series of rain water data at Pune, are given in Table 3-6. Significant positive correlation (r = 0.56) has been noticed only with calcium suggesting that it is this

Table 3-6: Correlation Coefficients of pH with Ionic Componentsin Rain Water at Pune

	SO_4	NO_3	NH_4	K	Ca	Mg
pH	-0.10	0.12	-0.01	0.27	0.56	0.43

ionic component which has influence on the observed pH of rain water. Khemani *et al.* (1985c) have shown an inverse relationship between hydrogen ion concentration and excess cations (calcium, potassium, and magnesium) in the rain water samples at Pune. They suggested that the excess cations in rain water are responsible for the observed alkaline pH values.

Sulfate and calcium are the dominating ionic components present in rain water in India. To illustrate the relationship of hydrogen ion concentration with calcium and SO_4, the data obtained for the rainy day during July 1965 at Delhi has been considered. Seventeen samples were collected in a shower which lasted for two and one-half hours on 31 July 1965. The variations in the concentrations of calcium and SO_4 along with hydrogen ion are plotted in Figure 3-5. No significant relationship has been noticed between SO_4 and hydrogen ion indicating that some of the SO_4 might be present in the form of salts. A weak correlation (r = 0.36) has been observed between SO_4 and hydrogen ion concentration. But calcium has shown significant (r with pH = 0.56) relationship with hydrogen ion,

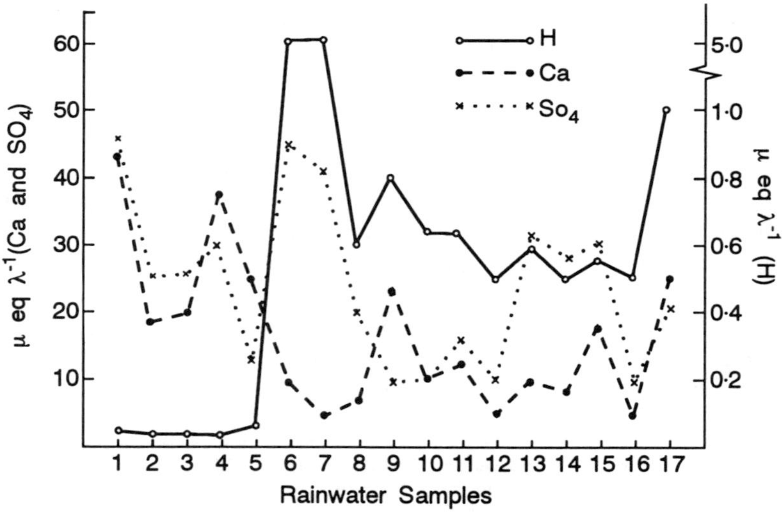

Figure 3-5: Variation of hydrogen, calcium, and sulfate concentrations in rain water samples collected on 31 July 1965 at Delhi.

Table 3-7: Average Concentrations of Anions and Cations in μeq L^{-1} and pH Values in Rain Water at Different Locations during 1974

Location	*Cl +*	*SO_4+*	*NO_3=*	*Σ^-*	*NH_4+*	*Na +*	*K +*	*Ca +*	*Mg=*	*Σ^+*	*$\Sigma^- - \Sigma^+$*	*pH*
Colaba	117	54	6	177	28	136	29	128	53	374	-197	7.2
Alibag	168	23	9	200	15	167	23	119	49	373	-173	7.1
Pune	48	46	13	107	28	60	14	72	30	204	- 97	6.6
Kalyan	111	108	31	250	21	102	26	92	38	280	- 30	5.7
Chembur	141	421	-	526	117	96	28	155	56	452	+110	4.8

indicating that this basic component reduces hydrogen ion in rain water rendering it alkaline. Similar results have been found in rain water samples collected at Pune (Khemani *et al.*, 1985). Soil is considered to be the main source of calcium and industrial pollution the main source of SO_4 particles in the atmosphere. The presence of these two components influences the pH of rain water as already pointed out.

This feature is further examined by calculating anion-cation balance of the rain water collected during the monsoon period of 1974 at five stations, Alibag, Colaba, Kalyan, Pune, and Chembur (Table 3-7).

The anion sum includes chlorine, SO_4, and NO_3, while the cation sum includes NH_4, sodium, potassium, calcium, and magnesium. Examination of the values of anions and cations indicates that cations are in excess at all the stations except at Chembur, a highly industrialized location, where the rain has been reported acidic (pH = 4.8). The main cause for alkaline rain is the presence of high concentrations of calcium and other soil-oriented elements which are alkaline in nature and neutralize the acidic effects generated by the presence of SO_4 and NO_3. However, the cause of acidic rain at Chembur is the presence of very high concentration of SO_4 (20.2 mg L^{-1}) which could not be neutralized by the available calcium (3.1 mg L^{-1}) and other alkaline elements. Nature has played its role in keeping the rain water in alkaline range in Bombay (Sequeira, 1976) in spite of the acidic nature of rain in a limited zone of two square kilometers at Chembur. The high concentration of calcium in rain water samples in Bombay has been attributed to the soil dust transported from the Arabian peninsula towards the west coast of India (Khemani *et al.*, 1985b). The alkaline elements in the soil dust play an important role in rainout and washout processes during the monsoon season and neutralize the acidic effects of pollutants released from the industrial complexes. It is to be noted that the Bombay region receives heavy rainfall during the southwest monsoon season.

The influence of alkaline particulates from soil dust on the pH of rain water has been reported in the United States and European countries. The analysis of rain water data for 1953–1954 and for 1977–1978, reported from the northeast United States (Stensland and Semonin), showed that rain water in that region became acidic because of the deficiencies of alkaline components like calcium and magnesium which were significantly less in the years 1977–1978 than in the years 1953–1954. Peters *et al.* reported an increase of hydrogen ion concentration in thirteen years of precipitation data (1965–1978) from nine collection sites in New York. Deficiency of soil-oriented ions (calcium and magnesium) and high concentration of acidic ions (SO_4 and NO_3) were considered to be the main reasons for the increase of hydrogen ion in the rain water reported from selected sites in North America and Scandinavia (Munger). The influence of soil particulates on pH of rain water can be clearly seen from the chemical analysis of rain water samples collected at Retz, a semi-arid region in Austria. The high concentration of calcium (6.06 mg L^{-1}) in rain water during 1978 neutralized the hydrogen ion concentration to some extent and increased the pH up to 5.4, which is very near to the boundary value of rain acidity (Cehak and Chalupa). Rain water at Retz, like India, could have been in the alkaline range, but high concentration of SO_4 ($\simeq$ 17 mg L^{-1}) transported from industrial sources in Europe overpowered the natural source and maintained pH towards the acidic range.

Hence, the main cause for alkaline rain water in India is the presence of high concentrations of alkaline aerosols (cations) dispersed from dusty soil and transported in the atmosphere, which are washed out by precipitation. The above influence is supported by the high pH value in rain water reported from ten background monitoring stations in India (WMO). Mukherjee and Krishna Nand (1981) have suggested another cause for higher alkaline pH values, close to 7.0 in precipitation in India. They have shown that higher atmospheric temperatures in tropics, lower solubility and dissolution rates of carbon dioxide, together with rapid cloud formation and precipitation processes result in higher pH values. Also, greater depth of saturated air (extending above aerosol layer) during southwest monsoon period may be another factor for higher pH.

Influence of Power Plants on pH of Rain Water

Divergent views are expressed on the neutralizing capability of fly ash particles released from the stacks of coal-fired power plants. Shannon and Fine have measured pH values of 9 to 11 from water extracts of fly ash. Lower concentrations of Na_2O and CaO in eastern coal fly ashes result in less alkaline ash from eastern coals (Ananth *et al.*). Likens and Bormann have suggested that the removal of alkaline fly ash emissions by industrial and utility electrostatic precipitators contributes to acid rain formation. On the other hand, Ananth *et al.* have indicated that the removal of fly ash may not be a contributor because the particles removed by the precipitator are probably larger in size than those remaining in the plume and, thus, if left uncontrolled, would settle more rapidly. They further indicated that there are insufficient data to determine whether the alkalinity of fly ash has a significant effect on the neutralization of acid rain.

During the field observational program, measuring gaseous and particulate pollutants around the thermal power plant at Sarni in May 1982, one rain water sample was collected which showed a pH of 6.8. Alkaline pH has been attributed to the washout effect of giant particles of fly ash released from the stacks of the power plant which is operating without electrostatic precipitators. Also, the lake water collected downwind of the power plant showed highly alkaline pH (8.2) which is due to the deposition of large amounts of giant particles of fly ash in the lake. It has been considered that alkaline fly ash emissions are capable of neutralizing precipitation acidity.

The above feature is corroborated by the reported alkaline pH value (7.4) in the rain water samples collected at the site of another thermal power plant, located in New Delhi, which was also operating without electrostatic precipitators (Subramanian and Saxena). It is of interest to find out what changes can occur in the pH of rain water if the particulates from the stacks of the power plant are allowed to escape along with gases. Fortunately, by the end of the year 1982, this thermal power plant at New Delhi was fitted with electrostatic precipitators and the collection of rain water samples was organized during the southwest monsoon season of 1985 at the site of the plant and also upwind and downwind of it. The samples at the plant were acidic on eleven occasions and the pH values varied between 4.2 and 5.3. The rainwater on the remaining sixteen occasions was alkaline and the pH varied between 5.7 and 7.8. However, the rain water samples in the upwind and downwind regions within 4 km from the power plant were alkaline and the pH varied between 5.7 and 8.2.

The above feature suggests that the fly ash particles are capable of neutralizing the acidity in precipitation. Also, it is confirmed that acid rain, wherever it has occurred in India, is a purely local phenomenon. Long-range transport of pollutants (SO_4 and NO_3) does not seem to be effective during the southwest monsoon period when the meteorological factors are favorable for quick dispersal of pollutants whose sources are weak.

Ionic Composition and pH at Cloud Base and Ground

Samples of rain water have been collected in the Pune region on four occasions both at cloud base (1.5 km) and at ground during the southwest monsoon of 1983. The values of ionic composition and pH are given in Table 3-8. On the average, the concentrations of ionic components at ground level have ranged up to four times those at cloud base level. The pH of rainwater at cloud base varied from 5.6 to 6.1 and were lower by 1.5 units than at ground level. This observation is in conformity with what has been reported on the island of Hawaii, namely, the pH of rain decreased with increasing altitude, averaging 5.2 at sea level and 4.3 at 2.5 km (Miller and Yoshinoga).

Examination of anion-cation balance at ground and cloud base indicated that the excess cations are three times larger at ground level than at cloud base. This feature suggests that excess cations (calcium and magnesium) are responsible for

Table 3-8: Concentrations of Major Ionic Components and pH Values in Rain Water Collected at Cloud Base (CB) and at Ground (G) at Pune

		Concentrations in $\mu eq\ L^{-1}$							
Date	*Level*	SO_4	NO_3	*Total*	*Ca*	*Mg*	*Total*	*H*	*pH*
07-14-83	CB	11.5	7.0	18.5	25.5	14.2	39.7	0.79	6.1
	G	21.8	30.3	52.1	94.5	25.0	119.5	0.05	7.3
08-05-83	CB	4.0	8.7	12.7	55.0	37.5	92.5	1.58	5.8
	G	5.6	15.8	21.4	212.0	41.7	253.7	0.02	7.6
08-28-83	CB	6.7	5.9	12.6	47.0	10.0	57.0	1.00	6.0
	G	18.5	21.0	39.5	156.0	36.7	192.7	0.10	7.0
09-05-83	CB	7.5	5.8	13.3	44.5	18.3	62.8	2.50	5.6
	G	18.0	–	18.0	179.0	23.3	202.3	0.04	7.4
Average	CB	7.4	6.9	14.3	43.0	20.0	63.0	1.45	5.8
	G	16.0	22.4	38.4	160.4	31.7	192.1	0.05	7.3

the observed increase of pH at ground. Non-marine magnesium also has been considered as a strong alkaline component, along with calcium which neutralizes the acid in rain water.

Peden and Skowron observed a remarkable reduction in hydrogen ion concentrations in precipitation samples and attributed it to the presence of particulate material in the samples. Also, Pack reported a significant decrease in hydrogen ion concentration with event samples collected from the ERRI network due to addition of basic particulate matter. Munger has reported a fourfold increase in hydrogen ion concentrations from west to east across northern Minnesota in the U.S.A., whereas in the same region alkaline dust (calcium and magnesium) decreased by two- to threefold. A considerable part of the hydrogen ion concentration in rain water in Hungary is neutralized mainly by soil-derived calcium components (Horvath, 1981). Recent chemical analysis of bulk rain water samples collected at a number of stations in Hungary has shown that calcium particles of soil origin play a role in the control of acidity (Horvath and Meszaros, 1984). In view of the above and in the light of the present study, it is confirmed that the alkaline properties of particulate matter in rain water are responsible for neutralizing the hydrogen ion concentration and consequently for the observed increase in pH.

Total Suspended Particulates (TSP)

The total acid or base potential of the material constituting the particulates is an important quantity because deposition of particulates can either lead to deterioration or an improvement of the soil that it lands on. Most acid soils are infertile and have lost much of their nutrient ionic materials, such as calcium and potassium by leaching of these soluble salts. It is of interest, therefore, to know the acidity of TSP measured at various locations. The calculations for sum of anions (Σ^-) and cations (Σ^+) for all the seven locations are shown in Table 3-9.

The hydrogen content in the TSP has been assumed to be characterized by the difference between anions and cations ($\Sigma^- - \Sigma^+$); that is, if this difference is positive the TSP are acidic and if this difference is negative the TSP are alkaline. If the principal ions have been accounted for, difference in the ion sums ($\Sigma^- - \Sigma^+$) are assigned to hydrogen ions (Hidy and Countess). The anion sum includes chloride, SO_4, and NO_3, whereas the cation sum includes NH_4, sodium, potassium, calcium, and magnesium. Examination of the values of anions and cations in TSP (Table 3-9) indicates that the cations exceed the anions at all locations. Hence, the ionic balance of TSP in different environments in India suggests that the TSP are alkaline in nature. The main cause for this feature is the presence of high concentrations of calcium and other soil-oriented elements (potassium and magnesium). Calcium has been found to be the dominant component in cations which neutralizes acid

Table 3-9: Average Concentrations in Nano Equivalents per Cubic Meter (neq m^{-3}) of Anion and Cation in TSP at Seven Locations in India

Location	*Cl* +	*SO_4*+	*NO_3* =	Σ^-	*NH_4* +	*Na* +	*K* +	*Ca* +	*Mg* =	Σ^+	$\Sigma^- - \Sigma^+$
Sea Region	205	21	38	264	23	163	8	37	38	269	- 5
Sinhagad	24	17	9	50	9	17	6	11	7	50	0
Raichur	14	19	17	50	48	17	11	44	24	144	- 94
Tuticorin	152	38	17	207	55	117	12	110	45	339	-132
Pune	38	26	20	84	25	34	9	51	17	136	- 52
Delhi	56	48	54	158	18	37	14	129	20	218	- 60
Sarni	19	15	21	55	18	21	10	80	24	153	- 98

generated by the presence of SO_4 and NO_3. Most of the cations are soil-originated and their excess concentrations in the atmosphere suggest that the soil is the major source of TSP in the atmosphere over India. Most of the anions are considered to be released by anthropogenic sources (SO_4 and NO_3) and their low concentrations in TSP suggest that pollution sources have not dominated the natural sources (soil and sea). Hence, the presence of excess cations are responsible for the absence of the acidity or the low concentration of hydrogen ions in the aerosols in India. The deposition of such aerosols should, therefore, give rise to increased soil fertility in India.

The effect of acidity in aerosols is observed in the pH of rain water. European rain is mostly acidic (Granat, 1972b). Rain in the northeastern United States is also reported to be acidic because aerosols are found to be acidic in nature (Hicy and Countess); but rain collected in interior sites in the United States is either neutral or slightly basic because of the large amount of dust components of the aerosols in that region (Gorham). The presence of alkaline aerosol is responsible for the alkaline pH in rain water discussed earlier. This is also corroborated by the reported alkaline pH values in rain water from different parts of India (Khemani, 1985; Khemani *et al.*, 1985c; Mukherjee *et al.*, 1986).

In view of the above, it can be generalized as follows: In arid areas and under strong sea influence, precipitation should be alkaline. In areas where soil is protected against wind erosion by soil moisture and vegetation and where sea salt contribution is small or normal, precipitation should be acidic. These features seem to be represented in the geographical distribution of precipitation pH-values over the northern hemisphere (Gravenhorst *et al.*, 1980; Georgii). In the arid regions of southwest United States and over the Indian subcontinent, relatively high pH values (pH > 6) occur. The low pH values (pH < 4) in the northeast United States, Canada, and central Europe are, however, not only due to the reduced alkalinity because of reduced sea- and soil-derived minerals in precipitation, but they also certainly include an effect related to anthropogenic acid-forming emissions.

Implications of Alkaline Particulates on Control of Acid Rain

Although the soil dust in India creates a great inconvenience to the inhabitants, it is a blessing in disguise. The alkaline properties of soil particulates has sustained alkaline pH in rain water so far and has been controlling acid rain in India. Deficiency of soil-oriented components (calcium and magnesium) and high concentrations of acidic components (SO_4 and NO_3) are considered to be the main causes of acid rain in the northeast United States, Europe, and Scandinavian countries.

In European countries, acid rain has reduced the fertility of the soils which have lost much of their nutrient ionic materials such as calcium and potassium. Already, farmers have started adding enormous quantities of lime to the soil to lower the acid content and to compensate for the loss of calcium. In the Netherlands, it has been calculated that more than 20 percent of the lime for agriculture purpose is used in compensating for the loss of calcium by leaching of the soil by acid precipitation (Vermeulen). It appears that larger quantities of lime will be required in coming years. A Danish report has suggested that lime be added to the acidified Danish lakes to make fish and other aquatic life possible again (Vermeulen). However, the phenomenon of acid rain poses no special problem in Asia, particularly in India, so long as the aerosol state of the air, which is presently alkaline, continues to remain so.

But the problem of acid rain has become alarming in the United States and in the European countries and requires immediate attention. Rain-bearing clouds in mid- and high latitudes can be seeded with water-soluble calcium salts for the control of acidity in rain water. Another alternative solution to the problem of acid rain lies in reducing the emission of sulfur and nitrogen oxides to the atmosphere. This objective can be achieved with the combination of improved control technology and the conservation of fuel.

Conclusions

The study of chemical composition of rain water and total suspended particulates at a few places in India has shown that concentrations of acidic components (SO_4 and NO_3) from the industrial sources are minimal as compared to alkaline components (calcium, magnesium, and potassium) which are released from the soil. Apart from sea salt, SO_4 and calcium have been found to be the dominant components in rain water.

The pH of rain water at different locations has been found to be much above the carbon dioxide-equilibrated value and has mainly varied between 6 and 9.

The correlations (r) worked out between pH and different ionic components from the long series of rain water data at Pune have shown that out of all the ionic components, only calcium has been found significantly correlated with pH. Hydrogen ion concentration has shown an inverse trend of variation with excess cations. This feature suggests that excess cations, which are alkaline in nature, are mainly responsible for the observed alkaline pH. Low pH values and reports of acid rain in India have been restricted to localized regions in highly industrialized cities like Chembur in Bombay.

Acid rain has been observed during the monsoon season of 1985 at the site of a thermal power plant when it was operating with electrostatic precipitators. Before the installation of electrostatic precipitators the rain water samples collected at the site of the plant were reported to be highly alkaline. However, the rain water samples collected within the distance of 4 km during 1985 in the upwind and downwind regions have shown alkaline pH. This feature indicates that long-range transport of pollutants does not seem to be effective in a tropical country like India during the southwest monsoon period when the meteorological factors are favorable for quick dispersal of pollutants whose sources are weak.

The pH values in rain water at Agra and Delhi during 1963 and 1965 respectively, were in the alkaline range and have remained in that range even after a period of two decades during which the cities developed industrially to a great extent.

A comparison of chemical composition and pH of rain water at cloud base with that at ground level has shown that the concentrations of cations decreased up to fourfold from ground level to cloud base level. The pH values of rain water at cloud base level varied from 5.6 to 6.1 and were lower by 1.5 units than the values at ground level. This feature suggests that increase in cations, especially calcium, is responsible for the observed increase of pH at the ground.

The cations exceeded the anions in the aerosols at the different locations studied. This result indicates that the aerosols are potentially basic in nature in India and are capable of rendering the pH of rain water alkaline.

The concentrations of soil-oriented components (calcium, potassium, and magnesium) in the atmospheric aerosols have been found to be higher and those of acidic components (SO_4 and NO_3) lower than those reported for the western countries, which are much more developed industrially. Hence, the alkaline properties of soil and fly ash particulates have sustained high alkaline pH in rain water and have been controlling acid rain in India. A deficiency of soil-derived components (calcium and magnesium) and high concentrations of acidic components in rain water are the main causes of acid rain in the northeast United States, Europe, and Scandinavian countries.

Acknowledgments

I express my gratitude to Dr. A.S.R. Murty for his valuable suggestions. I am also indebted to Mr. R.N. Chatterjee for organizing the collection of rain water samples at Delhi and Agra.

References

Ananth, K.P., J.B. Galeski, L.J. Shannon, F.I. Honea, and D.C. Drehmel. 1977. "Impact of particulates on SO_2 from coal-fired power plants on acid rain." In: *Proceedings of the Fourth Clean Air Congress*, the Japanese Union of Air Pollution Prevention Association.

Cehak, K. and K. Chalupa. 1985. "Observations of various chemical contaminants of the precipitation at a BAPMoN station in the Eastern Pre-Alpine region." *Archives for Meteorology, Geophysics and Bioclimatology* 35:307.

Cox, R.A. 1974. "Particle formation from Homogeneous reactions of sulphur dioxide and nitrogen dioxide." *Tellus* 26:235.

Dams, R., K.A. Rahn, and J.W. Winchester. 1972. "Evaluation of filter materials and impaction surfaces for non-destructive neutron activation analysis of aerosols." *Environmental Science and Technology* 6:441.

Das, D.K., G.K. Dev Burman, and A.L. Kidwai. 1981. "Chemical composition of monsoon rain water over Bhopal, Madhya Pradesh (India)." *Mausam* 32:221.

Eriksson, E. 1959. "The yearly circulation of chloride and sulfur in nature, meteorological and pedological implications, Part I." *Tellus* 11:375.

__________. 1960. "The yearly circulation of chloride and sulfur in nature, meteorological, geochemical and pedological implications, Part II." *Tellus* 12:86.

Galloway, J.N., G.E. Likens, W.C. Keene, and J.M. Miller. 1982. "The composition of precipitation in remote areas of the world." *Journal of Geophysical Research* 87:8771.

Georgii, H.W. 1982. "Global distribution of acidity in precipitation." In: *Deposition of Atmospheric Pollutants.* Dordrecht, Boston, London: D. Reidel Publication Company.

Gorham, E. 1976. "Acid precipitation and its influence upon aquatic Ecosystem, an overview." *Proceedings of the First International Symposium on Acid Precipitation and the First Ecosystem,* VSDA Technical Report, NE–23:425.

Granat, L. 1972a. "On the relationship between pH and the chemical composition in atmospheric precipitation." *Tellus* 24:550.

__________. 1972b. "Deposition of sulphate and acid with precipitation over northern Europe." *Meteorological Institute, Stockholm University Report* AC–2.

Gravenhorst, G. 1975. "Der sulfatantecil in atmospherischem aerosol uber dem Nordatlintik." Berichte des Institutes für Meteorologie und Geophysik der Universitat Frankfurt um Main, Nr.30.

__________. 1978. "Maritime sulfate over the North Atlantic." *Atmospheric Environment* 12:707.

__________, S. Beilke, M. Betz, and H.W. Georgii. 1980. "Sulfur dioxide absorbed in rain water." In: *Effect of Acid Precipitation on Terrestrial Ecosystem.* New York: Plenum Publication Corporation, 41–55.

Green, W.D. 1972. "Maritime and mixed maritime continental aerosols along the coast of southern California." *Journal of Geophysical Research* 77:5152.

Handa, B.K. 1969. "Chemical composition of monsoon rains over Calcutta, Part I." *Tellus* 21:95.

_________. 1973. "Chemical composition of monsoon rain water over Chandigarh in 1971." *Indian Geohydrology* 9:31.

_________, Adarsh Kumar, and D.K. Goel. 1982. "Chemical composition of rain water over Lucknow in 1980." *Mausam* 33:485.

Healy, T.V., H.A.C. McKay, A. Pilbeam, and D. Scargill. 1970. "Ammonia and ammonium sulphate in the troposphere over the United Kingdom." *Journal of Geophysical Research* 75:2317.

Hidy, G.M. and R. Countess. "Deposition both wet and dry." Acid Precipitation series 4, Butterworth, London, 1982.

Hirose, K., Y. Dokiya, and Y. Sugimura. 1983. "Effect of the continental dust over the North Pacific Ocean: time variation of chemical components in maritime aerosol particles in spring season." *Journal of Meteorological Society of Japan* 61:670.

Horne, R.A. 1969. *Marine Chemistry.* Wiley Inter Science.

Horvath, L. 1981. "A csapadekriz kemiai osszetetele Magyarorszagon (Chemical Composition of Precipitation over Hungary)." *Idojaras* 85:201.

_________ and E. Meszaros. 1984. "The composition and acidity of precipitation in Hungary." *Atmospheric Environment* 18:1843.

Junge, C.E. 1972. "Our knowledge of the physico-chemistry of aerosols in the undisturbed marine environment." *Journal of Geophysical Research* 77:5183.

_________. 1979. "The importance of mineral dust as an atmospheric constituent." In: *Saharan Dust.* New York: John Wiley.

_________ and G. Scheich. 1969. "Studien zur Bestimmung des sauregehaltes Von Aerosolteilehen." *Atmospheric Environment* 3:423.

Kapoor, R.K. and Bh.V. Ramana Murty. 1966. "Sulphate aerosols at Delhi in relation to the associated chloride component." *Journal of Applied Meteorology* 5:493.

Khemani, L.T. 1974. "Some aspects of atmospheric chemistry as applied to cloud physics." *M.Sc. Thesis.* University of Bombay.

_________, G.A. Momin, M.S. Naik, A. Mary Selvam, and Bh.V. Ramana Murty. 1980. "Thermal, microphysical and chemical conditions in an urban environment." *Proceedings of VIII International Conference on Cloud Physics,* 1980, 153.

__________, S.K. Sharma, A.S.R. Murty, and Bh.V. Ramana Murty. 1982a. "A simple gadget for collection of cloud/rain water from aircraft." *Proceedings of the Conference on Cloud Physics.* Chicago, November 15–18, 1982.

__________, G.A. Momin, M.S. Naik, R. Vijayakumar, and Bh.V. Ramana Murty. 1982b. "Chemical composition and size distribution of atmospheric aerosol over the Deccan Plateau, India." *Tellus* 34:151.

__________, M.S. Naik, G.A. Momin, R. Kumar, R.N. Chatterjee, G. Singh, and Bh.V. Ramana Murty. 1985a. "Trace elements in atmospheric aerosols at Delhi, north India." *Journal of Atmospheric Chemistry* 2:273.

__________, G.A. Momin, M.S. Naik, P.S. Prakasa Rao, R. Kumar, and Bh.V. Ramana Murty. 1985b. "Trace elements and sea salt aerosols over the sea areas around the Indian sub-continent." *Atmospheric Environment* 19:277.

__________, G.A. Momin, M.S. Naik, P.S. Prakasa Rao, R. Kumar, and Bh.V. Ramana Murty. 1985c. "Impact of alkaline particulates on pH of rain water in India." *Water, Air and Soil Pollution* 25:365.

__________. 1985. "Characteristics of atmospheric gaseous and particulate pollutants and their influence on cloud microphysics and rainformation." *Ph.D. Thesis,* University of Poona.

Likens, G.E. and F.H. Bormann. 1974. "Acid Rain: A serious regional environmental problem." *Science* 184:1176.

Miller, J.M. and A.N. Yoshinoga. 1981. "The acidity of Hawaiian precipitation: a preliminary report." *Geophysical Research Letters* 8:779.

Mukherjee, A.K. 1957. "Hydrogen ion concentration of monsoon rain water in Calcutta." *Indian Journal of Meteorology and Geophysics* 8:321.

__________. 1964. "Acidity of monsoon rain water." *Indian Journal of Meteorology and Geophysics* 15:267.

__________. 1978. "pH of monsoon rain water over the Bay of Bengal." *Indian Journal of Meteorology and Geophysics* 29:749.

__________. and Krishna Nand. 1981. "A physical reason for higher pH of tropical rain water." *Mausam* 32:231.

__________, Krishna Nand, B. Mukhopadhyay, and S.G. Ghanekar. 1986. "Rain water chemistry over Indian sea areas during monsoon season." *Mausam* 37:173.

Munger, J.W. 1982. "Chemistry of atmospheric precipitation in the North-central United States: Influence of sulphate, nitrate, ammonia and calcareous soil particulates." *Atmospheric Environment* 16:1633.

Murty, Ramachandra A.S. and Bh.V. Ramana Murty. 1969. "Chemical composition of rain water across the Western Ghats." *Indian Journal of Meteorology and Geophysics* 20:595.

Nagamoto, C.T., F. Parungo, R. Reinking, R. Puesohel, and T. Gerish, 1983. "Acid clouds and precipitation in Eastern Colorado." *Atmospheric Environment* 17:1073.

Naik, M.S. 1985. "Chemical characteristics of monsoon rainfall." *M.Sc Thesis*, University of Poona.

NEERI (National Environmental Engineering Research Institute). 1980. *Air quality in selected cities in India, 1978–1979.*

Pack, D.H. 1980. "Precipitation chemistry patterns." *Science*, Washington 208:1143.

Peden, M.E. and L.M. Skowron. 1978. "Ionic stability of precipitation samples." *Atmospheric Environment* 12:2343.

Penkett, S.A., H.F. Atkin, and M.H. Unsworth. 1979. "Chemical composition of the ambient aerosol in Sudan Gazira." *Tellus* 31:295.

Peters, N.E., R.A. Schroeder, and D.E. Troutman. 1982. *Temporal Trends in the Acidity of Precipitation and Surface Waters of New York.* Geological Survey Water Supply Paper 2188. Washington, D.C.: U.S. Government Printing Office.

Prospero, J.M. and T.N. Carlson. 1972. "Vertical and areal distribution of Saharan dust over the western equatorial North Atlantic Ocean." *Journal of Geophysical Research* 77:5255.

__________ and T.R. Nees. 1977. "Dust concentration in the atmosphere of the equatorial North Atlantic; possible relationship to the Saharian drought." *Science*, Washington 196:1196.

__________, R.A. Glaccum, and T.R. Nees. 1981. "Atmospheric transport of soil dust from Africa to South America." *Nature* 289:570.

Riley, J.P. and R. Chester. 1971. *Introduction to Marine Chemistry.* Academic Press.

Rao, Y.P. and B.N. Desai. 1973. "The Indian summer monsoon." *Meteorology and Geophysics Review* 4, India Meteorological Department 18.

Sadasivan, S. 1977. "Studies of some important trace constituents in the monsoon rains and their role in atmospheric process." *Ph.D. Thesis*, University of Bombay.

Sequeira, R. 1976. "Monsoonal deposition of sea salt and air pollutants over Bombay." *Tellus* 28:275.

__________ and D. Kelkar. 1978. "Geochemical implications of summer monsoonal rain water composition over India." *Journal of Applied Meteorology* 17:1390.

Shannon, D.G. and L.O. Fine. 1974. "Cation solubilities of lignite fly ashes." *Environment Science and Technology* 8:1026.

Stensland, G.J. and R.G. Semonin. 1982. "Another interpretation of the pH trend in the United States." *Bulletin of American Meteorological Society* 63:1277.

Subramanian, V. and K.K. Saxena. 1980. "Chemistry of Monsoon Rain Water at Delhi." *Tellus* 32:558.

Tsunogai, S., O. Saito, K. Yamada, and S. Nakaya. 1972. "Chemical composition of oceanic aerosol." *Journal of Geophysical Research* 77:5283.

Vermeulen, A.J. 1980. *Polluted Rain.* New York: Plenum Press.

Winkler, F.L. 1980. "Observations on acidity in continental and marine atmospheric aerosols and in precipitation." *Journal of Geophysical Research* 85:4481.

WMO (World Meteorological Organization). 1984. *BAPMoN-data for 1980.* National Climatic Data Centre, Federal Building, Ashville, N.C., U.S.A.

FOUR

IMPORTANCE OF BIOGEOCHEMICAL PROCESSES AND WEATHERING FOR NORWEGIAN STREAMWATER CHEMISTRY

Ivan Th. Rosenqvist

Surface water acidification and fish decline in modern times have been taken as examples of the harmful effect of the sulfur emission from the modern world industry. Especially in Sweden and in Norway, several thousand lakes have lost their fish populations in the twentieth century, mostly after 1950. However, many lakes have been said "always" to have been barren, and these lakes also have proven to be acid (Strøm). In other cases we know of, several lakes lost their fish populations in the nineteenth century (Helland; Dahl, K., 1921; Torgersen). In these cases we know that the reason for the fish decline was the chemistry of the lake water, mainly the low pH and content of aluminum, etc. In other parts of the world with no pollution or pyritic rocks, for instance in South American laterite areas with tropical podzols, very acid rivers at pH around 4 have been known for a long time (Sioli).

There seems to be general agreement that modern industrialized society emits to the atmosphere around 70 million tons of sulfur, corresponding to about 200 million tons of sulfuric acid per year from combustion of fossil fuels and from metal smelters. This is about one-third of the total sulfur precipitating from the global atmosphere. Thus, two-thirds of the total sulfur precipitation has natural origins, partly from sulfate salts in sea spray, which during stormy weather are blown up into the air. To a greater extent, airborne sulfur originates from decaying organic material, where the sulfur-bearing amino acids in the proteins from the flora and the fauna in the oceans and on land pass through complicated chemical processes into gases. Finally, these oxidize to sulfate and then fall back to the ocean and on land surfaces. Volcanos seem to contribute only a minor part of the sulfur emitted to the atmosphere, some 2–4 percent. It seems as if the Scandinavian problem of modern acidification of water courses has been taken as a typical example of the harmful influence of atmospheric pollution. Because thousands of lakes in southern Norway and Sweden have lost their fish populations, this massive fish decline has been taken as proof of acidification due to atmospheric pollution, mainly because it coincided in time with an increase in industry and the installation of high stacks after World War II, which further spread the flue gases from European industry.

It has been claimed that the acidity in rain has increased ten to one hundred times and that the acidity of surface water has followed this increase directly. Based on sulfur content in cores from ombrogenic peat, N.A. Sørensen (1983) concludes that the sulfur content of the rainfall in southern Scandinavia has increased only by about 70 percent from pre-industrial time, and not by a factor of ten, as claimed by some environmentalists. In addition to the lakes which have been scientifically reported to have lost their fish population in the nineteenth and twentieth centuries, we have in our folklore stories about streams and lakes which suddenly lost their fish population.

This was, in medieval time, taken as a divine punishment for the sinful behavior of individuals or of the whole community.

In all those cases where lakes or streams have become barren in this century, we observe that the water, episodically or permanently, has low pH values, mostly between 4 and 5. In most cases we have no reliable measurements of acidity priorto the fish declines. Investigation of bottom sediments from lakes may, in some cases, show quantitative changes in the diatom flora.

Although there are many factors of importance for algae growth, the diatom composition has been taken as a hypothetical indicator of acidity. According to these investigations, there seems to have been a decrease of less than one pH unit in modern time (corresponding to the upper 10 cm of the lake sediments). In an undated, probably less than 2,000-year old, sediment core, Berge, however, claimed to have found greater variations deeper down. According to the formula used, he calculated a pH value of less than 5 at 20 cm depth and less than 4.3 at 78 cm depth in the sediment. In our institute, Dr. Bjørg Stabell also has recorded considerable variations in the diatoms with depth in some sediment cores. This is following variations in the pollen examined by Dr. Helge Høeg. The acidity of the rainfall in the investigated areas of southern Norway and Sweden is relatively low and *fairly homogeneous*, corresponding to around 40 micro equivalents H^+ per liter, varying by 50 percent on an annual basis. (This corresponds to one small glass of fresh lemon juice dissolved in one ton of distilled water). On the other hand, H^+ in streamlets in gneiss-granitic areas varies within factors of 1:1000. (If gabbro areas are included, we have variations of 1:10000.) We know today that *there is no direct connection* between acid episodes in stream water and acidity in the rainfall prior to measurements of the stream water pH.

We know, however, that acid episodes in the streams always occur when the streams are in flood following heavy rainstorms or snow melts and that the most acid rivers are found in areas with the highest precipitation. In all cases, episodic rainstorms of *high or low pH, or low or high sulfate contents*, are followed by episodes with low pH in the river (Skartveit *et al.*). The pH, however, varies from streamlet to streamlet on the same type of bedrock.

Frequently, run-off water may be much more acid than rain, and in small creeks within a very limited area, with the same rain and same bedrock, we may have very strong variations in the acidity in the run-off water. At base-flow, all these rivers have low acidity, often only one-twentieth of the H^+ content of the same streamlet at storm-flow. The most important factors today seem to be the hydraulic run-off pattern and ion exchange processes in the top soil, combined with weathering processes in the deeper parts of the profiles.

All acid streams and lakes in Norway are found in areas with acid podzolic soil profiles. Frequently, we have off-shortened podzols containing only the mor (immature raw humus mostly in forest soil with low base saturation), O-layer and the A-layer. There are today sufficient data to say that a given amount of precipitation will give the same acidity, or H^+ concentration, in the run-off water from a given catchment, provided the ionic strength in the precipitation water is the same. Variations in the sulfate/chloride ratio do not lead to any noticeable differences.

By laboratory experiments, it has been shown that after they have percolated through topsoil from the watershed of an acid river, equivalently diluted solutions of sulfuric acid, calcium sulfate and calcium chloride give exactly the same pH values.

The acidity will, however, vary from field to field, depending on two factors—the run-off hydrology and the nature of the soil profiles.

Changes in vegetation and, consequently, changes in accumulation or oxidation of components in the biogeochemical environments are dominating factors. The organic material may occur as raw humus (mor) or mull (mature organic material in equilibrium with neoformation balancing natural oxidation—base saturation of ion exchange material is high). Mor will be active as a pH-buffer at low pH values whereas mull has much lower capacity and the pH of the soil water will give run-off water of higher pH.

After uncontrolled, unfenced goat and sheep grazing was forbidden in the first part of the nineteenth century in Denmark, the sandy west Danish calluna heath areas (without carbonates in the underground) underwent a transformation from mor to mull (since P.H. Møller examined those profiles in 1879 and 1884). Now these areas have been planted with leafwood—ash, elm and maple. The topsoil has totally changed. Fleming Juncker estimated that the pH has *increased* by more than one pH unit, in spite of the fact that the rain is considerably more polluted than in central Scandinavia.

In Norway and Sweden, however, we have had a change in the vegetation in the opposite direction. Coniferous forests of spruce, pine, and heather have taken over more or less naked land with low humus content, just in the areas where we today have the acidified rivers. (These areas were investigated in 1854 by the Prussian chief forest inspector, von Berg.) They were reexamined by Tryggve Troedesson. The changes have led to formation of thick layers of acid mor during these 130 years. In Sweden, Per A. Melkerud investigated an area which had been grazing land, at least from the beginning of the seventeenth century, but from 1820 the area has gradually been reforested by spruce.

It appears well documented that we have had in this case a strong chemical weathering and acidification after the first, second and third generation of spruce trees. This is followed by neo-formations of smectite minerals from the early mica and feldspar-sandy soil.

In podzol profiles in Sweden and Norway, Gunnar Jacks and I have independently found there is a reservoir of exchangeable H^+ in the acid soil profiles, corresponding sometimes to many hundreds, or even a thousand years of the present polluted rainfall. This H^+ will, by ion exchange with electro positive ions in the water, be liberated provided there are sufficient anions. In all these acid areas, the profiles are *out of equilibrium*, so that the amount of organic material formed from the litter from the trees is greater than the oxidation, whereas, in the natural forests, before the influence of man (some four thousand years ago), there must have been an equilibrium—so that the oxidation of organic material corresponded to the neo-formation.

As mentioned, we have no acid brooks, except in areas where we have acid topsoil. On the other hand, there are areas with acid topsoil and deep sandy feldspar-bearing subsoils. In these cases we have no acid run-off.

During the International Biological Program, the amount of annual growth of sphagnum mosses was found to be between 100 and 300 grams dry material per m^2 per year in the subalpine parts of Norway. In these mosses the dry material contained polyuronic-acid-pentosane-aggregates corresponding to *from 18 percent to 35 percent* polyuronic acid. That means we have a neo-production of 18 to 105 grams per m^2 per year. As the equivalent weight of polyuronic acid is 176 grams, this means we have a neo-formation of between 100 and 600 milli-equivalents of exchangeable H^+ per m^2 per year. This may be compared to a 1,000 millimeter rainfall with a pH of 4.3, which corresponds to 50 milli-equivalents H^+ per year.

Most of the newly-formed polyuronic acid cannot contribute to the acidification in surface water as there are not sufficient amounts of anions in the rain. The material is then either oxidized and returned to the atmosphere as carbon dioxyde and water, or it may form peat layers.

In some peat moors the amount of exchangeable protons corresponds to several kilometers of the present polluted rainfall. The importance of the exchange processes, as the determining factor of the run-off water, is beautifully illustrated by Richard Wright and his co-workers (Wright and Gjessing) in the so-called project RAIN. In this case, two mini-catchment areas in Risdalsheia, in southern Norway, were covered by transparent roofs in the spring of 1984. The water from one of these roofs was first de-ionized and ocean water was then added in order to give the same level of chlorine as in the other field. This means about 80 micro-equivalents chlorine per liter. Then, the treated low-sulfate high-pH water was spread on the catchment area under the roof. The other mini-catchment also was covered by a transparent roof, but the rain water was directly spread under the roof.

In the first case, the artificial rain water had a pH of 5.05 and 7 micro-equivalents of sulfate per liter. Whereas, the rain in the field had a pH of 4.3 and 64 micro-equivalents sulfate per liter. The total run-off in the first case, for the period 13 June, 1984 to 14 November, 1985, had a pH of 4.07, whereas, in the case where the untreated precipitation was spread under the roof, the pH was 4.03. In both cases the run-off water was more acid than the applied "rain." The difference between the two fields was, however, insignificant.

We have to be further aware of the fact that under the first roof, nitrate and ammonium have now been removed and we will probably, in a few years, observe a change in the flora in the field with the de-ionized water. This may increase the pH.

If we pass along a sloping field during heavy rainfall and rake samples of the surface water which is running off, we frequently observe that the H^+ content varies within a factor four, or more, over a distance of 100 meters, and the run-off water may be ten times more acid than the rain water. This shows that the ion exchange between the cations in the rain and the cations (mainly H^+) in the mor is a very quick process, which always leads to acid run-off. Deeper down in the profiles, the weathering of the feldspar and other minerals will again neutralize the acid formed in the upper layer. This is the reason for the much higher pH in water courses at base-flow.

The acidity of the run-off water from the topsoil depends mainly on the nature of the mor and the content of mobile anions—such as sulfate, chloride or nitrate. For each small area, there exists at *high intensity* of run-off, a linear relationship $H^+ = A\Sigma\ (Cl^-, NO_3- SO_3-) + B$. (A and B vary from area to area.)

In a soil profile, the O-layer and the bleached A-layer are nearly pure cation exchange material whereas the B-layer may at the same time be cation and anion exchange.

In acid soil profiles, the cation exchange layers are often more than 90 percent in the proton state—and the ion exchange capacity is of the order of one gram equivalent per kilogram.

Also, the B-layers are chiefly cation exchanger, but especially polarizable anions like sulfate and phosphate may here partly exchange on the (OH) of hydroxides and clay minerals and be retained.

A. Henriksen (1979) presented correlations between alkalinity (mainly HCO_3-) and SO_4- and bivalent cations (calcium + magnesium) in lake water. He gave the following statements:

"The acidification of fresh water is a continuous process analogous to the titration of a bicarbonate solution with sulfuric acid"; further, "A consistent characteristic of acidified waters is the replacement of bicarbonate by sulfate while little, if any, change in the relative concentrations of cations has occurred."

That these statements are wrong is obvious to anyone who has worked with ion exchange processes and hydrology in the field.

He could take, as we have, one liter of the Oslo drinking water from the tap in his office and 10 grams of the raw humus from the reforested spruce area outside the building—then he could rinse the mor material with deionized water to get rid of all water soluble acids and subsequently filter the tap water through the soil. The analysis would give the following results:

Tap water				*Filtered through mor*			
pH 6.4				pH 4.26			
H^+	=	0.4	micro eqv./liter	H^+	=	55	micro eqv./liter
Ca^{++}	=	145	"	Ca^{++}	=	25	"
Mg^{++}	=	46	"	Mg^{++}	=	10	"
SO_4++	=	150	"	SO_4++	=	n.d.	"
HCO_3	=	43	"	HCO_3-	=	0	"

N.A. Sørensen (1985) was probably the first Norwegian scientist who stressed the fact that the rivers and lakes where salmonide fish had high mortality were characterized by very low calcium content. This does not necessarily depend upon extremely low calcium content in the precipitation or the bedrock and regolith, but upon the fact that calcium is fixed in a biogeochemical ion exchange material which is *out of equilibrium* with the chemical environment in which it is situated. This is illustrated by the above listed change in chemistry due to the actions of the fresh, raw humus.

The importance of the organic layer is demonstrated in three places examined in South Norway (Rosenqvist, 1985). In these three localities, extensive forest fires in the dry season had burned off the total organic layer 6 to 20 years before the examination. In all three cases, run-off from these areas had acidities 1,000 to 4 times less than in the areas just outside the burned fields. On the other hand, the content of magnesium and calcium was found to be considerably higher in the run-off from the area which had burned 6–10 years before the investigations as compared to the run-off outside the area. This difference also existed, but to a minor degree, in the two re-vegetated areas, which burned 20 years before. This has led some people to say that the bedrock must have been richer in calcium and magnesium in the burned area than outside—and perhaps the waters always had had a higher pH in the burned area. Besides the low probability that forests are more inflammable at high pH than at low pH, we have investigated the effect of ion exchange on such waters. Kjell Hegna has, from 1982 to 1985, examined the limnology of an area which burned in 1976 and compared it with a neighboring area which did not burn and had very acid water.

In our laboratory, my colleague, Turid Winje, carried out the following experiment:

To an artificial solution of chlorides, sulfates, and bicarbonates of magnesium, calcium, sodium, and potassium, corresponding to the neutral run-off from the field which burned 10 years ago, 0.1 – 0.05 percent humus (mor) was added from outside the burned area and the filtrate also was analyzed.

Original sol. micro eqv.				*Filtrate micro eqv.*			
	pH 6.1				pH 4.5		
Mg^{++}	20.6	SO_4^{--}	70	Mg^{++}	3.5	SO_4^{--}	73
Ca^{++}	85	Cl^-	50	Ca^{++}	8	Cl^-	50
Na^+	27	HCO_3^-	28	Na^+	53	HCO_3^-	0
K^+	11.8			K^+	25		
H^+	1			H^+	30		
	+ 145.4		– 148		+ 120		– 123

We can thus see that the ion exchange, mainly of the calcium and magnesium ions, liberated the univalent ions Na^+, K^+, and H^+. The H^+ first reacted with the bicarbonate ion and gave, at the lower pH, carbon dioxide and water. After having neutralized the bicarbonate, further ion exchange lowered the pH. By increased contact with one gram humus per liter, magnesium dropped to 3, calcium to 6.5, sodium to 50, and potassium increased to 27.5—and pH dropped further.

The average composition of the quartz-bearing magmatic and metamorphic areas in southern Norway is close to:

SiO_2	70.0%
TiO_2	0.5%
Al_2O_3	14.5%
Fe_2O_3	1.5%
FeO	2.0%
MnO	0.1%
MgO	0.8%
CaO	2.0%
Na_2O	3.5%
K_2O	4.0%

The relatively high percentage of magnesium and calcium in the weathering rock is generally not found in the acid run-off water, in spite of the fact that plagioclase, biotite, and amphibole are *more* attacked than the other silicate minerals.

This clearly demonstrates the fact that the bivalent ions have the highest capacity to exchange and liberate H^+ from the ion exchange material. Thus, it is the fixing of Ca^{++} in the catchment which is the reason for the low calcium content in the run-off, and at the same time the reason for the high H^+ content.

The sulfur and chloride in southern Norway seem to be higher than the continental crust average, particularly in the rocks of the so-called Bamble Formation. The neutralization capacity of rock and regolith, by weathering, may be calculated. Different methods have been used.

The precipitated CBD (citrate, bicarbonate, dithionite) soluble part of the B-layer in one postglacial podzol weathered until Rosenqvist *et al.* (1980) calculated a minimum value of consumed H^+ of 140 milli-equivalents per m^2 per year.

In the Cascade Mountains, U.S.A., Reynolds and Johnson calculated 930 milli-equivalents H^+ per m^2 per year.

The enrichment of zircon in the tills of upper Numedal in southern Norway, compared to the zircon content in the Precambrian rock from which the till originated, indicate that half the silt fraction must have gone into solution (Stiberg). If this had taken place in 10,000 years (post Weichselian) more than 300 milli-equivalents H^+ per m^2 must have reacted by weathering per year.

If we compare these figures with the H^+ content of 1,000 mm of rainfall of pH 4.3 = 40 milli-equivalents per m^2 per year, we understand that if it had not been for the "rejuvenation" of the acidity due to the increasing raw humus (mor) by exchanging cations suchas Mg^{++} and Ca^{++} "from the weathering water" against H^+, no acid surface water would *exist in these areas* by the relatively low degree of pollution. In fact, nobody has ever been able to point to any area in Scandinavia where the H^+ of acid rain makes up as much as 15 percent of the total H^+ budget—e.g., even in the areas with the most acid run-off. Also, in areas with little regolith cover, the weathering on the barren granitic rock, has been of the same magnitude—e.g., on the order of 3–10 mm per year, that is, 10-340 grams per m^2 per years since the Ice Age.

Although the emission of sulfur in the United Kingdom (UK) was more than 50 percent higher in 1972 than today (UK Department of Environment) and the other western countries, including Scandinavia, have reduced their emissions to similar degrees, the number of acid rivers is steadily increasing in Norway.

Near intense local emitters, such as the nickel smelter in Sudbury (Dillon *et al.*), we have very much higher acidity in the precipitation and dry deposition than in Scandinavia, but the lakes are rarely as acid as the Scandinavian barren lakes. We had, from 1880 to 1946, an important nickel smelter in Norway at Evje. In a narrow valley, this smelter emitted, every winter, more sulfate than now falls over the surrounding 10,000 km^2, but the lakes in the district had plenty of fish. Some fifteen years after the smelter was closed down, some of the surrounding small lakes became barren, together with the general fish decline in southern Scandinavia.

One of these lakes is now being thoroughly examined by many methods, thanks to the British, Swedish, and Norwegian SWAP program.

The hydrology of a catchment area in the acid areas may be schematically illustrated by a model consisting of two types of material: (A) ion exchange materials, mainly in the H^+ state, and (B) rock-forming minerals. This is schematically illustrated by Figure 4-1.

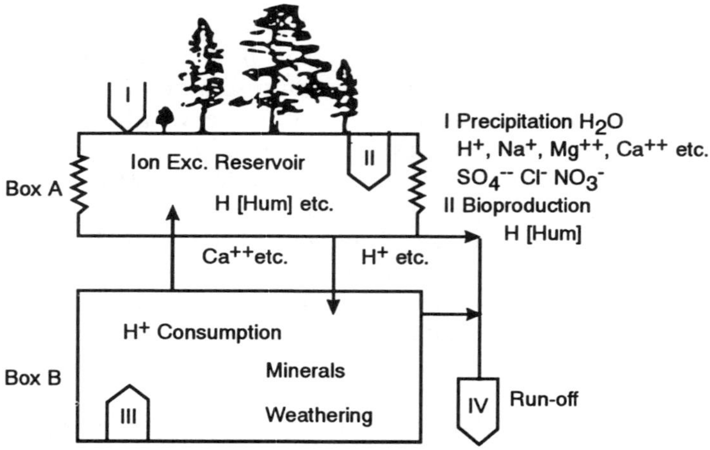

Figure 4-1: Schematic of ion exchange reactions in the hydrology of a catchment area consisting of two types of material: A - ion exchange materials, mainly in the H^+ state, and B - rockforming minerals, which undergo the following types of reactions:

Box A: $Na^+ + An^- + H(Hum) \rightarrow Na(Hum) + H^+ + An^-$

Box B: $2H^+ + Ca\ Silicate \rightarrow Ca^{++} + H_4SiO_4 + Al(OH)_3$

Box A: $Ca^{++} + 2Na(Hum) \rightarrow Ca(Hum) + 2Na^+$

$An = SO_4^{--}, Cl^-, NO_3$

On its way to the water courses, rain water will pass both types of material. Depending upon variations in permeability and intensity of precipitation, water may first pass through material (A) then through material (B), and again be pressed up in material (A) and so on. During rain storms, the influence of material (A) may dominate. At base-flow, material (B) will be *the dominating material* factor. If the material of (B) is of gneiss granitic composition, as in the Scandinavian podzols, weathering processes will tend to neutralize the percolating water, but this will again turn acid by the passage through and upon material (A).

But tropical podzol material, as represented in areas with acid rivers (Sioli), (B) is lateritic and not capable of neutralizing acids.

If material (A) is in equilibrium or absent, because it has burned off, material (B) will dominate and the run-off will have high pH. Only if material (A) is increasing (out of equilibrium), as in the present acid areas of Scandinavia, will any rain result in acid run-off.

Conclusion

If you ask the right question, Nature will give you the correct answer.

References

Berge, F. 1975. "pH Forandringer og Sedimentasjon av Diatomeer i Langtjern (pH Changes and Sedimentation of Diatoms in Lake Langtjern)." SNSF Project Internal Report 11/75, Oslo-Ås.

Dahl, K. 1921. "Undersøkelser over ørretens ut døen i det sydvestlige Norges fjellvand (Investigation of the extinction of trout in mountain lakes of SW Norway)." Norvk Jeger og Fisker Forening Tidskrift. 1921, 4.

Dillon, P.J., N.D. Yan, W.A. Scheider, and N. Convoy. 1977. "Acid Lakes in Ontario, Canada: Their Extent and Responses to Base and Nutrient Additions." Jubilee Symposium on Lake Metabolism and Lake Management. Uppsala.

Hegna, K. 1986. "Sammenligning av Vann og Sedimentkjemi mellom et 6-9 År-Gammelt Skogsbrannområde og et ikke Brent Skogsområde i Telemark (Comparison of Water and Sediment Chemistry between an Area Subjected to Forest Fire 6–9 Years Earlier and an Unburned Area in Telemark)." Thesis, Limnological Institute, University of Oslo.

Helland, A. 1900. *Norges Land og Folk* (*The Country and the People of Norway*). Bratsberg County, Norway.

Henriksen, A. 1979. "Alkalinity is the equivalent sum of bases that are titrated with strong acids." *Nature* 278:542.

__________. 1980. "Acidification of Freshwaters—A Large Scale Titration." *Proceedings of an International Conference on the Ecological Impact of Acid Precipitation.* 68–74. SNSF Project, Oslo-Ås.

Jacks, J. 1984/85. "Liming of Soil as an Amendment for Acid Ground Water." Research Activity Catalogue. National Swedish Environmental Protection Board Bulletin 1984/85.

Juncker, F. 1985. *Dynamisk Okologi Humus* (*Dynamic Ecology. Humus*). Copenhagen: Gyldendal.

Melkerud, P.A. 1984/85. "Dynamics and Structure of Pedogenically Formed Smectite in Soil Sequence below Stands of 1st, 2nd and 3rd Order of Coniferous Forests." Research Activity Catalogue, National Swedish Environmental Protection Board Bulletin 1984/85.

Reynolds, R.C. and N.M. Johnson. 1972. "Chemical Weathering in the Temperate Glacial Environment of Northern Cascade Mountains." *Geochemical and Cosmochemical Acta* 36:537.

Rosenqvist, I. Th. 1977. *Sur Jord—Surt Vann* (*Acid Soil—Acid Water*). 123pp. Oslo: Ingeniorforlaget.

__________. 1981. "Betydningen av Sur Nedbor og Sur Jord i Innsjokjemi (Importance of Acid Precipitation and Acid Soil for Lake Chemistry)." *Vann* no. 4.

__________. 1985. "Acid Rain, Acid Precipitation and Acid Soil in Fresh Water Chemistry." *Land-use Policy* 1985(Jan.):70.

__________, P. Jørgensen, and H. Rueslåtten. 1980. "The Importance of Natural H^+ Production for Acidity in Soil and Water." *Proceedings of an International Conference on the Ecological Impact of Acid Precipitation.* 210–242. SNSF Project. Oslo-Ås.

Sioli, H. 1977. "Studies in Amazonian Waters." Atas to Simposio Sabre a diata Amazonia. *Limnologia* 3:9.

Skartveit, A., B. Halsvik, and E. Meisinset. 1980. "Nedbørtilførsel av Sjøsalter og Sure Komponenter og Avrenning av Joner i Nedborsfelt på Vestlandet (The Input of Marine Salts and Acid Components in the Runoff of Ions in Precipitation in Western Norway)." SNSF Project Internal Report 63/80. Oslo-Ås.

Stiberg, J.P. 1984. "En Undersokelse av Tungmineral Inholdet i Bunumorener fra Numedalstagens Nedslagsfelt (An Investigation of the Heavy Mineral Content in the Basal Tills from the Numedal-Rivers Precipitation Area)." Thesis. Institute of Geology, University of Oslo.

Strøm, K.M. 1925. "pH Values in Norwegian Mountain Lakes." *Nyt Magasin Natur* 62.

Sørensen, N. A. 1983. "De Tre Forsuringsteser (The Three Theses for Acidification)." *Miljøvern Teknik* no. 4:5.

__________. 1985. "Vedrorende det Langsiktige Behov for Kalking av Norske Vassdrag (Concerning the Long Lasting Need for Liming of Norwegian Surface Waters)." *Vann* no. 4:276.

UK Department of the Environment. 1984. "Acid Deposition in the United Kingdom." Warren Spring Laboratory, Department of Trade and Industry.

Torgersen, H. 1934. "Forsøk med Opparbeidelse av Sure, Fisketomme Vann (Experiments in Reclamations of Acid-Fishbarren Lakes)." *Stangfiskeren* 1934.

Troedsson, T. 1984. "Brukas eller Forbruces Skogsmarken (Is the Forest Soil Used or Consumed)?" *Räcken Skogen (Do We Have Sufficient Forests)?* 32–39. Stockholm: Svenske Forskningsrådnämnden.

von Berg, K.H.E. 1855. "Ueber die Vegetations-Verhaltinisse in Skandinavi besonders in forstlicher Beziehung (On the Vegetation Conditions in Scandinavia, Particularly in Relation to Forests)." *Tharander Jahrbuch* 11.

Wright, E.F. and E. Giessing. 1986. Acid Rain Research Report 9/1986. Oslo: Norwegian Institute for Water Research.

FIVE

FOREST CONSERVATION IN TROPICAL ASIA

THE INDIVIDUAL DIMENSION

Warren Y. Brockelman

Forest Destruction: Global or Individual Problem

Tropical forest destruction is one of the most pressing problems facing the world today. Reports are written year after year summarizing various implications of the problem: wood shortage, loss of thousands of animal and plant species, changes in weather, increased soil loss, more severe flooding, and uglier landscapes.

A few commentators (e.g., Simon and Kahn, 1984; Simon, 1986) have disputed the accuracy of available statistics and the importance of the problem at the global level, but no one can deny the overall trend of forest loss in tropical countries. No one can fail to be concerned about the seriousness of losing thousands of species from the earth in such a short time.

Is forest destruction truly a global problem of human behavior? "The individual dimension" in my title is intended to echo a dilemma. Although the problem is in some sense "global" in scale, our continuing preoccupation with its global importance may distract us from designing effective solutions to it. All of our problems relate in some way to the actions of individuals. We have largely failed to come to grips with the problem at this end of the scale in our efforts to reverse the trend of forest destruction.

It is a bit like trying to steer or reverse the direction of a large column of tropical termites when we have no control over the behavior of a single termite. We may have an elegant model of the flow of the whole colony as a system, but to little avail. We just have not accounted for what is in the mind of the individual termite.

The seriousness of tropical forest destruction, however, can only be appreciated from a global perspective. This perspective was most suddenly and dramatically provided by the radio beams from the first U.S. Landsat (NASA) satellite during 1972–73. Satellites will now be a permanent source of information and are increasing in resolution and, hence, in usefulness. In many tropical countries, including Thailand, the availability of satellite images made it impossible for the government to exaggerate the actual extent of forest resources. Thailand's forest area dropped overnight from 51 percent of its land area to under 40 percent (Boonyobhas and Klankamsorn). This immediately stimulated press accounts, policy statements, and plans for increased forest protection (e.g., *Bangkok Post,* 1975, 1978; *The Nation Review*). This concern has continued to this day, while the forests have continued to decline (Royal Forest Department).

Global analyses of the problem show that rapid deforestation is occurring in virtually all countries of the tropics (UNFAO; Myers, 1980; USCEQ; USITFOTF; World Resources Institute). Each year, forty to fifty thousand km^2 of forest is logged, nearly all of which (90 percent) becomes converted to cropland.

Shifting cultivation also permanently reduces forest area: about 145,000 km^2 of forest fallow is cleared each year, and 23 percent of this becomes converted to permanent cropland, due to increasing human populations (World Resources Institute).

Another major cause of destruction of forests is direct conversion to rangeland. This is especially tragic and shortsighted in Amazonia (Myers, 1984), where peasants have been encouraged by special incentives to clear forests to raise beef cattle. The major beneficiaries are people in the U.S. who, as a consequence, enjoy hamburger meat at lower prices.

Fuelwood collection is also an important cause of forest deterioration throughout the non-industrial tropics. The vast majority of rural poor depend on fuelwood (including charcoal) for household cooking, and it is already acutely scarce in the more populous poor countries (World Resources Institute; Worldwatch Institute, 1975, 1979). It is difficult to assess the impact of fuelwood collection on forests because it is highly dispersed and not reflected in trade statistics, and tends to cause general deterioration of forests rather than a visible reduction in area. More than any other cause of forest destruction, fuelwood collection concerns only the individual resident and his immediate family, and their direct needs for survival.

The most prevalent causes of forest destruction and deterioration thus relate to the maximization of individual and household benefits. In contrast, the problems caused by forest loss most often appear to be global in scale. This is reflected in the following passage: "Tropical forests provide humanity with a cornucopia of benefits. Tropical forests are living museums and laboratories that have yielded only a tiny fraction of their treasures to scientific study. Many useful chemical products essential for making medicines and pharmaceuticals originate in tropical plants. More than 50 percent of modern medicines come from the natural world, many from tropical forests . . . Industrial products derived from tropical moist forests include volatile essential oils, gums, resins, latexes and other exudates, steroids, waxes, rubber, fibers, dyes, tanning agents, turpentines, edible oils, rattans, bamboo flavoring, spices, and pesticides" (Myers, 1984: 111; World Resources Institute).

The benefits from all these products rest largely on the existence of the science and technology of the industrialized nations, as well as marketing and consumers in developed countries, trade agreements between developed and developing countries, international shipping arrangements, capitalized entrepreneurs in developing countries, a market for collecting and buying the raw products from forest areas, and the availability of local harvesters. "Middlemen" who oversee the collection and pay the laborers are also likely to be involved. Thus, poor residents in or near the forests of tropical habitat countries will share at best only a trivial part of the wealth derived from a given forest product. In fact, marketing of a tropical forest product is likely to remain profitable only if this long global chain from forest plant to consumer can be cut short by synthesis of the product directly in the industrialized country. While the tropical forest with its diverse cornucopia of species may have provided the world with the initial knowledge of the product, the product no longer depends on the forest for yielding its benefits to man.

An exception is rubber, once an uncommon tree in its native forest, which is still profitably managed and harvested on a commercial scale in the tropics. Coffee, tea, and cocoa are further examples. Ironically, however, all these plants are raised in locally owned plantations which compete with natural forests for growing space. The natural forests are losing ground.

It is true that the preservation of the mother forest is necessary if we are to continue to discover exciting new products of benefit to us (Myers, 1984); this is a global concern. The individual persons whose actions determine the fate of the forest, however, will never share in significant benefits unless management and product development are drastically changed. Ironically, then, the mother forest which yields new products to the world is all too often destroyed by these same products, unless they can be synthesized elsewhere. In that case, there is no local incentive for either saving or destroying the forest.

Numerous other products of known use are harvested directly from natural forests and marketed locally. People in Western countries are unaware of most of these. For example, villagers in southeast Thailand harvest the orange-yellow sap of the tree *Garcinia hanburyi* which is dried and concentrated into cakes and made into paints and dyes (Kanchananaga; Brockelman and Srikosamatara, 1978). Other tree species of the genus *Aquilaria* in southern and Southeast Asia produce a special core wood which, when decayed by fungi, has aromatic properties useful in producing food seasoning and incense (Gianno). Groups of villagers roam through all moist forests in south and southeast Thailand felling these trees to look for the valuable wood (Brockelman and Sophasan, 1981), and shoot wild animals for subsistence. The sap collectors also hunt animals to survive in the forest. Other resins used for fuel are widely collected from holes chopped in the trunks of large *Shorea* and *Dipterocarpus* trees.

In these examples, as in many others that could be described, there is no proper management of the forest resources, even though local individuals are obtaining some benefits. This is largely because none of the people who harvest these minor products own any of the forests, and in most cases now are collecting illegally in protected forests.

Another whole class of benefits from tropical forests is often termed ecological services, or maintenance of the "life support system" (IUCN). These services include effects on regional and local climate, protection of soils and water quality in watersheds, flood abatement, reduction of silt in rivers and reservoirs, etc. These benefits are usually regional or even global; few actually accrue to residents in or at the edge of the forests. The benefits depend on fairly large areas of intact forest; maintenance of forest in one's own lot will not make much difference in the level of benefit received.

Therefore, whose problem is maintenance of ecological services? We have little knowledge of how such services relate to forest area and how far away they carry. These effects on the life support system present us with perhaps the most difficult dilemma of all.

The problem of forest destruction thus consists of a wide array of problems which extend from the farmer's back yard to the community, local province, nation, region, and to the globe as a whole. The causes and effects are often complex and poorly understood, both ecologically and economically. One thing seems certain, however, and that is that trees are being cut down by individual people, and usually by the poor. In some cases, decisions are being made by company executives, or government officials; these are also individual people whose personal motivations need to be understood.

International Responses to the Problem

To what extent have our responses to the problem of tropical deforestation been based on an accurate understanding of human causes of the problem, in addition to its ecological effects? This is difficult to answer because responses have been myriad and varied, just as are the basic causes.

Recommendations of foreign governments or international agencies rarely address the "individual dimension" of the problem in enough detail to be very useful because they usually concern broad policy objectives or the possibility of collective international action (such as the international timber consortium). Moreover, how individuals are most effectively incorporated into conservation planning depends greatly on the local culture, social organization and patterns of national administration.

Most tropical countries experiencing rapid deforestation have free-market economies where the rural residents have considerable freedom to do as they please, though often with limited resources and options for making a living. Even in centrally planned economies such as Burma and Viet Nam, the central government may exert little control over inhabitants of remote forested areas who are often ethnic minorities.

The most widely known international policy document on conservation, the *World Conservation Strategy* (IUCN), is probably the most enlightened. It emphasizes the need for public participation in conservation action and points out that poverty and lack of adequate rural development is probably the major cause of conservation problems. One section of the *Strategy* deals with the need for "conservation based rural development," and calls for such measures as credit and marketing incentives to help poor farmers adopt conservation practices. It is also recognized by IUCN that protected areas need to serve local people in material ways to be effective in conserving forests and wildlife (McNeely and Miller).

What is lacking even in IUCN policy statements, however, is an appreciation of how national conservation agencies are to bring the benefits of conservation to rural residents. Conservation agencies are simply not responsible for integrating conservation with development, but with traditional protection and management of conservation areas and wildlife as taught in the western countries where many officials in developing countries were trained. The cross-sectorial approaches to conservation and development cannot be carried out without fundamental changes in the organization of the bureaucracies of developing countries. Most conservation officials in Thailand, for example, have a highly negative attitude towards rural villagers who often violate national park and wildlife laws, and no responsibility for serving them.

The World Resources Institute has formulated recommendations for global action: "The task force concluded that many proven responses to land misuse and deforestation exist and that the lack of political and, consequently, financial will to support them stems from a poor understanding of the consequences of failing to act" (World Resources Institute). A large increase in investment is called for by both public and private sectors.

In Thailand, at least, it is not clear how large increases in investment would have been spent by the government. Awareness of the seriousness of the deforestation problem and a desire to do something about it have existed in the Thai government for over a decade. Contrary to WRI's assertion, what seems to have been lacking for the most part is a successful project model for preventing deforestation. Part of the problem is that forest destruction is due largely to poverty and agricultural problems and policies (such as the promotion of cassava exports) over which the Forest Department has little or no responsibility. Another major problem is the lack of profitability of investment in forest management where slow growing hardwood species are concerned.

In 1980, the U.S. Interagency Task Force on Tropical Forests (USITFOTF) issued a plethora of recommendations as part of a proposed strategy and program for the U.S. Government. These cover virtually every aspect of the problem, but one inconspicuous sentence under the "applied research" recommendations (p. 47) strikes a chord: "What are the effective incentives for individuals, communities and enterprises to engage in sound forest management practice?" Without proper incentives, all the rest is to no avail.

Many excellent recommendations have been prepared by the Interagency Task Force on the Conservation of Biological Diversity (USITFOBD), but there is little focus on tropical forests per se, perhaps to avoid covering the same ground as the previous task force. Virtually no recommendations deal with studying the motivations or options of the people who are most engaged in destroying biological diversity.

The most recent recommendations of the United Nations Food and Agricultural Organization (Rao) on reversing forest destruction cover virtually all aspects of the problem, both local and international, but emphasize the incorporation of forestry into rural economic development. Whether or not these measures would significantly reduce deforestation is uncertain, however, because nearly all community forestry schemes involve the establishment of plantations of fast-growing species, and do not concern directly the problem of squatters encroaching in the natural forests. For reasons explained further below, most forest management schemes will become viable only after all available natural forests have been destroyed.

National Responses in Thailand

Traditionally, all forest area in Thailand belonged to the King; since the country became a constitutional monarchy in 1932, these lands have been administered by the government. To further tighten control over forest use, the National Forest Reserves Act was enacted in 1964. This act established extensive reserved forests within which cutting, burning, settlement and other forms of exploitation were prohibited. Logging concessions within the area must be approved by the Ministry of Agriculture and Cooperatives.

Later, two other land classifications were established: national parks and wildlife sanctuaries, within which all forms of exploitation, including hunting, are prohibited. These protected areas now constitute about twelve percent of the area of the kingdom and are still being gazetted.

In practice, the forest reserve system has been a failure in protecting forests because of encroachment by landless farmers and unregulated logging, both legal and illegal. Sustained yield logging, a concept originating in European and U.S. forestry, has not been achieved (Mungkorndin) and forest utilization has constantly been eating away at the capital and not the yield.

In contrast to the reserved forests, the parks and wildlife sanctuaries have permanent resident officials and armed guards and these areas have been relatively well-protected against deforestation. Poaching by local villagers continues to be a serious problem in these areas, however, as relations between protected area staff and local villagers are typically poor and the amount of patrolling carried out is inadequate to prevent violations by villagers, who normally have little respect for forestry laws (Brockelman, 1986; 1987).

It has been estimated that there are at least a million slash-and-burn farmers squatting in the reserved forests (Ruangpanit), a number which the Forest Department is incapable of dealing with, and these forests are at least half destroyed. Several attempts to begin community forest villages have been initiated since 1967 to promote reforestation and prevent continued destruction, but the attempts have been too small in scale and not highly successful for a variety of reasons to have much impact on the overall problem (Ruangpanit; Ramitanondh).

In order for social or community forestry to succeed, several conditions must be satisfied:

- Firewood or wood products must be in short supply.
- There must be access to outside markets to sell wood and buy needed supplies.
- Trees should be fast-growing species to permit a rapid return on investment.
- Villagers must have some capital to invest or outside financial support.
- There must be some form of land ownership.

If wood is not in short supply, people will not plant trees for their own use, and will not be able to find markets for fuelwood or construction wood. As the reserved forests are essentially commons, poor farmers unable to obtain land in more fertile areas settle in them and take advantage of the "free" land and wood. Some trees may be cut for lumber merchants and some made into charcoal, but the vast majority are simply burned on the spot to put more nutrients into the soil. It is difficult to persuade these settlers to move elsewhere and begin planting trees, as it makes no economic sense to them; they often must be resettled by force, which is also troublesome and rarely attempted. The government's policy of promoting social forestry villages seems to be catching on best in areas completely denuded and almost totally devoid of trees.

The requirement for using fast-growing, usually exotic species (Vergara) also precludes management of valuable native hardwood timber species. It also fails to address the problem of how to manage the natural forests for sustained yield. There seems to be no economically profitable way of managing native tropical forests for sustained yield; indeed, although many Thais deplore the destruction of these forests throughout the kingdom, most people have little idea of how to manage them, are afraid to enter them, and see economic progress only in terms of cutting them down and replacing them with agricultural or tree crops.

Another difficulty with applying social forestry is that it must carefully take into account the social and political structure of the village (Ramitanondh). The government usually works through the existing administrative and power structures, with the result that benefits remain concentrated in the hands of a few influential or relatively wealthy people.

In view of these problems, the government's most recent policy on forests (Royal Forest Department, Thailand) places most emphasis on plantation forestry (including rubber) in achieving its 40 percent target for forest area coverage for the nation. Sustained yield management of natural forests is hardly mentioned; the target will now be met by increasing completely protected forests (parks, wildlife sanctuaries and protected watershed units) to 15 percent, and commercial plantation forests to 25 percent. Private sector management is to be promoted. Indeed, in lowlands far from remaining reserved forests, plantations of *Casuarina* are being profitably established in rice-growing areas.

Conservation and the Individual

Government tends to approach problems such as these "from the top down," beginning by formulating broad policies from crude "macro" analyses. National targets are set and the line agencies are then given general directions to meet the targets. Regulations and policies are passed down. These may consist of laws specifying what people can or cannot do, or where they can or cannot settle, fish, hunt, or cut trees. As enforcement powers are very limited in frontier areas, individuals will more often than not behave in a manner which suits their best interests, regardless of the law. Often, due to limited budgets, the law is not widely promulgated and most rural residents have little or no knowledge of it unless attempts are made to enforce it.

Planning and policy-making must be done at the top of the bureaucracy, but they must be based on experience and information from the "grassroots" level. This is often lacking, especially since government officials often lack appropriate grassroots experience. Policy and administration must also take place at the village level. This usually does not happen where government officials seldom go into rural villages except on inspection or ceremonial visits.

When rural villagers fail to follow the law or use wise conservation practices, the problem is often treated as an educational one: if people understood the full consequences of their actions and how nature works, they would behave in a more ecologically sound manner. Therefore, we must educate them about nature conservation.

My own experiences convince me that this is seldom the case. People usually know what benefits them the most, and they understand what is happening to the environment. Their world is limited, however, and their options (given their lack of capital and technical support, etc.) are limited. Slash-and-burn farmers know that soon the land will all be exhausted, the forests gone and the wildlife shot out. Their time horizons are very short; they must produce a harvest this year to feed their family and pay their debts to the local merchant who buys their rice, corn, or cassava at too low a price and charges extortionate rates of interest (customarily 5 percent per month). There is no money for investment and no time to experiment.

Let us look at some of the benefits wise conservation practice is supposed to bring to the individual farmer. We all know that forest cover protects watersheds: they prevent the loss of soil, siltation of reservoirs, excessive flooding; they help recharge of underground aquifers and improve water quality. However, all but the first of these benefits accrue to farmers downstream, not the upland farmers cutting the forest. With regard to soil fertility, it makes little sense for a poor farmer to try to maintain it by terracing, adding fertilizer or growing bush or tree crops that he cannot market. He has no capital and does not own the land. It is far easier to move on and clear a new patch of forest.

Hunting of wildlife will always be beneficial to the farmer because poorly protected parks, wildlife sanctuaries, and reserved forests are, in effect, common property resources. If a hunter does not shoot an animal when he has the opportunity, someone else will. Rural residents will tell you that the military and government officials also hunt, so why shouldn't they? Somehow there is a mismatch between costs and benefits of conservation. We all know, however (at least those of us who have been "educated"), that our world could be in a sorry mess unless measures to conserve the global resources are implemented. But implemented by whom? There is something inappropriate about reckoning global costs and benefits. While such a reckoning does point up the global problems, global costs and benefits are not the economically relevant ones. The solution is in the province of microeconomics, not global economics and ecology. The individual is not yet convinced.

Some Beneficial Programs

The welfare of the individual farmer or forester is crucial to the conservation of forest resources. Not all is gloomy for individual farmers, however. Some programs have benefited them and have been sustained to some degree. The following are some schemes that have worked:

Charcoal Production

In Kanchanaburi Province, western Thailand, a village is situated near a reserved forest. The local headman in this village is an enterprising and respected man. He sells plots in the forest to influential people, who hire local residents to cut trees and make charcoal for sale in the towns. After the land is cleared, tractors are brought in to plow the soil and cassava is planted. Local farmers work the fields and get a percentage of the profit. They are all happy.

Incense Wood Collection

Valuable *Aquilaria* trees, mentioned above, are scattered through forests in eastern Thailand. Local villagers are supplied with axes, carving tools, guns, and other supplies by certain merchants, who purchase the wood from the locals. If the collectors happen to be caught by the park guards and put in jail, there is no great sacrifice. The merchant will usually cover their bail and any lost supplies.

Forestry and Agricultural Development in the Southeast

Southeast Thailand has rich hardwood forests on fertile, gently rolling terrain. Settlers move in from more populous parts of the east and are hired to cut trees in the reserved forests by influential people. They are encouraged to clear the land and plant cassava or maize which are purchased by merchants (though at well below the government's minimum price). When the land is devoid of all trees, wealthier farmers somehow acquire it and plant large areas in sugarcane, pineapples, and rubber trees. The poor farmers are hired to work on these holdings.

All of these examples, of course, illustrate highly illegal activities. They do show, however, that methods of helping poor farmers do exist, with a little imagination. All we need to do is devise some globally beneficial (and legal) method of helping the rural people.

This will not necessarily be easy. Enforcement of forestry laws will have to go hand-in-hand with individual incentives. Why is this so difficult? Another example will help to illustrate. One official in the Forest Department told a story something like this:

"One of my friends was a good, honest forestry official and once tried to stop an illegal logging operation. He went out in the forest one day and was caught and beaten to a bloody pulp. It was horrible. They made an example out of him. I know because I carried him out of the forest—he died within a day.

"Sure, conservation is great. I'm all for it. I try to do a good honest job; we all should. But look, there is a limit. Conservation is fine, but there are other things in life; like caring for your family and relatives."

Are there really that many bad guys? Obviously, there are some, but I am not sure any more who is a bad guy. I have yet to run across a bad guy or even an unkind person on my treks through areas devastated by slash-and-burn farmers in search of gibbons and other wildlife (Brockelman, 1982). These farmers, in fact, are the most helpful Thais I have run across. They often invite you to sleep in their thatch-roof houses. But they must look out for their own interests.

Not all examples of individual benefit do violence to conservation, fortunately. Numerous examples of projects could be cited which illustrate good conservation in watershed management, forest planting, stabilization of hill tribe agriculture, etc. Most of these examples are still relatively limited in scope and require much extension work. I will discuss one promising project which is familiar to me.

Khao Yai National Park, in central Thailand, has about two thousand km^2 of excellent moist tropical forest covering hills, deep valleys and mountains and is well known internationally as an important preserve for wild elephants, tigers, gaur, gibbons, hornbills, silver pheasants, and many other forest species. The park is now an island of forest surrounded by agricultural areas, and problems of poaching and encroachment have increased in severity in recent years.

In 1981, I decided to try to survey the remote northwest part of the park for gibbons with some other naturalists. As no park personnel were familiar with the area, we sought the help of local villagers. The headman of one village, Ban Sap Tai, located virtually on the park boundary, invited us to stay overnight and willingly supplied us some carriers and guides.

The villagers were poor and had run out of forest land to clear outside the park. Some had encroached inside the boundary, and we heard shots of hunters in the park. The dirt road out to the nearest market was very poor and impassable during much of the rainy season. One asset of the village, however, was its very scenic setting in the foothills, and the beautiful views from mountain bluffs just inside the park. But no visitors ever visited this part of the park. It immediately occurred to us that the area had great tourism potential for trekking over wild elephant trails and camping in the forest.

Soon we began mapping out a strategy for developing trekking for the benefit of the villagers. On our succeeding hikes, we always paid them high wages, at a level we considered suitable for tourist guides, and did not bargain them down. This helped generate good will and a readiness to help us any time we showed up. Thus, we never had to convince them that they might be able to make money from visiting trekkers.

Out of these first efforts was eventually born World Wildlife Fund Project 3001: "Elephant conservation and protected area management." The project was submitted by the National Environment Board of Thailand on behalf of the Royal Forest Department. This three-phase project involved preparation of a management plan for the park, study of the elephant population and its distribution in the park (carried out with village helpers), and initiation of a trekking program to benefit the village.

Trekking offers only sporadic or part-time employment, however, and does not benefit the majority of villagers. More substantial benefits were created when the Population and Community Development Association (PDA), in collaboration with the local NGO, Wildlife Fund Thailand, were invited to initiate a "conservation and rural development" project in the village. With a grant from the German group, Agro-Action, PDA carried out a socio-economic survey, established an economic cooperative to provide cheap loans and help in marketing, and provided agricultural and family planning help. In order to become members of the cooperative, villagers had to vow not to encroach or hunt in the park any more. The majority have joined.

PDA is an autonomous agency in Thailand that has been extremely effective in promoting family planning in villages throughout the country. It has recently incorporated diverse rural development activities into its programs (Stoeckel *et al.*). PDA is extremely capable of generating public support and organizing village level help—in this regard it bears the imprint of its founder and secretary-general, Dr. Meechai Viravaidya, whose name has become synonymous with birth control in Thailand.

For the duration of the project, the PDA has had a staff extension worker living in the village to guide the project and provide daily help to the villagers. This individual has worked with a level of dedication and good will seldom seen in civil service workers, who at any rate (except for school teachers) are seldom assigned to live in rural villages.

What are the effects of the Ban Sap Tai project? Poaching and encroachment by the villagers have virtually ceased, and gun shots are seldom heard in the surrounding hills. Several households have had to move outside the park boundary without any apparent serious grudges. It is still too early to call the projects in Ban Sap Tai unqualified successes, though, because several long-term problems remain unsolved: (1) how to maintain a stable cooperative after the PDA ends active involvement; (2) how to establish a more permanent trekking program; and (3) how to initiate similar programs in other villages with a lower level of initial investment.

Defining the Conservation Problem

The conservation problem is best seen as a problem of motivating people to behave in a manner which benefits the ecosystem in the long run. Efforts to motivate poor people must concentrate primarily on their individual welfare and material benefits. For more well-to-do people, any number of other factors might be important in motivating appropriate behavior.

In a developing agricultural country, arguments about the "intrinsic value" of nature or living things are ordinarily of little use (Callicott). The same applies to arguments that there are strong ethical reasons for conserving living things. My personal inclination (and that is perhaps the most valid guide) is to reserve ethical arguments for contexts which involve more-or-less direct interactions between human beings. Of course, we should be concerned with indirect environmental effects on others, and with future generations, but determining and evaluating such direct and indirect effects becomes extremely complex and is more an intellectual and technical exercise than an ethical one.

We should not exaggerate the material and economic benefits of conservation, either (Ehrenfeld). This often happens when gaps in our technical knowledge become filled, perhaps subconsciously, with a measure of faith and belief. Hypothetical benefits or future benefits are important in motivating government decision-makers but have little value in motivating rural farmers. They will only respond to perceived actual benefits, so our goal must be to actualize the benefits of conservation to the fullest extent possible.

Rural residents, of course, are not the only persons who cause conservation problems and who must receive benefits. They are simply the most numerous, and most often neglected. The limiting problems vary from country to country and from local area to area. In some areas, influential businessmen are the culprits; in others, government leaders or national policies are at fault. The policy of a foreign country may exacerbate tropical forest destruction.

In every case, the benefits of conservation must be appropriately tailored to motivate the party responsible. It does no good to talk about national or global benefits to a farmer, but when one is attempting to motivate a high government official, one may raise arguments about national aggregate indices, trade balance, tourism revenue, per capita income or its distribution, etc. As private enterprise has important roles to play also, we must devise environmentally sound and sustained ways of earning profits.

One of the most common forms of neglect of individual benefits occurs when protected conservation areas are set aside in rural regions without any consideration of how they will affect, and be affected by, local residents. This has often led to abuse of the parks or refuges by disgruntled locals. Humans are in some way a part of every ecosystem; if animals and habitats are to be protected from further exploitation, new ways of allowing humans to participate and receive non-exploitative benefits must be devised.

This principle is emerging out of direct experience in many countries where protected area systems are encountering trouble, on all continents. At the Second (1982) World Parks Congress in Bali, representatives from many nations of Asia discussed new approaches to the problem of integrating protected areas into local economies, and allowing increased local participation (Child; Saharia; Misra; Soekiman *et al.*). Unfortunately, this has not yet become official policy in Thailand, but it is being increasingly discussed.

A parallel can be found between the conservation movement and the population control movement. When the rationale and strategy of human population control was first being debated, some experts pointed out that family planning was not a population control strategy and could not be expected to bring birth and death rates into balance (e.g., Davis). Among population control techniques discussed were such things as tax structure, costs of higher education, retirement benefits, housing costs, etc. (e.g., Berelson). These things, however, have very uncertain and indirect connections with individual reproductive behavior. The most relevant questions to the individual are: How many children do you want? Why? It turned out, fortunately, that most couples desired somewhat fewer children than they would have had without birth control technology. Even though the human population is still not stabilized, family planning efforts have completely carried the day in population control efforts. In Thailand, family planning programs are administered by the Ministry of Public Health, and family benefits are stressed.

Just as population control, as such, is no one's individual concern, neither is tropical deforestation. Conservation strategists must identify and focus on the individual decision-making process just as population planners have come to focus on the reproductive decision-making process.

My rather pragmatic approach to conservation should not be construed to imply that I have a lack of respect for ethical or "intrinsic value" attitudes toward nature. I, myself, have a vague feeling that natural communities have intrinsic worth and beauty. I am not motivated by want of food and shelter. For an increasing number of Thais, also, the appreciation of religious or intrinsic value is a motivating force to be reckoned with. This was borne out recently in an attempt by the government to put a cable car lift in Doi Suthep (mountain) National Park near Chiang Mai City in northern Thailand. The project was withdrawn in the face of spirited opposition by the public, including many Buddhist monks, who argued that the cable car would desecrate a sacred temple on the mountain, destroy some of the forest, be aesthetically offensive, and lead to over-development within the national park. Individual benefits are not always utilitarian; this took the more pragmatic government officials and developers, who based their arguments on the economic benefits of tourism, completely by surprise.

We will continue to be taken by surprise as societies develop and change all over the world.

Acknowledgments

The ideas in this paper were stimulated by field work supported by the Charles A. Lindbergh Fund and the New York Zoological Society. I thank Marion Clawson and Adam Gretton for useful comments and suggestions.

References

Bangkok Post. 1975. "The destruction carries on." Supplement, Dec. 31, 87, 89.

__________. 1979. "Forests are shrinking fast." July 11, 3.

Berelson, B. 1969. "Beyond family planning." *Science* 163:533–544.

Boonyobhas, C. and B. Klankamsorn. 1976. "Application of ERTS-1 imagery in forestry." Bangkok: Royal Forest Department, mimeo.

Brockelman, W.Y. 1982. "In search of the elegant ape." *Animal Kingdom* 85(6):34–39.

__________. "Wildlife conservation in Thailand: a strategic assessment." *Journal of the Bombay Natural History Society, Centennial Symposium* (in press).

__________. 1987. "Nature conservation." In *Profile of Natural Resources of Thailand.* Bangkok: Thailand Development Research Institute.

__________ and Samaisuk Sophasan. 1981. "Tree poaching in Khao Yai." *Bangkok Post,* June 7, 24.

__________ and Sompoad Srikosamatara. 1978. "Surveys for pileated gibbons in Khao Chamao and Khao Srabap National Parks, Thailand." Unpublished Special Report to the National Park Division, Royal Forest Department. Bangkok.

Callicott, J.B. 1986. "On the intrinsic value of non-human species." In B.G. Norton, ed. *The Preservation of Species*. Princeton: Princeton University Press, 138–172.

Child, G. 1984. "Managing wildlife for people in Zimbabwe." In J.A. McNeely and K.R. Miller, eds. *National Parks, Conservation, and Development: The Role of Protected Areas in Sustaining Society*. Washington, D.C.: Smithsonian Institution Press, 118–123.

Davis, Kingsley. 1967. "Population policy: will current programs succeed?" *Science* 158:730–739.

Ehrenfeld, David. 1976. "The conservation of non-resources." *American Scientist* 64 (Nov.–Dec.):648–656.

Gianno, Rosemary. 1986. "The exploitation of resinous plants in lowland Malayan forest." *Wallaceana* 43 (March):36.

IUCN (International Union for the Conservation of Nature). 1980. *World Conservation Strategy*. Gland, Switzerland: IUCN.

Kanchananaga, Suraphong. 1973. *Resources and Products of Thailand*. Bangkok: Siam Communications Ltd.

McNeely, J.A. and K.R. Miller, eds. 1984. *National Parks, Conservation and Development: The Role of Protected Areas in Sustaining Society*, Washington D.C.: Smithsonian Institution Press.

Misra, H. 1984. "A delicate balance: tigers, rhinoceros, tourists and park management vs. the needs of the local people in Royal Chitwan National Park, Nepal." In J.A. Mcneely and K.R. Miller, eds. *National Parks, Conservation and Development: The Role of Protected Areas in Sustaining Society*, 197–205.

Mungkorndin, Sompetch. 1980. "Sustained yield management in Thailand." In *ASEAN Seminar on Management of Tropical Forests: ASEAN Forests for Social Welfare. Chiang Mai, Thailand*. Bangkok: Royal Forest Department of Thailand.

Myers, Norman. 1980. *Conservation of Tropical Moist Forests*. Washington, D.C.: U.S. National Academy of Sciences.

__________. 1984. *The Primary Source: Tropical Forests and Our Future*. New York: W.W. Norton.

The Nation Review. 1979. "Concessions for logging stopped." Jan. 10, 3.

Ramitanondh, Shalardchai. 1984. "Socio-economic benefits from social forestry: for whom? (The case of northern Thailand)." In Y.S. Rao, N.T. Vergara, and G.W. Lovelace, eds. *Community Forestry: Socio-Economic Aspects.* Bangkok: FAO Regional Office for Asia and the Pacific, 371–384.

Rao, Y.S. 1986. "FAO's tropical forestry action plan—a summary." *Tigerpaper* 13(2):12–16.

Royal Forest Department, Thailand. 1985. "*Policy on National Forests.*" Bangkok: Royal Forest Department.

Ruangpanit, Niwat. 1984. "Manpower and technological constraints in community forestry projects and programs in Thailand." In Y.S. Rao, N.T. Vergara, and G.W. Lovelace, eds. *Community Forestry: Socio-Economic Aspects.* Bangkok: FAO Regional Office for Asia and the Pacific, 371–384.

Saharia, V.B. 1984. "Human dimensions in wildlife management: the Indian experience." In J.A. McNeely and K.R. Miller, eds. *National Parks, Conservation and Development: The Role of Protected Areas in Sustaining Society.* Washington, D.C.: Smithsonian Institution Press, 190–196.

Simon, J.L. 1986. "Disappearing species, deforestation and data." *New Scientist,* May 15, 60–63.

__________ and H. Kahn. 1984. *The Resourceful Earth,* New York: Basil Blackwell.

Soekiman, A., L. Daryadi, J. MacKinnon, and P. Hillegers. 1984. "National parks and rural communities." In J.A. McNeely and K.R. Miller, eds, *National Parks, Conservation and Development: The Role of Protected Areas in Sustaining Society.* Washington, D.C.: Smithsonian Institution Press, 237–244.

Stoeckel, J., A.A. Fisher, M. Viravaidya, and R. Na Pattalung. 1986. "Maintaining family planning acceptance levels through development incentives in northeastern Thailand." *Studies in Family Planning* 17(1):36–43.

UNFAO (United Nations Food and Agriculture Organization). 1982. *Tropical Forest Resources.* Forestry paper no. 30. Rome: FAO.

USCEQ (U.S. Council on Environmental Quality (CEQ) and Department of State). 1979. *The Global 2000 Report to the President.* Vol. 1. Washington, D.C.: U.S. Government Printing Office.

USITFOBD (U.S. Interagency Task Force on Biological Diversity). 1985. *U.S. Strategy on the Conservation of Biological Diversity.* Washington, D.C.: U.S. Agency for International Development.

USITFOTF (U.S. Interagency Task Force On Tropical Forests). 1980. *The World's Tropical Forests: A Policy, Strategy and Program for the United States.* Washington, D.C.: U.S. Department of State.

Vergara, N.T. 1984. "Expanding populations and shrinking resources: the economic setting and development potential for social forestry." In Y.S. Rao, N.T. Vergara, and G.W. Lovelace, eds. *Community Forestry: Socio-Economic Aspects.* Bangkok: FAO Regional Office for Asia and the Pacific, 3–12.

The World Resources Institute and The International Institute for Environment and Development. 1986. *World Resources 1986.* New York: Basic Books, Inc.

Worldwatch Institute. 1975. *The Other Energy Crisis: Firewood.* Worldwatch Paper 1. Washington, D.C.: Worldwatch Institute.

__________. 1979. *Planting for the Future: Forestry and Human Needs.* Worldwatch Paper 26. Washington, D.C.: Worldwatch Institute.

SIX

AFFLUENCE VERSUS EFFLUENTS—MARINE WATER POLLUTION CONTROL IN THE USA AND IN THE DEVELOPING COUNTRIES

Harvey F. Ludwig

Introduction

One of the most fundamental questions of concern in evaluating man's impact on "Spaceship Earth" is the extent of degradation of ocean waters caused by discharge into the seas of wastes from man's activities, and whether such degradation will likely lead to irrevocable loss of precious ocean resources essential for man's continuing well-being. Because the bulk of the waste discharges into the seas have occurred over the past century, since the advent of accelerated population growth and large-scale urbanization and industrialization, it is timely now to review and assess the accumulated experience to attempt to determine the merits and demerits of ongoing waste disposal practices, and to develop conclusions for guiding future practice.

Unlike the problem of waste disposal to inland fresh waters, which is regarded as a localized in-country affair, because the oceans are international they tend to be considered to some extent as common domain, as evidenced by the recent efforts to formulate a universally acceptable "Law of the Seas." In the spectrum of international concerns about the ocean, the pollution parameter is one of the major issues, along with rights for navigation, aquatic fisheries, and seabed mineral extraction.

The present evaluation is based on review of the literature on marine waste disposal as affecting marine environment and ecology, most of which has been published over the past three decades, plus experiences gained from monitoring effects of waste disposal on marine ecology for marine waste disposal systems which are of major size or which discharge into especially sensitive marine ecological zones.

Evolution of Waste Disposal Technology in USA

Sewage Treatment Technology Prior to 1960

Treatment of wastes prior to discharge to environment (to receiving waters in all but a few cases) is a modern practice, barely a century old, which came into being in the western world when the growth of cities and the resulting volumes of sewage collected by municipal sewerage systems reached magnitudes that seriously depreciated downstream beneficial water uses. This included streams or other confined waters to which the wastes were discharged, including confined coastal waters and bays.

The entire technology of sewage treatment which evolved was thus related to control of pollution from man-generated wastes discharges to confined waters (water bodies of limited volume). Such treatment became necessary when the amount of waste discharges exceeded the natural assimilative capacities of the receiving waters, thus resulting in pollution, that is, degradation of water quality due to discharges exceeding the assimilative capacities. Under such conditions it was found necessary to use treatment measures, not to remove all of the waste substances, but to reduce the amount to levels within the assimilative capacity. It was found that protection of these waters to sustain their beneficial uses (water supply, fisheries, swimming and recreation, and environmental aesthetics) required essentially two treatment steps, (1) removal of sludge materials and of floatable materials, such as greases, to prevent deposition of these materials on shorelines and the associated nuisances and public health hazards, and (2) removal of the remaining organic materials in the waste flow, namely colloidal and dissolved organics, sufficient to reduce the concentrations of these organics to levels which would not deplete the dissolved oxygen concentration of the waterway to very low or septic levels, which would turn the entire waterway into a "sanitation mess" unfit for practically all beneficial water uses (except for their uses for waste disposal and for navigation). Thus, the design technology which evolved, primarily by engineers in the USA and western Europe, in addition to removal of settleables/floatables to prevent sludge bank formation in the waterways, was based essentially on use of the parameter of dissolved oxygen (DO), to reduce the DO demand of the degradable organic matter in the effluents discharged to the confined waterways, to be within the capabilities of the waterways to absorb the degradable organics without reducing the waterway DO to levels below those considered necessary for the particular beneficial uses of value. The salient parameter for matching treatment needs with environmental effects was the DO content of the waterway, and the key parameters for measuring the DO loading to be imposed for the discharged effluent were the biological oxygen demand (BOD) and the suspended solids (SS) of the effluent.

A common requirement, up to about 1960, was to procure a "20/30" effluent, meaning 20 mg/1 BOD and 30 mg/1 SS. This was accomplished by a combination of "primary treatment" or removal of settleables and floatables by gravity separation, followed by "secondary treatment," meaning use of some biological oxidation process such as activated sludge or trickling filtration, the two together being called "complete treatment," although the actual removals of organics were usually of the order of 90 percent or less. Both primary and secondary treatment resulted in production of sludges with considerable organic content, which were further processed by biological stabilization (usually anaerobic digestion), resulting in a fairly stable sludge product which could be readily de-watered and thus made suitable for disposal in rural areas (usually into farmlands where it was of value for its soil conditioning properties plus a small but significant concentration of nutrients).

In addition to the above processes, the overall sewage treatment system also included chlorination of the effluent, to reduce its content of pathogens to acceptable levels, and use of the coliform determination to measure the probable level of pathogenic hazard in the receiving water before and after chlorination.

The above practices represented standard means for treating municipal sewage to about 1960, except that, following World War II, the characteristics of the municipal sewage were significantly altered by the increasing discharge to the system of effluents from industrial manufacturing plants. At first, the industrial loadings were largely organic, hence, their impact was simply to require more capacity in the municipal sewage treatment systems. But as industrial processing steadily expanded, their effluents contained increasing amounts of hazardous substances, including toxic, inflammable, explosive, corrosive, and other materials which could either cause physical damage to the sewage system, or impair its biological treatment processes, or pass through the treatment system to damage downstream beneficial water uses. As a result, regulations were developed by the governmental agencies to require individual industries to remove hazardous substances from wastes before discharging them to the public sewage system (Ludwig, 1978b). The hazardous sludges from such in-plant treatment systems, together with solid hazardous wastes produced by the industry, were usually disposed of by landfilling and sometimes by incineration.

The landfilling was performed at sites where presumably no groundwater pollution hazards were involved or by using specially-designed fills with liners and other measures which presumably afforded suitable protection of groundwater from contamination by hazardous substances leaching from the filled areas.

During this period, up to 1960, in a relatively few cases the sewage effluents were discharged into the open sea, or into unconfined marine waters, and for these the sewage treatment processes employed were essentially the same as those planned for discharge into confined waters. The engineers who planned them hardly thought otherwise.

The Present Technology

During the decades of the 1960s and 1970s the basic concepts for appropriate treatment of liquid wastes were markedly changed, as part of the general trend toward much better attention to protecting natural resources which stemmed from the Environmentalist-Conservationist Movement (ECM) beginning in the 1960s. As is well-known, this period was characterized by the emergence in the USA of the ECM with so much power that it pushed the regulatory requirements for waste disposal ahead very rapidly, so that by the 1970s and ever since, the regulatory pressures for water pollution control (emanating mostly from U.S. EPA) have far transcended the technology. That is, the pressures for clean-up reached levels where the traditional engineering approach (how to apply scientific knowledge to solve real world problems in an economic manner) was more or less abandoned in order to meet preconceived desired levels of environmental quality more or less regardless of costs.

This was clearly evident by the push by EPA for "tertiary treatment," generally meaning use of expensive physical/chemical (rather than biological) processes to "work over" the effluents from secondary treatment, to achieve further reductions of BOD and SS together with reductions in nutrients substances, such as phosphorus and nitrogen, and reduction in the other "refractory" substances remaining in secondary treatment plant effluents in low concentrations, including residuals of heavy metals and synthetic organics. While tertiary treatment had a glamorous appeal (it was also termed "advanced waste treatment"), it was very costly per unit of removal obtained, virtually doubling the total treatment costs just to obtain the small extra removals. Subsequent monitoring of downstream water quality has not, in most cases, turned up evidence of downstream improvements to justify the high extra costs (Ludwig, 1971).

The lesson learned is that the relationship between extra treatment process removals and improvement in downstream ecology is much more complex than had been assumed in the implementation of the tertiary treatment program. An excellent example of this is the current, very confused picture of Chesapeake Bay as reported by recent U.S. EPA publications, where the famed striped bass fishery has been markedly depreciating over the past decade despite the application of tertiary treatment and other measures for reducing total pollutant discharge into the Bay down to levels believed to be sufficient as judged by all known water quality parameters. It is now realized that, despite these protection measures, the "balance" between physical/chemical constituents in the Bay waters may, nevertheless, have shifted in some subtle manner which has depreciated certain bottom grasses (in favor of other plant species, presumably) which are vitally important to the striped bass. Thus, the treatment/ecological impact relationship is not at all well understood, this despite the tremendous magnitude of EPA's research program on such relationships over the past two decades.

Another very significant change in the nature of the national water pollution control program in the USA, again as a result of the "big push for accelerated cleanup" sponsored by the ECM, was the shift of administrative management of the program from the U.S. Public Health Service, where the program was essentially engineer managed, to the new U.S. EPA where control has shifted almost totally from engineers to attorneys. Ever since, the concept of efficiency in cleanup, wherein costs of treatment are related to and justified by improved beneficial water uses, has not been of primary importance but has been replaced by what may be called the "legal approach," whereby the "decision makers" (attorneys) manage the entire program to achieve a "simplistic uniformity" regardless of cause-and-effect relationship between treatment costs and measurable benefits accrued.

In this context, the changes in downstream water quality, as measured by the established parameters, are simply deemed worthy of the cost, and benefit/cost analyses are conducted accordingly. One of the most basic elements of the "SUA" (simplistic uniform approach) was the upgrading of the engineer's earlier credo, that all wastes should have at least primary treatment (to avoid sludge bank formation), to the new credo that all wastes must have no less than secondary treatment regardless of place of disposal; that is, regardless of the oxygenating capacity of the receiving waters. While the first credo makes engineering sense, the second credo represents the actual opposite of engineering, and it is the second credo that has led to great controversy in the USA in the field of marine waste disposal.

Marine Waste Disposal in California

The Engineer's Era

With respect to disposal of waste into the marine environment, the West Coast of the USA has been the "action center" because there the deeper unconfined ocean waters are readily accessible, whereas in the east the gradual slope of the continental shelf places the unconfined waters, in most cases, beyond the zone of economic accessibility. On the West Coast, the earliest recognition of the actual resources of the unconfined seawaters for absorbing wastes took place in California, especially in southern California with the pioneering work of Mr. A.M. Rawn, chief engineer of the Los Angeles County Sanitation Districts. This led to construction of a major submarine outfall system discharging into unconfined waters offshore of White Point, with treatment limited to conventional primary treatment, and to a similar system for the Orange County Sanitation Districts discharging offshore in the vicinity near the city of Santa Ana.

Both of these systems included rigorous continuing monitoring programs for determining whether any of the discharged sewage materials were being returned to the shoreline zone and, in the case of the Orange County outfall, whether return of coliform organisms was sufficient to exceed the state's established standard for beach waters of 10 total coliforms/ml. Both systems operated very economically and quite satisfactorily, based on the records of the monitoring programs which employed all of the known pollution indicating parameters.

In 1950, water pollution control regulatory authority in California was transferred to the State Water Resources Control Board. Based on studies conducted for the state by E.A. Pearson of the University of California at Berkeley, the board took the pioneering step of requiring comprehensive monitoring of the effects of all significant waste discharges to marine waters, including both confined estuarine/bay areas and unconfined open seas, and including use of all the conventional water pollution parameters, including DO, BOD, SS, pH, major cations and anions, nutrients (nitrogen, phosphorus, potassium), heavy metals, oil/greases, and others, plus analyses for benthic biota and for accumulation in bottom sediments of heavy metals and petroleum compounds.

Ten years later a vast backlog of monitoring data had been accumulated, and a special study was sponsored by the board for evaluating the significance of the accumulated data. The results of this study (Ludwig and Onodera, 1962) concluded that (1) the conventional sewage treatment concepts, based essentially on the DO parameter, while applicable to confined estuarines/bays, had little if any application to waste discharge into unconfined seawater; (2) the only meaningful parameters relating to disposal to unconfined waters were those for coliforms, plus a new parameter termed "floatables," that is, all materials which rise to the surface and thus can be returned to the shore by onshore winds; (3) the oxygenating and diluting capacity of the unconfined seawater are virtually unlimited, hence, oxygenation of degradable organics using secondary treatment is of virtually no value for wastes discharged to such zones; (4) with the exception of materials settling out in the vicinity of the end or discharge points of the submarine outfall pipe system, all other constituents are rapidly diluted and cease to exist in terms of detectability, including heavy metals and other toxics; (5) the increase in levels of such materials in the open ocean environment, considering all of the liquid waste discharges reaching the oceans around the world, should be insignificant, i.e., should not result in any significant increment which would degrade marine ecology; (6) the natural organics contained in the waste effluent exert a positive influence on marine ecology by furnishing valuable nutrients which are normally scarce in the open sea environment; (7) while the bottom ecology in the vicinity of the outfall will certainly be degraded, the total affected area is virtually of no statistical influence; (8) for disposal into the unconfined ocean waters the primary treatment process alone should be suitable, given an outfall reaching sufficiently offshore to reach the unconfined water zone, except that the removal of floatables accomplished by conventional primary treatment should be improved (given greater attention); and (9) for such disposal it would be unnecessary to remove the settleable sludge from the waste, in view of the ocean's virtually unlimited waste absorption capacity. Because sludge disposal represents a big part of overall treatment plant costs, the most appropriate treatment system would be one not removing the settleable solids at all but one focused on removal of floatables, with a sufficiently long outfall. In some cases not even floatables need be removed; that is, they would be carried out to sea and eventually degraded and stabilized.

Following the pattern set by the Los Angeles County and Orange County systems, in the 1960s the cities of San Diego and Los Angeles both constructed and operated similar systems, again with rigorous monitoring which showed excellent protection of the environment. Another submarine outfall system was constructed for discharge of waste clay slurries from mining of high quality beach sand, in the open waters offshore from the famed Seventeen Mile Drive zone near Monterey, California. This represents one of the most sensitive, fragile marine ecological zones in the world, and rigorous monitoring showed the system gave excellent performance in protecting this resource (Engineering-Science, 1963). Still another similar major system was planned and implemented for Honolulu/

Oahu in the 1970s, based on detailed evaluation of treatment level alternatives versus needs for protection of the precious and fragile recreational values of the marine ecosystem (Engineering-Science, 1972).

In addition to the comprehensive monitoring of the waste disposal systems discharging to the ocean, a supplementary and complementary comprehensive research and development project was initiated in the 1960s, again with Los Angeles County Sanitation District leadership, with mostly EPA funding. This project, known as the Southern California Coastal Water Research Project (SCCWRP), which is administered jointly by the local governments operating marine waste disposal systems in southern California, is still continuing. Its scope, depth and magnitude have transcended by far all other studies of impacts of wastes on marine environment, and its publications include a series of reports which have virtually rewritten the technology in this field (SCCWRP). With the large funding levels made possible by both strong local governmental support, plus that from EPA, and under the excellent and stimulating leadership of Mr. Willard Bascom, SCCWRP has furnished a solid and extensive data base of facts on effects of wastes on marine en- vironment and ecology. It is of interest to note that SCCWRP's findings and conclusions have essentially confirmed all of the findings and tentative conclusions referred to above, first made in 1962.

Another important contribution by the State of California Board was preparation of a comprehensive manual on the technology of marine waste disposal monitoring (CSWRCB). This pioneering contribution is still appropriate and is widely used.

The Attorney's Era

The pattern of waste disposal developed and required by EPA for disposal to confined waters, despite its uniformity, nevertheless makes sense for most inland disposal situations, because for most of these the DO resources are definitely limiting. Moreover, EPA's push to upgrade practices for inland disposal of digested sludges has also been constructive, (1) in recognizing that the sludges may contain sufficient pathogens to be a significant communicable disease hazard, especially to animals grazing in lands where the sludges are commonly discharged, and (2) in leading to use of the forced-air BARC composting process, the first economic composting process which comprehensively destroys pathogens, for processing of sludges prior to disposal in farmland areas (Ludwig, 1983). In addition, EPA's "Superfund" program for management of hazardous wastes, while very expensive and highly controversial as to its cost effectiveness, at least has focused attention to the need for much better management of hazardous wastes and, inter alia, prohibits any discharge of these materials to the marine environment.

The EPA's position on disposal of liquid wastes into unconfined ocean waters, unfortunately, seems to make no engineering sense at all, that is, neither scientific nor economic sense, and uniformity seems to be the "name of the game." Despite strong protests and prolonged congressional hearings on the subject (USCSOWR), despite the excellent records of performance of submarine outfalls/primary treatment systems on the West Coast, and despite the conclusive findings of SCCWRP, all of the major ocean disposal systems of the West Coast are being converted to secondary treatment systems at huge construction costs, to say nothing of the extra very high annual power and other operations costs which presumably will continue on forever. One of these situations, for the city of San Francisco, is discussed in a recent issue of Civil Engineering (Murphy and Eisenberg), which notes that secondary treatment was required even with a submarine outfall 23,400 feet long, one of the world's longest. This article describes San Francisco's situation as "lucky," in that even higher levels of treatment were not required.

Importance of Estuaries

Another resultant of the ECM's push for better understanding of impacts of wastes on ecology is the large-scale program of research on the role of estuaries in the overall world ecosystem, leading to firm conclusions which recognize these zones as the most precious of the earth's areas for aquatic reproduction purposes, thus quite reversing the earlier concept that such swamps represented mosquito-borne communicable disease hazards which properly should be "reclaimed" by draining or filling. In addition, the U.S. Corps of Engineers, as the primary agency of the USA engaged in dredging of coastal areas, including disposal of spoils into coastal areas including swamps, sponsored a major and pioneering research project for developing technology on how to conduct dredging/spoils disposal with minimum adverse impacts on ecology and with use of measures for offsetting unavoidable residual adverse effects (U.S. Corps of Engineers).

These various research and development projects furnished a substantial data base for evaluating the relative role of estuarine waters, the interface between the rivers and the oceans, as compared with the rivers and ocean waters as appropriate receiving waters for waste disposal purposes.

Marine Waste Disposal in Developing Countries

Bare-Bones Engineering

As described above, the present practice of marine waste disposal in the USA corresponds to an era of affluence, possible and affordable, in a country with enough affluence to be willing to allocate high costs for protecting environmental resources which had hardly been of public concern only a few decades earlier. The change came about following World War II when a very sizeable percentage of the adult population had enough affluence to spend vacation periods in the outdoors, engaging in fishing, skiing, hunting, swimming, boating, and other outdoor sports, and to participate as active members in wildlife and conservationist organizations in such numbers as to establish a very powerful lobby which clamored for and got action on the part of the government.

No such affluence exists in developing countries (DCs). Hence there is no significant popular movement dedicated to environmental protection. Under the circumstances, the case for environmental protection must be made by appeal to the social consciousness of the decision-makers (DMs) of the government, who are usually forced by circumstances to concentrate their efforts on solving the immediate problems resulting from the advent of rapid population growth, rapid urbanization, and rapid industrialization. Under these circumstances, it is not easy to gain the interest of the DMs in supporting environmental protection programs, but neither is it impossible. Rather, it is possible and feasible if the proposals are presented with solid justification on economic as well as socio-economic grounds.

This means not only use of the engineering/economic-based approach, but adaptation of it to prepare projects requiring minimum costs and which are clearly cost-effective and devoid of the expenditures intended for preserving environmental resources which are clearly outside the limits of the DCs financial and economic resources. This also means abandonment of the ongoing USA concepts of appropriate environmental standards, and means setting environmental standards levels consistent with the status of the local economy. As shown in Figure 6-1, environmental standards are not absolute but depend on levels of affluence; hence, for the DCs the appropriate

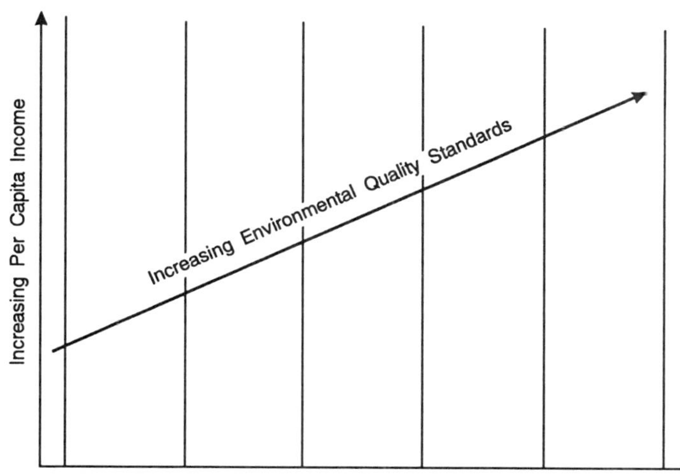

Figure 6-1: Schematic drawing showing increases in environmental standards in the USA over the decades.

pollution clean-up targets will not match those of U.S. EPA today but will correspond to those practiced in the USA in earlier eras (Kiravanich et al., 1985).

Case Examples

A considerable number of very successful marine waste disposal systems have been implemented in the DCs over the past two decades, following the design concepts of 1962 noted above, based on use of a submarine outfall extending far enough offshore to reach unconfined waters, with provision for treatment only for removal of floatables, to be furnished as a second step provided that monitoring of impacts of the system on marine environment shows this to be needed. An outstanding example is the submarine off-fall disposal system serving the entire city of Rio de Janeiro (Brazil), which extends several kilometers offshore from the famed Ipanema Beach area and which has provided excellent protection of Ipanema, Copacabana, and the other renowned beaches in the region. While the basic plan for this system included provisions for removal of floatables, if needed, the rigorous monitoring program (which commenced some ten years before the outfall was built) has shown this not to be necessary (Engineering-Science, 1967).

Other examples where this same design basis has been very successfully utilized (and in which the writer has participated), include the sewage disposal systems serving the metropolitan/urbanizing areas of Accra (Ghana), (Engineering-Science, 1965), Montevideo (Uruguay) (Ludwig, 1960), and the coastal cities of Puerto Rico (Engineering-Science, 1971). Examples in the planning stage include systems for serving coastal communities and industries in the eastern seaboard region of Thailand (Ludwig, 1982) and the metropolitan capital area of Abidjan (Ivory Coast) in Africa (Ludwig, 1982). For the case of the Abidjan system, which involves assistance from the World Bank, the plan will eliminate the present drainage of sanitary and industrial wastes into the extensive coastal lagoon region in the midst of which Abidjan is located, and which is the capital area's most precious environmental resource, for purposes of recreation and environmental aesthetics as well as fisheries.

With respect to the disposal of industrial wastes per se, as compared to sanitary sewages, the advantages of disposal to unconfined ocean waters is even greater than for sanitary sewages. Figure 6-2 shows, for disposal to confined waters, that the required treatment processes can be very complex and costly because of the diverse types of pollutants, including toxics, which are commonly found in industrial wastes. For disposal to unconfined waters, the problem is relatively simple, and especially if the industrial effluents can share with sanitary wastes in use of a common submarine outfall (Ludwig, 1978a,b).

An excellent summary of the use of the technology of appropriate use of unconfined marine waters for waste disposal, as described above, is given in the proceedings of the Regional Conference on Marine Waste Disposal held at Trieste (Italy) in 1972, sponsored by the International Association on Water Pollution Research (IAWPR). Three of the conference's conclusions are quoted as follows:

> "The effective use of the sea or marine waters from the final disposal of adequately treated wastes is an economic social and ecological necessity. With adequate source control of objectionable materials, plus treatment to remove those objectionable materials not removable at the source, an adequate dilution dispersal system can produce minimum adverse ecological effects and possibly even beneficial effects on sea or ocean ecology. Such disposal systems will provide greater protection to the regional ecology than other alternatives.
>
> It appears that for open sea waste water disposal, the minimum cost and optimum benefit system would entail a primary treatment plant giving special attention to the floatables removal, plus an adequate submarine outfall to achieve sufficient initial dilution of the waste into the sea water (at least 70 to 100 parts of sea water per part of waste water) and a considerable transport time to the beach areas (or other areas) needing a high degree of protection.
>
> For open sea waste water disposal systems there appears to be little if any justification for costly waste water treatment processes (such as secondary or tertiary treatment). Contrary to popular belief, secondary treatment plants, while effective for reducing BOD and suspended solids, do not achieve much extra removal of coliforms nor of refractory materials such as organic toxicants and biostimulants.

Significant Parameters for Treating Wastes From Various Industries for Discharge to Confined Waters
(After Point Source Publications of US/EPA)

Treatment Parameter / Industry	Inorganic Chemicals	Organic Chemicals	Bauxite Refining	Aluminum Primary Smelting	Aluminum Secondary Smelting	Feedlots	Fruit/Vegetable Canning	Plywood	Iron Smelting	Cement	Asbestos Building Materials	Cane Sugar Refining	Fiberglass Insulation	Meat Processing	Grain Processing	Soap and Detergents	Synthetic Resins	Steam Power	Pulp and Paper	Brewery	Copper Wire
Chlorinated Hydrocarbons																					
Coliforms (fecal)						●	●							●							
Sulfate			●							●											
Flouride				●	●			●					●								
Phenols		●							●		●										
Cyanide	●	●							●												
Heavy Metals											●										
Aluminum					●																
Potassium										●											
Zinc								●									●				●
Manganese									●												
Arsenic								●													
Mercury	●																				
Copper					●			●													●
Lead	●																				
Chromium	●							●	●								●				●
Surfactants																●					
Nitrogen		●			●	●		●			●	●	●	●	●						
Phosphorus		●				●		●			●	●		●	●		●				
Grease/Oil		●						●					●	●		●					
Color		●											●						●		
Temperature								●		●	●	●	●	●	●			●			
TOC	●	●									●										
COD	●	●			●	●		●			●	●	●	●	●	●	●		●		
BOD		●				●	●	●			●	●	●	●	●	●	●		●	●	
TDS		●	●			●		●		●	●	●	●	●	●		●				●
TSS	●	●	●	●	●	●	●	●	●	●	●	●	●	●	●	●			●	●	●
Acidity/Alkalinity		●	●					●		●		●		●	●	●					
pH	●	●	●	●	●		●	●	●	●	●	●	●	●	●	●	●		●		●

Figure 6-2: Chart showing significant parameters for industrial waste effluents discharged to rivers, lakes, estuaries, and other confined waterways.

In fact, secondary treatment plants actually introduce additional biostimulatory factors."

One of the salient points brought out in this summary relates to the relative degrees of adverse effects of disposing of waste effluents into inland rivers or to estuaries or other confined waters as compared to use of a submarine outfall for reaching the open sea. A common "mis-notion" on the part of marine ecologists is that waste effluents should be discharged to the inland rivers or to the estuaries, following proper treatment (meaning primary/secondary treatment), thus avoiding damage to the marine environment. Aside from not including due consideration of the relative ecological values of the riverine and estuarine environments compared to the ocean, this notion hardly appreciates the fact that primary/secondary treatment processes do not remove much of the toxics in the wastes (the processes are intended to remove degradable organics), hence, these toxics will, in any case, flow into the sea and on the way will also permeate the riverine and estuarine zones which generally represent much more fragile and valuable ecologies than the ocean area of concern.

Major Capital Areas in Southeast Asia

Of the various capital areas of southeast Asia, both Manila and Kuala Lumpur have metropolitan sewage systems underway. The system of Manila, largely constructed over the past several years with World Bank assistance, when completed will discharge the waste effluents via submarine outfalls into the offshore waters of Manila Bay. The system at Kuala Lumpur (Ludwig, 1986), now being considered for upgrading with assistance from the World Bank, will depend for treatment mostly on use of plastic-media trickling filters, with the effluents discharged to the river system which flows downstream some 20 km before reaching the coastal part zone at Kelang (Engineering-Science, 1985). In this case, the urbanizing and port areas downstream of Kuala Lumpur are also planning to implement their own sewage systems to prevent the downstream zone from turning into a sanitation mess. The downstream treatment problem will be compounded by the fact that Kuala Lumpur will use up most of the dissolved oxygen capacity of the Kelang riverine system which drains the Kelang Valley region.

The situations at Bangkok and Jakarta lag far behind those of Manila and Kuala Lumpur.

Jakarta. The problem of many major capital cities is the non-affluence of the bulk of the population, who are mostly former farm families recently migrated to the capital, where, poor as they are, they are better off generally than their relatives who remain in the country. Under these circumstances, there is no hope in the foreseeable future for affording construction in their living areas of a comprehensive sewage system (such as that at Kuala Lumpur). Instead, it will be necessary to continue to depend, for excreta disposal, on use of individual subsurface leaching units for most of the non-affluent areas, with sewers limited to the commercial, affluent residential and industrial zones. Although a comprehensive plan of sewage and sanitation was prepared for Jakarta in the 1970s, implementation of the sewage system will be a slow process, and only the first phase of construction, covering about ten percent of the city, has been funded (Ludwig, 1984). As a result, the river system passing through and draining through the city is usually septic/black in the dry season due to inflows of both sullage and solid wastes. Other than for their uses for navigation, practically all beneficial uses of the Jakarta rivers have been lost, including fisheries, water supply, recreation, and bathing and washing.

In addition, due to contributions from industries, the pollutants reaching the Bay of Jakarta contain appreciable amounts of heavy metals or other toxics which are accumulating in the bottom sediments. To prevent continuing accumulation to levels which would contaminate the bay fisheries, consideration is now being given to establishing a project for systematic collection and safe disposal of the hazardous wastes produced by industries throughout the metropolitan area.

Bangkok. Bangkok is located a short distance from the Gulf of Thailand, on both sides of the Chao Phya River, which is Thailand's major river draining the main central/northern zone of Thailand. Along with the Chao Phya, three other major rivers also drain into the Upper Gulf, and the continuing urban growth in the zone (now about six million population) is progressively transforming the Bangkok/Upper Gulf region into a megalopolis, which will likely extend the urbanizing complex all the way to the sea (Ludwig, 1973). The four rivers, in their lower reaches passing through the urbanizing region, are heavily polluted in the dry season and no longer support any beneficial uses except navigation. Moreover, their discharges into the Upper Gulf have markedly reduced the very lucrative Upper Gulf fisheries. Unfortunately, no comprehensive plan of water pollution control has ever been prepared for Bangkok or the overall urbanizing region. The urgently needed first step in bringing the situation under control is to prepare such a plan (Kiravanich and Ludwig, 1979), so that an appropriate regional system can be gradually implemented on a step-by-step basis. As it is, without the plan, it is not possible to delineate the appropriate components for the regional system nor to give them meaningful priorities.

Summary and Conclusions

Summary

The majority of major cities of the world are located on or near the oceans and, hence, the non-managed pollutants from these centers drain to the marine waters, passing into and down the river system, through the estuarine zones, and then into the unconfined ocean waters where (except for floatables) they are diffused and diluted to non-detectable levels. Except for navigation, the beneficial uses of the rivers are lost. Moreover, before the waste materials are diffused into the open sea, they greatly damage the very valuable fishery reproduction in the shallow estuarine zones. While these shallow estuarine waters make up a negligible portion of the total marine water volume, they are all-important in terms of beneficial uses of interest to man.

Because of rapidly increasing pollution of rivers, estuaries, and near-shore coastal waters due to increasing population and industry, sewage treatment practices have evolved over the past century for processing liquid wastes prior to discharge, essentially to reduce their organic pollution loads. This technology has been oriented to suit the limitations of shallow/confined waters into which the treated effluents have been commonly discharged; hence, the treatment processes have focused on removal of settleable/floatable solids which can form sludge banks in the shallow waters, and on removal of degradable organic materials which deplete the oxygen resources of the receiving waters. This technology has been engineer-oriented, with the type and degree of treatment chosen to match the receiving water conditions so as to protect the aquatic environment as well as make use of its waste absorbing capacities.

Beginning in the 1960s it has been found that when the wastes can be directly discharged to the open or unconfined seawater by means of a submarine outfall, the receiving water conditions are markedly different so that the conventional parameters of BOD, etc., used for conventional design of treatment systems, are not of significance and, instead, the only parameters that do appear to be of significance are floatables and coliforms, both of which can be returned by onshore winds/currents to enter into the zone of shallow water to be protected. For most systems discharging into the open seas, it appears that only floatables removal may be required.

In the USA, the U.S. EPA has, since the 1960s, taken over the responsibility/authority for regulating waste discharges to the environment, and the control of such regulation has progressively passed from engineers to attorneys and, as a result, the use of engineering/economic solutions for solving waste management problems has been re-oriented into an overly simplistic system of uniformity whereby the same minimum level of treatment is required for all wastes, regardless of the capabilities of the receiving waters. As a result, the communities and industries in the country located along the West Coast and elsewhere (including the Gulf of Mexico), which can readily and economically build submarine outfalls reaching into the unconfined offshore waters, are being required to upgrade their levels of treatment from primary to secondary at the cost of hundreds of millions of dollars, with little, if any, real measurable gain in environmental protection. This is the opposite of good engineering practice.

All of the available scientific data indicate that the optimal disposal system, where the non-confined seawaters can be feasibly reachable by use of a submarine outfall, is to utilize such an outfall together with treatment only for removal of floatables. Thus, the wastes do not enter into the shallow waters of sensitive ecology but instead are by-passed for direct discharge into the unconfined waters. The data indicate that the build-up in the ocean waters and bottom sediments of the constituents reaching the open sea will be negligible and of no significance in altering the overall ocean physical/chemical/biological balances, even with discharges indefinitely into the future.

In developing countries, where the affluence of the USA is non-existent, and where most people are quite poor, the expensive non-cost-sensitive environmental control standards of the USA have little, if any, application. Instead, if water pollution control systems are to be approved and funded by DMs, they must be clearly cost-effective. In this context, marine waste disposal systems have been and are now being planned and built in DCs which follow the logical submarine outfall/floatables removal concepts that are noted above. Probably the most interesting of these is the system serving Rio de Janeiro, where a submarine outfall extending several kilometers offshore affords excellent protection of the beaches in the vicinity (Ipanema and Copacabana), with no treatment of the waste at all (because of the favorable offshore current patterns).

Conclusions

The conventional technology for design of systems for treating liquid wastes from municipalities and industries, which is based on essentially the dissolved oxygen parameter, plus removal of settleable/floatable solids which can form sludge banks in rivers and other confined receiving waters, has virtually no scientific basis for design of treatment systems which discharge to open or unconfined marine water via a submarine outfall. For such discharges the applicable design parameters are removal of floatables and coliforms, which might be returned to the near-shore zone of marine water which needs to be protected for swimming, fisheries, and other beneficial uses.

Discharge to the open sea is advantageous, when the open

sea can economically be reached by means of an outfall, not only because of much lower costs but because this affords much better protection of the regional ecological values and moreover, furnishes valuable nutrients to support marine ecology. When waste effluents are discharged to rivers or confined estuaries/bays, their toxic constituents (which are only partially removed by conventional/economic treatment) thus permeate the ecologically sensitive confined waters while passing on to the relatively insensitive open sea zone.

The current U.S. EPA practice of requiring secondary treatment for effluents discharged to the open sea and which prohibits discharge of organic sludges to the open sea, represents the opposite of good engineering practice and is causing urban centers on the West Coast of the USA and elsewhere to waste hundreds of millions of dollars.

There is no real evidence that the toxic or pathogenic constituents contained in waste effluents discharged to the open sea will result in any harm to the marine environment. When diffused into the sea, the resulting additions to the ocean's basic levels of these constituents are insignificant, even for discharges over hundreds or even thousands of years.

An urgent need which should be addressed by the U.S. EPA is to produce a "Manual of Appropriate Design Practice for Disposal of Sewage and Industrial Wastes from Coastal Communities," which evaluates the costs and levels of protection to the regional ecology for three alternative places of discharge, (1) to the inland river, (2) to the confined estuary or bay, and (3) to the open sea, based on the known scientific facts relating to effects of wastes upon sensitive aquatic ecology. This would upgrade the use of alternative (3) from its present status, whereby it is considered ecologically less desirable, to its proper status as the most ecologically protective. The presently available data indicate that alternative (3) will be the most economic and will afford the best ecological protection, whenever the open sea is economically reachable by use of a submarine outfall.

Another task which should be considered by the U.S. Agency for International Development, which would be extremely valuable to developing countries in promoting their environmental pollution control efforts, would be to sponsor a program by the U.S. EPA for adding an additional chapter to the EPA's many very valuable manuals on pollution control technology. As now prepared, these manuals present the basic background data. Then, in the final chapter, these data are interpreted to develop conclusions on appropriate treatment practices and standards for USA conditions. The purpose of the extra chapter would be to interpret the basic data to develop similar conclusions appropriate for developing country conditions. Thus, at very small investment, USAID could greatly enhance its program for assisting the DCs in solving their very difficult problems of environmental pollution control.

References

CSWRCB (California State Water Resources Control Board). 1972. *Guidelines for the Preparation of Technical Reports on Waste Discharges to the Ocean and the Effects of Waste Discharges on the Ocean.*

Engineering-Science, Inc. 1963. *Report on Submarine Outfall Clay Slurry Disposal System at Seventeen Mile Drive on Monterey Bay.* Prepared for San Luis Obispo Regional Water Quality Control Board by Engineering-Science, Inc., Arcadia, California.

__________. 1965. *Report on Metropolitan Sewerage System for Accra-Tema Urbanizing Region.* Prepared for UNDP/WHO/Government of Ghana by Engineering-Science, Inc., Arcadia, California.

__________. 1967. *Proposed Comprehensive Plan of Sewage Disposal for Rio de Janeiro.* Prepared for State of Guanabara, Rio de Janeiro.

__________. 1971. *Report on Master Plan of Sewage Disposal for Coastal Cities of Puerto Rico.* Prepared for Government of Puerto Rico by Engineering-Science, Inc., Arcadia, California.

__________. 1972. *Proposed Water Pollution Control Plan for Oahu.* Prepared for City of Honolulu by Engineering-Science, Inc., Arcadia, California.

__________ and Minco. 1985. *Reappraisal and Review of the Kuala Lumpur Master Plan for Sewerage and Sewage Disposal.* Prepared for City of Kuala Lumpur by Engineering-Science, Inc., Arcadia, California.

IAWPR (International Association on Water Pollution Research). 1972. *Summary Report of Regional Conference on Marine Water Pollution of International Association of Water Pollution Research.* Trieste (Italy), Pergammon Press.

Kiravanich, P. and H. Ludwig. 1979. *A Proposed Action Program for Water Pollution Control for Bangkok/Inner Gulf Megalopolis Region.* National Environment Board, Bangkok.

__________, P. Sirithan, and H. Ludwig. 1985. *Environmental Quality Standards and Criteria, the Problem in the Developing Countries.* National Environment Board, Bangkok.

Ludwig, H. 1960. *Report on Proposed Metropolitan Sewage Disposal Plan for Montevideo, Uruguay.* Prepared for WHO/Pan American Health Organization.

__________. 1971. *The Case for Advanced Waste Treatment, Circa 1971.* Water and Sewage Works.

__________. 1973. *Water Pollution Control in Metropolitan Bangkok.* Prepared for National Research Council and USAID, Bangkok.

__________. 1978a. *Guidelines for Preparation of Ordinances for Regulating Sanitary and Industrial Waste Discharges for Municipalities in Indonesia.* Prepared for UNDP/Institute of Hydraulic Engineering, Bandung.

__________. 1978b. *Design of Marine Waste Disposal Systems for Industrial Wastes.* Regional Conference of International Association for Water Pollution Research, Sorrento (Italy), Pergammon Press.

__________. 1982a. *Report on Recommended Plan of Water Pollution Control for Metropolitan Abidjan (Ivory Coast).* Prepared for Government of Ivory Coast and World Bank, Abidjan.

__________. 1982b. *Environmental Assessment for Eastern Seaboard Development Planning Project.* Prepared for National Environment Board, Bangkok.

__________. 1983. *Environmental Impact Assessment for Agriculture Waste Recycling Project.* Prepared for Asian Development Bank, Manila.

__________. 1984. *Report on Jakarta Sewerage and Sanitation Project.* Prepared for World Bank.

__________. 1986. *Report on Review of Proposed Metropolitan Sewerage System Improvements for Kuala Lumpur.* Prepared for World Bank.

__________ and Ben Onodera. 1962. *Scientic Parameters of Marine Waste*

Discharge. International Conference of International Association on Water Pollution Research, London, Pergammon Press.

Murphy, G. and Y. Eisenberg. 1985. *San Francisco Outfall, the Champ?* Civil Engineering, American Society of Civil Engineers.

OCBS (Orange County Board of Supervisors). 1946. *Report on Sewage Disposal Program for County of Orange.* Prepared for County of Orange by Board of Consultants, Santa Ana, California.

SCCWRP (Southern California Coastal Water Research Project). Annual Reports for years 1970 to 1986.

USCSOWR (U.S. Congress, Subcommittee On Water Resources). 1978. *Modification of Secondary Treatment Requirements for Discharges into Marine Waters.* Congressional Hearings 24 May 1978, Washington, D.C.

U.S. Corps of Engineers, 1979. *Final Summary, the Dredged Material Research Program.* Vicksburg, Mississippi.

SEVEN

HORMESIS AND NURTURE WITH IONIZING RADIATION

T.D. Luckey

Introduction

"From the evidence of various experimenters, it is apparent that considerable more investigation will be required to determine what the physiologic effects of x-rays are, or even whether any such effects are produced by them at all. The testimony on this subject is very conflicting, even taking only that of the highest authorities." (Anon., 1896)

Hormesis is the stimulation from low doses of any agent. This broadened concept came from Southam and Ehrlich who used it for a microbial toxicant.[1] Hormesis is equivalent to older terms which convey a more limited or similar meaning; the Arndt-Schultz law is one example. Nurture is the sum of environmental effects upon an organism; it includes essential agents, as oxygen, nutrients and gravity, and nonessential agents such as starlight and helium (Figure 7-1). The concepts of hormesis and nurture with ionizing radiation will be examined to provide understanding of the physiologic effects of this important agent of our environment.

The three major models for the effects of different amounts of ionizing radiation upon organisms are presented in Figure

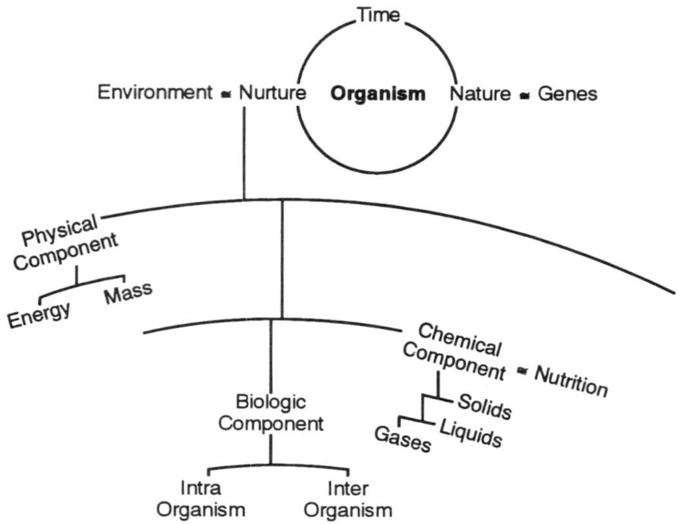

Figure 7-1: Nature and Nurture. Each individual is a composite of his nature, the genetic potential which directs his ontogenetic development, and his nurture, his total environment, from conception to death.

7-2. The lower line represents any of the linear models which suggest that all radiation is harmful. No variation of this model allows a threshold, a point at which there is no difference between the irradiated and control organisms. The horizontal line indicates that a threshold has been established with low doses of ionizing radiation. The threshold model indicates there is no difference between lightly irradiated and control organisms. The latest BEIR report accepted the threshold model for all physiologic functions except mutation and cancer induction.[2] The hormesis model projects diametrically opposite responses to high and low doses of ionizing radiation. The hormesis model provides a "point" threshold, the zero equivalent point (ZEP), and decreases the significance of any of the non-threshold models.

Ionizing radiation is a component of our nurture. Many radioactive elements are in our bodies, food, air, clothes, homes, furniture, tools, and machines. Such information abounds in

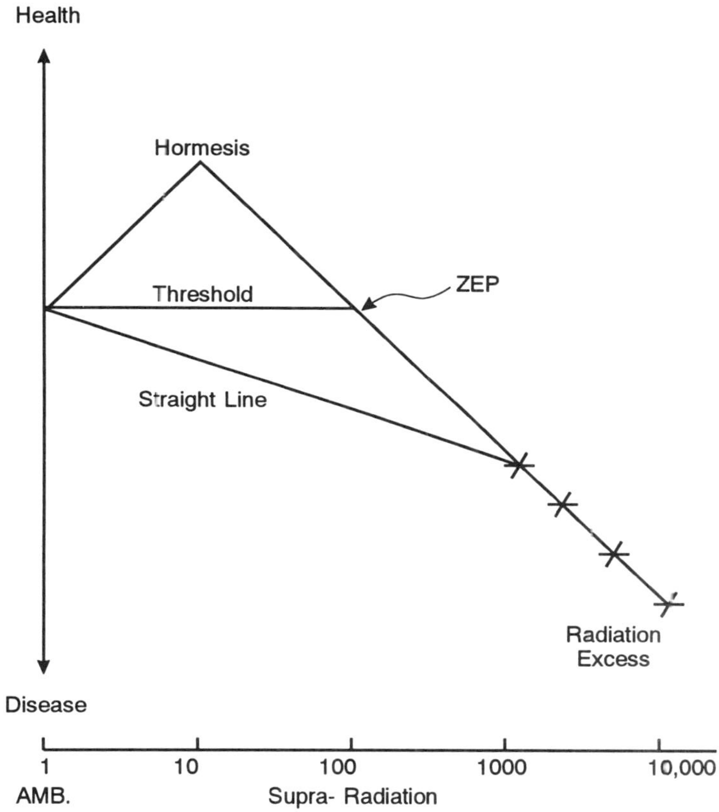

Figure 7-2: Dose-response Models For Ionizing Radiation. The three major dose-response models for the effects of ionizing radiation are the linear model with its several modifications, the lower curve which has several variations, the threshold model and the hormesis model in which the curve crosses the threshold at the zero equivalent point (ZEP). The abscissa provides an arbitrary log scale for either acute or chronic (rad/y) exposure for mammals.

texts and articles about background radiation. Each of us has ten thousand measurable events per second caused by ionizing radiation from earth and cosmic sources.[3] Unless otherwise stated in the following, radiation means exposure to x- or gamma rays. Hormesis with ionizing radiation has been noted throughout this century and has recently been reviewed generally,[4] in immunity[5] and as the expression of an essential agent for protozoa.[6] This review will emphasize radiation hormesis in cancer mortality.

A short summary of the effects of low doses of ionizing radiation in bacteria, plants, and invertebrates support the evidence provided by animal experiments and human experiences. A brief overview of radiation hormesis in vertebrate growth, reproduction, and resistance is background for radiation hormesis with cancer. The evidence of radiation hormesis in cancer induction and mortality will be examined in more detail to show that whole body exposure studies provide evidence of an hormetic model for cancer incidence and mortality following low dose irradiation. Exposures of tissues and cell cultures are not considered since these are not representative of an integrated organism. The evidence that low levels of ionizing radiation provide greater health and increased average longevity raises vital questions.[7] Is ionizing radiation essential for optimum physiologic performance? for life?

When studied with microbes both questions were answered in the affirmative. The positive microbic results combined with the extensive evidence for radiation hormesis in animals and humans suggest the need for changed attitudes, recommendations and practices in government, industry, medicine and public health regarding ionizing radiation. Discussion of the consequences of these concepts includes evaluation of the scientific data, new areas for critical research, public health implications, and a peek into the quagmire of the provision of a set of recommended allowances for increased background levels of ionizing radiation.

Radiation Hormesis in Physiologic Functions

Bacteria

As early as 1896 x-rays were found to kill bacteria[8] and soon bacterial growth was found to be increased by exposure to low doses of x-rays.[9] Protozoan growth and survival were increased by low levels of radium.[10] Gager found that fermentation was doubled by appropriate exposures of yeast to radium rays.[11] Uranium salts were reported to stimulate yeast[12] and bacterial growth.[13] These early reports have been amply confirmed and summarized.[14] One example is illustrated in Figure 7-3. Note the consistency in these two experiments which were performed two years apart.

Plants

There are well over a thousand references regarding radiation hormesis with plants in the reviews of Berezina *et al.*, Kuzin, Simon and Bhattacharya, and Luckey.[15] Excellent work at the Hannover Institute of Radiation Botany provided good examples of radiation hormesis in plants (Figure 7-4). The experiments with strawberries provide reliable results where the use of clones precluded genetic variability.[16] The experiments with duckweed, a small two leafed plant which floats on

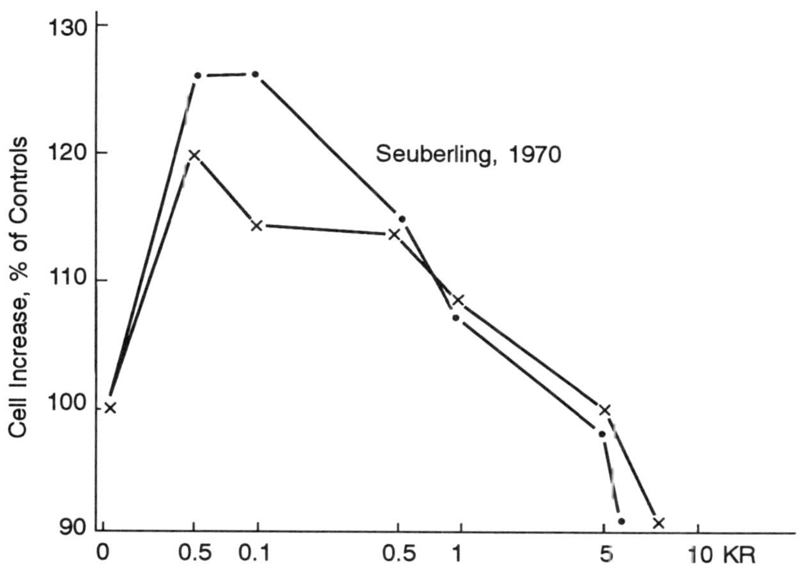

Figure 7-3: Effect of a Single Dose of X-ray upon *Euglena* Growth. The rate of growth, cell division, of *Euglena gracilis* is given on the ordinate as percent of control growth. The cultures were illuminated at 25°C for 24 hours. The abscissa gives dose rates in kilorads.[17]

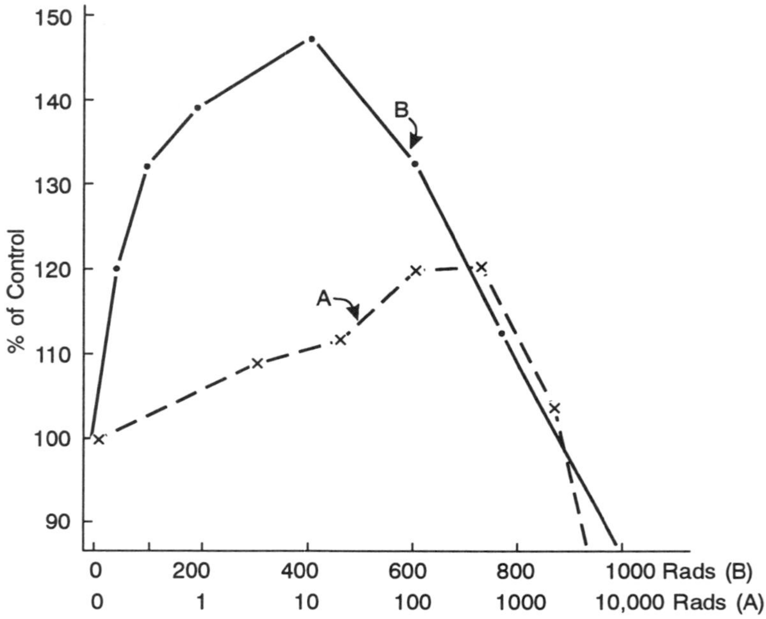

Figure 7-4: Radiation Hormesis in Plants.

A. X-ray Exposure of Strawberry Clones. Exposure of young strawberry plants, *Senga precosa*, increased early and total yield, ordinate, when irradiated with one dose of x-rays, abscissa A.[18]

B. X-rays Stimulate Growth of *Lemna minor*. The irradiated plants weighed more than controls with six hours of light at 26°C. The ordinate is percent of control and the abscissa indicates the acute doses administered at the start of the experiment.[19]

water, give an easily repeatable experiment with consistent results.[20]

Endpoints studied with plants include amount and time of germination, pollen tube growth, root growth and development, stem and leaf growth, branching, number and timing of flowers, and yield at harvest. The harvest may be measured as total plants, total grain, fruit or vegetable, yield per unit of land, uniformity of harvest, timing of harvest and nutritive value of the product. Zelles and Ernst records the increased growth of the vegetative nucleus, the generative nucleus and the cytoplasm within a single cell, the pine tree pollen.[21] He noted that exposure had to be completed within one hundred seconds to be effective.

Russia[22] and Hungary[23] have commercial ventures using mobile irradiators for seeds. However, formidable variables seem to have precluded their general commercial acceptance and large-scale use on farms in the western part of the world. The variables include both exogenous variables, such as dose rate, total dose, type of radiation and several environmental factors, and endogenous factors such as water content, maturation storage and strain. Results are generally more readily obtained under suboptimum conditions. The temperature change immediately following irradiation seems to be a synergistic factor. Timing of the exposure within the photoperiod was found to be very important in the example given; irradiation early in the light cycle had the greatest effect. Many experimental papers involve continuous irradiation of the whole plant, the green portion, the growing tip or the roots. Another problem has been interpretation using lumped data; obviously, important components of the dose-response curve may be missed if different doses are combined. In spite of these difficulties the potential for commercial use of radiation hormesis is being examined in several countries.

Invertebrates

The results from microbes and plants are amply confirmed by data from invertebrates. Much of the work utilized the easily handled flies, moths or beetles using fecundity or lifespan as endpoints. The effect of a single dose of gamma rays upon the reproduction rate in the flour mite is illustrated in Figure 7-5.[24] At the end of the six-week experiment there were 10,000 mites/gm of flour in the control colony and 16,000 in the colony which received 5.3 krad. Review of the fourteen reports confirming this result include increased number of eggs per adult per day, increased total progeny, increased total population, increased parthenogenesis and decreased generation times.[25] These results appear even when radiation reduced the fertility of part of the cohorts.

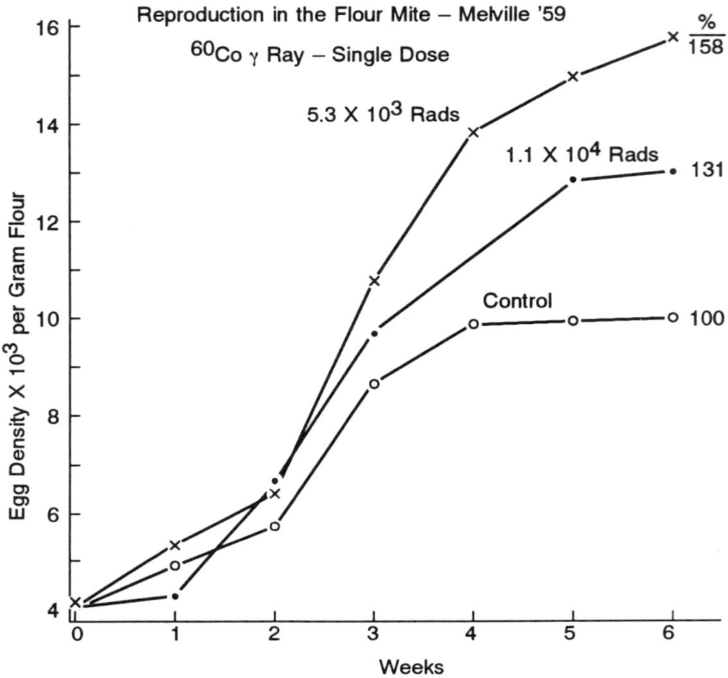

Figure 7-5: Radiation Hormesis in Insect Reproduction. Increased population in flour mite colonies was obtained following acute gamma irradiation.[26]

The same reviews list 35 reports showing increased average lifespan in invertebrates. Sometimes this is noted in one sex and not the other and sometimes it is noted in both sexes. While most find the increase in lifespan of irradiated insects is about 120 percent of the control values, Strehler found irradiated flies had a lifespan twice that of controls.[27]

The results seen in microbes, plants, and invertebrates provides a unified concept about the effectiveness of radiation hormesis which may be anticipated in vertebrate data. The unity of biochemistry, physiology and nutrition throughout the phylogenetic spectrum of life appears to be amply confirmed by the extensive data showing radiation hormesis in vertebrates; this includes epidemiology and exposure data from humans. In the same manner the data from microbes, plants, and invertebrates provide both background and confirmation of the effects reported for animals, including humans.

Vertebrates

Radiation hormesis in vertebrates was first considered seriously when Lorenz and co-workers reported increased growth rates and average lifespan in rats, mice, rabbits, and guinea pigs.[28] Confirmation of this result (Figure 7-6) by Morris and 22 others suggests a practical use in farm animals.[29]

The biopositive effects of low doses of ionizing radiation

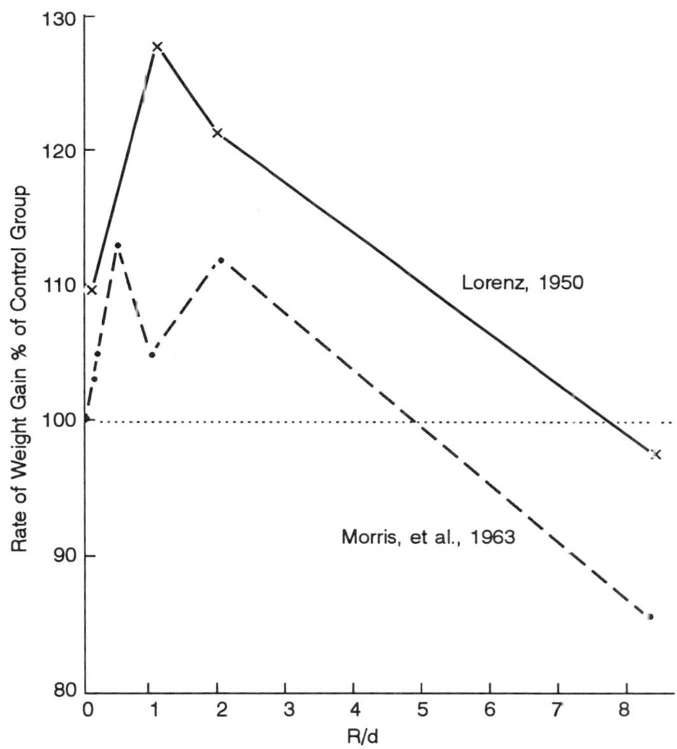

Figure 7-6: Radiation Stimulates Mouse Growth Rate.[30, 31] Growth of mice, ordinate, was significantly increased ($p<0.01$) by acute, daily exposure to x-rays, abscissa, for three weeks.[32]

upon neurologic functions (Table 7-1) include several examples of increased excitability of nerve tissues and acuity in hearing or sight. The studies of McDowell suggest that both learning and memory may be increased by low doses of whole body irradiation.[33] Obviously, more systematic study is needed.

Increased fecundity following light irradiation of vertebrates was found in 23 reports.[34] Muramatsu *et al.* found the mean litter size of mice was greater ($p<0.05$) in irradiated mice than in controls (Figure 7-7).[35] The total population was greater after several generations of continuous irradiation due to decreased sterility, increased sperm viability, increased number of embryos and newborn with decreased embryo mortality.[36] Kaplan reported irradiation to be helpful in the treatment of human infertility.[37] The specter of increased mutation was not evident in 82 generations of lightly irradiated mice[38] nor in the children born to exposed Japanese parents.[39]

Immune competence appears to be one key to understanding radiation hormesis. Increased wound healing, greater radiation resistance, and decreased infection and cancer morbidity

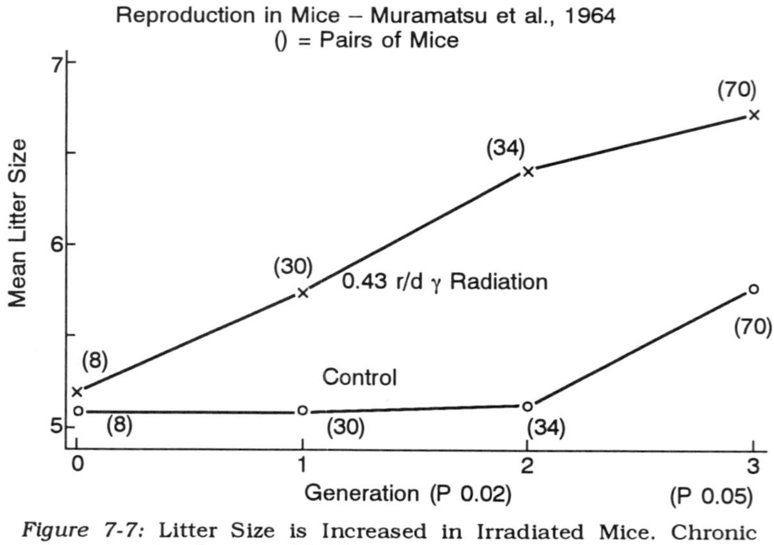

Figure 7-7: Litter Size is Increased in Irradiated Mice. Chronic gamma ray irradiation of 0.43 rads per day increased the litter size of mice, ordinate, for three successive broods.[40]

Table 7-1: Radiation Hormesis of Neurologic Function[41]

Date	*Authors*	*Observations*
1903	Zhukovskii	Dog Cortex Excitability
1933	Girden & 1935	Auditory Acuity of Dogs
1937	Brogden *et al.*	Dog Acoustic Acuity
1952	Galkovskaya	Nerve Regeneration
1956	Lomonos	CNS Excitability
	Rugh	Neonatal Mice
1958	McDowell & 1960	Monkey Learning
1960	Lott	Rat Nerve Excitability
	Livanov *et al.*	Rat Brain Excitability
1962	Dawson *et al.*	Frog Nerve-Muscle Prep.
	Hunt *et al.*	Rat Arousal
1963	Garcia *et al.*	Sensitivity of Brain Receptors
	Smith *et al.*	Moth Flight Induced
1965	Kimaldorf	Shrimp Eye Reaction
1968	Pora	Rat Brain Activity
1969	Grigeryev	Rabbit Vestibular Excitability
1975	Semaginin	Cerebral Cortex Development
1978	Carpenter *et al.*	Pacemaker Neuron Stimulated

provide evidence of increased immune competence as the main basis for increased average lifespan.

Wound healing was found to be accelerated over fifty years ago using a simple, readily observed system.[42] Small incisions were made in rabbit skin; then the rate of healing was observed both macroscopically and histologically in previously irradiated and control animals. Similar results were found with ulcerated epithelioma[43] and callus in bone fracture.[44]

Radioprotection may be provided by previous exposure to ionizing irradiation. Increased resistance to lethal doses of ionizing radiation was found by several investigators, as listed in Table 7-2. The increased radiation required for an LD[50] was quadrupled in two papers.[45] This suggests that evaluation of human responses to radiation should consider previous exposure of each individual.

Table 7-2: Radiation Hormesis in Radioresistance[46]

Date	*Authors*	*Observations*
1950	Cronkite *et al.*	Three weekly exposures of 144 rads with 1 month rest.
	Betz	Similar data.
1954	Lorenz *et al.*	Used continuous low level (1.1 rad/d) pretreatment.
1960	Maisin *et al.*	Acute (20 day) or chronic irradiation of fetal mice quadrupled radioresistance. Young mice given 5 rads showed double radioresistance over controls.
1963	Mezentsev and Gubin	Pre-treatment with 43 rads increased radioresistance.
1964	Vkasenko and Shakhnazaryan	Pre-treatment prolonged life.
1970	Nishio	Drinking 0.1 uCi strontium 90 and 0.4 uCi cesium 137/ml quadrupled radioresistance.
1978	Stienon-Smoes *et al.*	Pre-exposure of mice increased LD[50] about 50 percent.
1979	Spalding and Holland	Early irradiation gave dogs increased survival times.

Increased immune competence and probably increased resistance to infection are major effects of radiation hormesis. The infection protection must be inferred because radiobiologists rarely design challenge experiments which could provide good evidence with limited numbers of animals in a short time. A delightful exception is the early controversy about the biopositive effects of ionizing radiation. It involved Professor William Schrader of the University of Missouri who provided evidence to contradict the statement by Professor J.J. Thomson of Cambridge University that ionizing radiation was not bactericidal.[47] Schrader's *in vitro* studies showed these new x-rays could kill bacteria. He also noted that the irradiated Guinea pig survived infection with diphtheria bacilli while the control died within 24 hours with histologic evidence of diphtheria. Increased resistance to diphtheria toxin injection was reported by Gerhartz and Bisgard *et al.*[48] The increased resistance to infection in exposed invertebrates[49] and rodents[50] gives limited direct support to this concept. The concept of radiation hormesis in infection is well supported by studies which show irradiated animals have increased serum immunoglobulins (Table 7-3) with some measurements of specific antibodies. Increased amounts of other immune parameters are noted on the lower part of the table. Liu (1986) and Liu *et al.* have confirmed the increased numbers of plaque forming cells following mutagen administration.[51] They also report the increased production of interleukin 2 in spleenocytes from lightly irradiated mice (0.1 rads acute or 0.54 rads/d chronic whole body irradiation) when compared with control mice; thymus of exposed mice showed increased reactivity to interleukin 1, and spleen showed increased sensitivity to interleukin 2. Recent findings of increased immune competence in Japanese bomb survivors when compared with controls[52] and in Chinese living with high natural radioactivity[53] seem to confirm the animal reports. Immune studies with carcinogen induced cancer in mice support the concept of increased immune competence following low-level irradiation.[54]

Hormesis in Radiation-Induced Cancer

Accepting its heterogeneity of types as a single entity, cancer is an obvious pathologic state which develops three to forty years following exposure to ionizing radiation. Although most studies involved low energy transfer (LET) radiation, e.g., x- and gamma rays, the same general response pattern was found for high LET radiation, e.g., neutrons.[55] These particulate beams act as molecular bowling balls; they literally knock out parts of molecules, as DNA, and produce free radicals comparable with the products of low LET radiation. Excess free radicals can cause mutations which may lead to cancer. This summary of hormesis with ionizing radiation provides a concept which was ignored or considered lightly by agencies and committees which have suggested guidelines based upon the assumption that all radiation is harmful. A good example is the acceptance of the threshold model (Figure 7-2) for all physiologic functions, but not for mutation or cancer, by the committee on the Biologic Effects of Ionizing Radiation.[56] This BEIR III report did not accept the linear hypothesis, did not intercalate (often

Table 7-3: Radiation Hormesis of Immunoglobulins[57]

Antibody

1913, Manoukhine
1920, Hektoen
1921, Kaznelson and Lorant,–Typhoid Toxin
1937, Thompson *et al.*
1950, Burrows *et al.*–Cholera toxin-transient
1963, USEAC
1964, Taliaferro and Taliaferro
1975, Gras *et al.*
1976, Servant *et al.*
1977, Macedo; Gorini *et al.*–Antibody affinity
1978, Onletz; Shubin

Hemolysins and Bacteriolysins

1913, Manoukhine

Total Amount of Lymphoid Elements

1915, Murphy and Morton
1979, Tanaka and Sakai

Lymphocytosis–Medium doses gave no effect

1919, Thomas *et al.*
1929, Murphy

1951, Pape
1954, Lorenz *et al.*

Granulocytosis-transient

1960, Sacchetti *et al.*

Lymphocyte Stimulating Factor

1959, Metcalf

Phagocytosis

1962, Buzine
1975, Vorbrodt

Spleen Lymphocyte Enzymes and Bactericidal Activity

1976, Paul *et al.*

Spleen Follicle Development

1951, Pape

Spleen Plaque Forming Cells—Prior to antigen administration

1973, Zaalberg *et al.*

Immune factors are increased by low doses of radiation.

called extrapolate) for doses less than ten rad, and did not speculate on the health effects of background levels of ionizing radiation. If either the hormesis or the threshold models is correct for radiation induced cancer, present guidelines are counterproductive to public health, industry and government activities.

There are few reports of hormesis in mutation following low doses of ionizing radiation. The results of both Mulholland and Connolly and Olivieri *et al.* suggest that mutation rates are decreased in lightly irradiated cells.[58] A practical threshold in radiation induced mutation is indicated by the lack of phenotypic changes seen in mice chronically exposed to x- and gamma rays for 82 generations, equivalent to two thousand years for humans.[59] Although chromosomal aberrations appear, no radiation-induced mutations have been found in children of the Japanese survivors of the atom bomb according to good, sensitive methods such as electrophoretic patterns of serum proteins.[60] Thus, the most serious late effect of atomic bomb radiation appears to be limited to cancer.

The complete dose-response curve for the effects of ionizing radiation upon cancer induction appears to be a complex with six components (Figure 7-8); the top three curves represent acute exposures and the lower three are chronic exposures. There is little question about the first two. Excess ionizing radiation kills the individual before there is time to develop cancer (Figure 7-8a). Smaller doses allow subjects to live long enough for cancer to develop (Figure 7-8b); leukemia is the classic model for radiation induced cancer. The next four components are novel.

The V-shaped curve (Figure 7-8c) was elucidated by Hellström and Hellström who suggest a mechanism involving one class of antigen-presensitized radiosensitive T suppressor cells.[61] These thymic lymphocytes tend to suppress messages to other T cells which then cannot mount an active response to foreign material. When the activity of suppressor T cells is decreased with 400 rads of whole body irradiation, other immune cells become more active to repress tumor growth. This is called the "Hellström effect." Sometimes total regression of tumors may be obtained. North has suggested these radiosensitive T cells

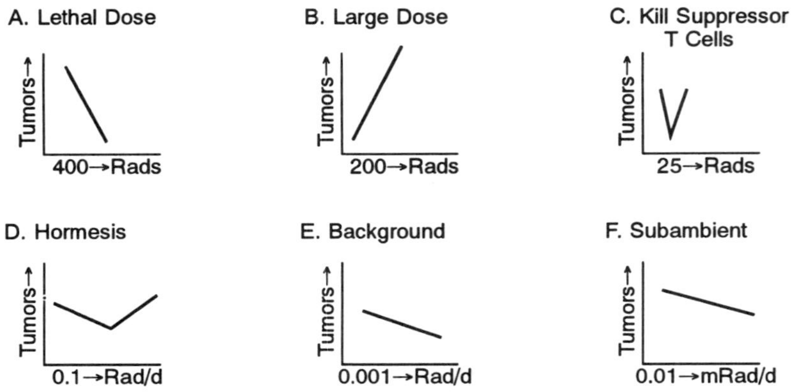

Figure 7-8: Six Dose-Response Curves for the Different Components of Radiation Induced Cancer.

may be manipulated to aid tumor therapy.[62] This concept was reinforced by Anderson *et al.* who implanted tumor cells into mice following whole body irradiation.[63] Tumor growth was slowed in mice exposed to 25 rads; repeated exposure to 25 rads was especially helpful in augmenting the host response. Data from Liu support this concept.[64] Although clinical application remains nebulous, these results help to explain some of the anomalies noted in the Japanese data.

Hormesis in radiation induced cancer (Figure 7-8d) occurs below doses used in the Hellström effect with acute doses and may be part of the background effect for chronic doses (Figure 7-8e). These may blend imperceptibly together. Curves 8d and 8e relate to the inverse correlation of cancer incidence and mortality with background radiation levels. No data is available to examine cancer induction in conditions of subambient levels of ionizing radiation (Figure 7-8f). Information from hormesis and background levels of radiation plus microbic data suggest this curve is a reasonable, if hypothetic, possibility.

The complete dose-response curve for radiation induced cancer (Figure 7-9) is a composite of the six components (Figure 7-8) with the acute and chronic effects differing in the area

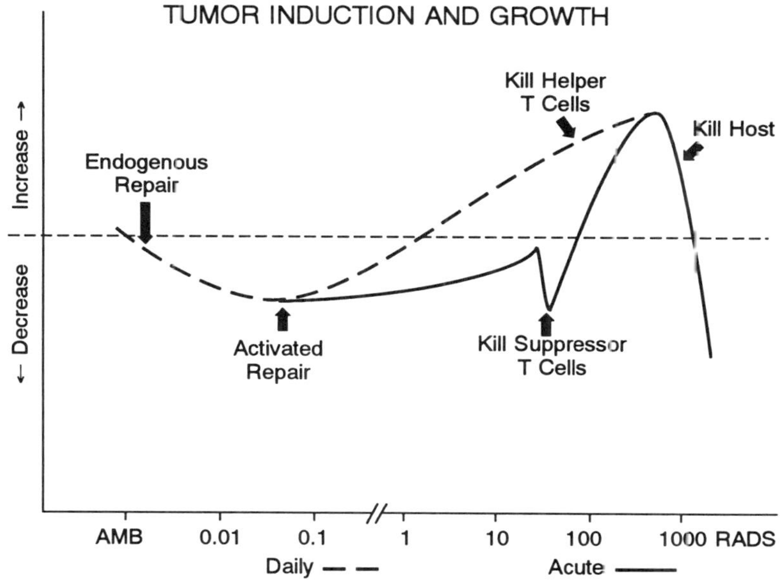

Figure 7-9: Complete Dose-Response Curve for Radiation Induced Cancer. This combines the six separate components shown in Figure 7-8.

of the Hellström effect. Of course, the effects for any given acute and chronic dose would be quite different, depending on the species exposed. For example, 1,000 rads acute radiation quickly kills most mammals; however, guinea pigs have survived 12,000 rads of chronic radiation.[65]

Studies on whole body radiation including cancer induction and mortality provide data which fit the concept established for many physiologic functions: i.e., high and low doses of ionizing radiation produce diametrically opposite results. Usually experiments which do not exhibit hormesis did not involve low enough doses. Animal experiments which cover a good range of doses are reviewed. The experience of humans exposed to radiation higher than background levels, 1 mrad/day, and human epidemiology studies appear to involve appropriately acute doses or chronic levels of exposure. The following information supports the hypothesis that low doses and levels of radiation do not induce cancer (the threshold model) and may decrease the incidence of cancer (the hormesis model). General acceptance of either would have a dramatic effect upon our present policies.

Animal Experiments

Experimental evidence for the hormetic model and a threshold in radiation-induced cancer in animals both support the concepts obtained from human medical and accidental exposures, ecologic experiments and epidemiology. Accepting biologic variability which makes each individual different from all others, there remains a remarkable biochemical and physiologic unity in living organisms.[66] Biochemical and physiologic responses of laboratory animals to ionizing radiation are often comparable with those in humans. Adjusted time scales are helpful for comparisons: Grahn suggested that one year for a man is equivalent to ten days for a mouse.[67] Since low doses of ionizing radiation often result in increased average lifespan, there is more time for cancer initiation in the older population. Thus, age specific data provide a reasonable basis for comparisons of irradiated and control cohorts.

Challenge experiments are concise and provide more information within a short time than the expensive "irradiate and watch" experiments conceived by most radiobiologists. The classic experiment of Murphy and Morton gave clear results with relatively few animals (Table 7-4).[68] They excised spontaneous tumors from mice and regrafted them into the groins of the original host following one of three treatments: (1) a "stimulation dose" of x-rays to the mouse only; (2) no irradiation to either mouse or tumor; and (3) a "stimulation dose" of x-rays to the tumor before replacing it into the host. They found whole

Table 7-4: X-Rays Stimulate Resistance to Cancer

Group	*No.*	*Irradiation* *Mouse*	 *Tumor*	*% With Tumor Resorption*	*% With New Tumors*
1	52	+	–	50	21
2	29	–	–	3	48
3	10	–	+	0	40

body irradiation induced host resistance to both the original tumor and to the subsequent appearance of new tumors. They presumed the cause involved the increased lymphoid tissue which they noted following the whole body irradiation, a conclusion remarkably close to present evaluations.

This concept was extended by Lisco *et al.* with carcinogen-induced tumors (Figure 7-10).[69] Young mice were given two large doses of x-rays with a few weeks rest after each. Then control and irradiated mice were injected with a carcinogen, methylcholanthrene. After 69 days five tumors were palpable in the control mice and none in the irradiated mice. The decreased susceptibility of irradiated mice to chemically induced tumors was significant ($p<0.01$) from day 96 through 129 following the injection; during this time about twice as many

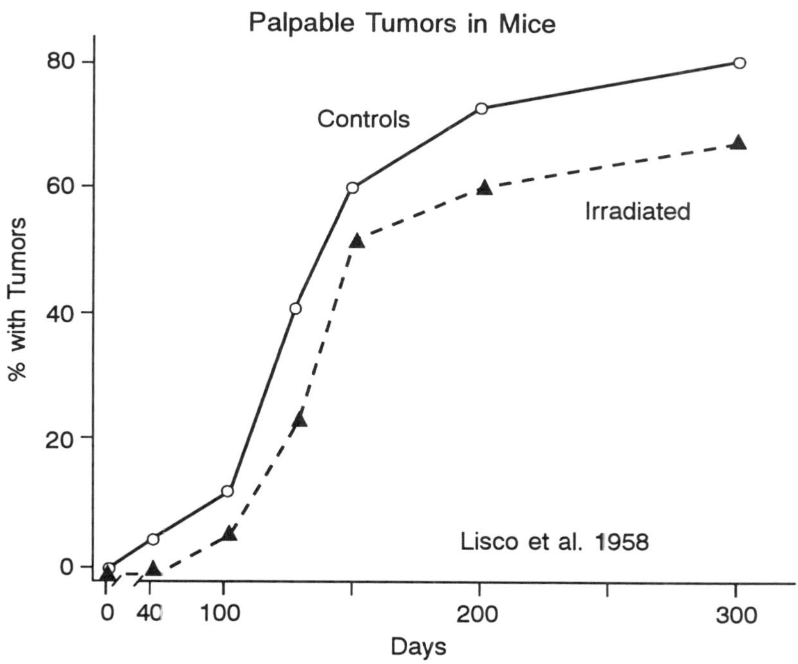

Figure 7-10: Reduced Tumor Incidence in Irradiated Mice. Prior to the administration of methylcholanthrene, the mice had recovered from two 450 rad doses of x-rays.[70]

chemically induced tumors were found in control as in the irradiated cohort.

The above result was confirmed in mice given drinking water with 0.1 uCi cesium 137 and 0.4 uCi strontium 90 per ml through several generations. The irradiated mice were found to be more resistant than unirradiated controls to implantations of Erhlich acites tumor.[71] However, cancer induction was enhanced when methylcholanthrene was administered immediately following or concomitant with irradiation.[72] These decisive results make it difficult to understand why challenge experiments are generally ignored by radiobiologists.

Oak Ridge National Laboratory has contributed two decades of research to the study of the incidence of cancer in irradiated mice. Dose-response curves presented by Upton *et al.* relate cancer incidence in mice to single and daily exposures of either gamma rays or fast neutrons.[73] The high incidence of neoplasms in control mice should provide a good test of hormesis; however, the low numbers allowed variation to preclude serious conclusions. Excepting a small initial increase with the lowest doses, no consistent pattern was established for either males or females. This may be due to the smallest dose being 25 rads.

An extensive study involving thirty thousand mice provided more definitive results from this laboratory.[74] When ten rads of gamma radiation were administered at 45 rad per minute, both sexes exhibited hormesis for all cancers studied with the exception of cancer of the Harderian gland, the third eyelid. When fifty rads was the minimum dose, chronically irradiated mice showed hormesis in radiation induced cancer. However, an acute dose of fifty rads appeared to exceed the threshold for several types of cancer. Neutrons at 25 rads gave a threshold for leukemia incidence and appeared to exceed the threshold dose for solid tumors. Since the relative biologic effectiveness (RBE) for neutrons is usually estimated to be four to twenty times that of gamma and x-rays, it is unfortunate that lower doses were not utilized. Less extensive work with a different strain of mice was reported; the lowest dose of fifty rads of high LET radiation exceeded the threshold for most tumors.

Other less extensive studies have found hormesis or threshold responses to low doses of ionizing radiation. "There is also evidence for threshold dose or dose rates below which early tumors are not induced."[75] Low doses of ionizing radiation resulted in a five-fold reduction in mouse leukemia.[76] This group also reported that low doses of ionizing radiation increased leukemia and lung cancer while all other cancer types were decreased (Figure 7-11).[77] Fewer reticulo-endothelial cancers were found in mice exposed to fast neutrons than in controls.[78] The incidence of non-lung cancer in mice following plutonium oxide inhalation was only 25–30 percent as great as that of untreated controls.[79]

The increased danger of low doses of ionizing radiation to embryos is not evident in the following studies. When male mice were exposed to 6.6 rads from deuterium water *in utero* for 22 days, they subsequently developed only 56 percent as many tumors as the unexposed controls; the females showed a threshold, 103 percent of controls.[81] This study confirmed

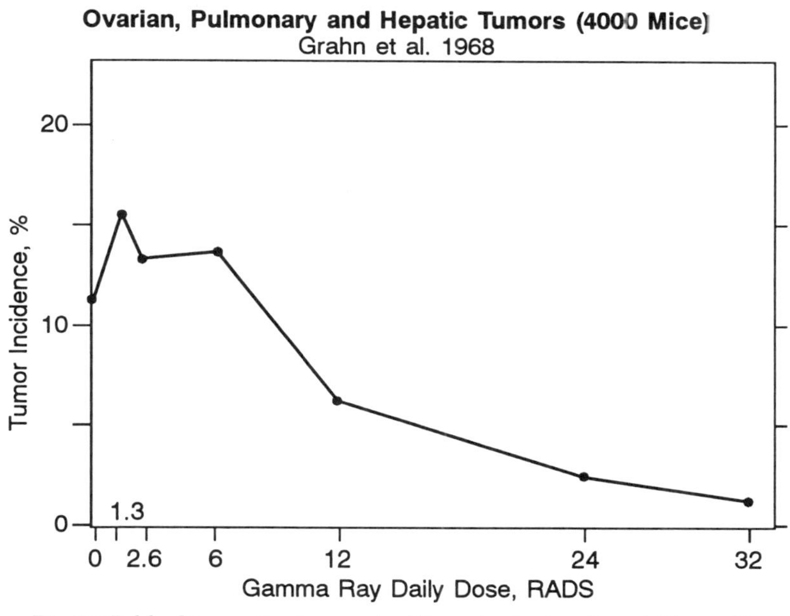

Figure 7-11: Cancer Incidence in Chronically Irradiated Mice. The incidence of lung cancer and leukemia were increased while "all other cancers" decreased.[80]

previous results showing a lower incidence of tumors found in mice and rats which had been irradiated with x-rays *in utero.*[82] These results appear to substantiate the conclusion of Webster that there is no established risk of childhood cancer following *in utero* low dose irradiation.[83]

Human Exposures

Although the use of ionizing radiation in medicine will not be reviewed generally, clinical data suggest the existence of hormesis or a threshold. Webster and Boice and Land provide general reviews of the medical aspects of ionizing radiation in cancer induction.[84] "New radiation-associated tumors have been predicted for one in one hundred children receiving one thousand rads or more."[85] This is a surprisingly low incidence for such a high dose. Webster could find no adequate explanation for the failure to demonstrate excessive leukemia in women treated for cancer of the cervix.[86] The narrow beam doses were five hundred to one thousand rads with a major portion of the bone marrow exposed to 100+ rads. Only nine leukemia cases were observed while the BEIR III (1980) prediction was 42–85 for this group.[87]

Leukemia, especially acute lymphocytic leukemia, is readily induced by high doses of ionizing radiation; low doses do not evoke the same reaction, according to Gunz and Atkinson (1964) and Linos *et al.*[88] The latter group questioned the interpretation of several reports involving clinical studies which suggested a positive association between low doses of ionizing radiation and leukemia. When diagnostic exposures were correlated with two carefully selected controls for each of 138 leukemia cases found in one county over a period of twenty years, Linos *et al.* state: "No statistically significant increase was found in the risk of developing leukemia after radiation doses of zero to 300 rads (3Gy) to the bone marrow when these amounts were administered in small doses over long periods of time, as in routine medical care." In fact, their extensive and thorough examination for doses of one to fifty rads suggest that the risk of either acute or chronic lymphocytic or myelocytic leukemia is lessened by low doses of ionizing radiation. This fits the hormesis model.

Atomic Bomb Survivors

High doses of ionizing radiation result in the production of cancers comparable with those induced by other carcinogens or those which occur spontaneously.[89] The 83,000 survivors from Hiroshima and Nagasaki had an average whole body dose (Kerma) of 27.2 rads, revised dose, and provided a cohort of 527 million person-year-rads.[90]

Recent revision has lowered dose estimates for both neutron quantity and energy in Hiroshima, with a concomitant increase in the gamma ray/neutron ratio, while leaving the radiation estimates for Nagasaki at about 98 percent gamma rays. Kerma adjustments from these changes have little effect upon the general shape of the dose-response curves.[91]

Refinements in this data base should include identification of prior radiation experience for each subject,[92] better estimates of radiation from fallout,[93] and separation of the control cohort into zero and 0.1–0.9 rad subgroups to correspond with those of Robinette *et al.*[94] Close routine inspection of all households plus determination of the cause of all deaths provide unusually good data for comparing exposed with control populations. Over seventy percent of all cancers were confirmed by autopsy. About 99 percent of any excess cancer mortality is expected to end at finite times following acute whole body radiation: 27 years for leukemia and 45 years for other cancers.[95]

The observed mortality data from individual types of cancer at Hiroshima-Nagasaki were used to calculate mortality per 10,000 persons and expressed as percent of control.[96] The dose response curves for leukemia and "all other cancers" (Figure 7-12) suggest hormesis and a threshold for cancer induction at about eight rads; doses less than eight rads gave no increased leukemia or "all other cancers." There was decreased incidence for most of the thirteen specific cancer categories (Table 7-5). Note the low incidence of "other cancers" in the high radiation doses; statistical validation of this concept is lacking in the Japanese data as well as much of the animal data. Note also the generally decreased cancer mortality at 345 rads compared with that at 244 rads; no mortality was found in four categories. Confirmation of this effect in animal experiments suggest that

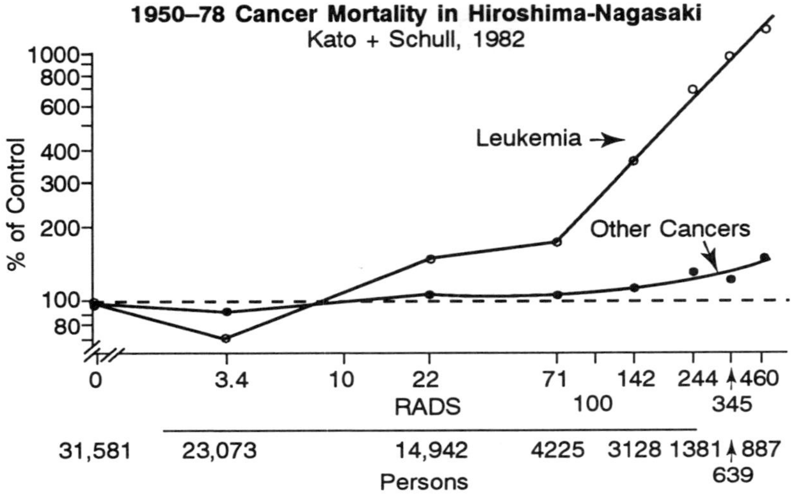

Figure 7-12: Cancer Mortality in Japanese Bomb Victims. Hormesis was noted in the dose-response curves for both leukemia and "all other cancers" in both cities.[97]

study of survivors indicate they retain increased cellular immunity when compared with controls.[99]

Leukemia. Special attention has focused upon leukemia as the most prominent radiation-induced cancer in atomic bomb survivors. High doses induce increased incidence of leukemia within thirty years following exposure. However, the low dose cohort had only 71 percent of control leukemia mortality (Table 7-5). Detailed examination of the low dose data makes it clear that the data do not fit any version of the non-threshold models as well as they do the hormesis model (Figure 7-13).

Esophagus. When compared with controls, esophageal cancer mortality exhibited a decreased incidence in the lowest dose cohort, a threshold at approximately fifteen rads, and unusually low incidence at 345 rads (Table 7-5). No increment in the incidence of esophageal cancer was noted in the Nagasaki data however, the total numbers involved were small.[103]

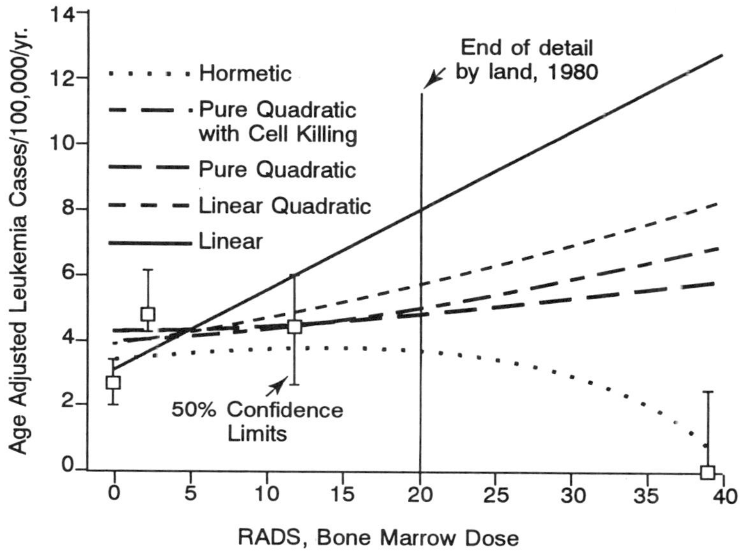

Figure 7-13: Radiation Induced Leukemia Mortality in Nagasaki.

13a - Comparison of low dose irradiation induced leukemia in Nagasaki according to Land and Pagnementa with 50 percent confidence limits displayed.[100] Note that the point at 38 rads is ignored in all but the lowest curve, the hormesis curve.

13b - Computer model of the dose-response curve for radiation induced leukemia mortality at Nagasaki.[101]

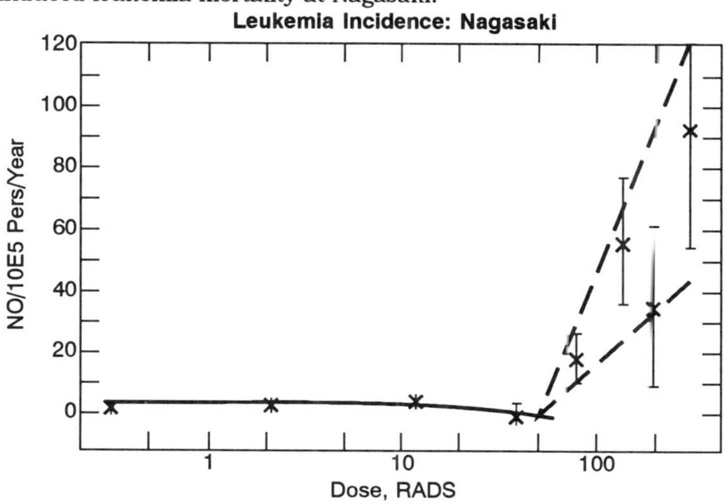

Table 7-5: 1950–78 Cancer Mortality in Atomic Bomb Subjects[102]

Rads	0	1-9	10-49	50-99	100-199	200-299	300-399	400+
Ave Rads	0	3.4	22	71	142	244	345	526
Persons	31,581	23,073	14,942	4,225	3,128	1,381	639	887
	controls per 10,000	% of control						
All Cancer	575	94	107	111	123	156	182	190
All Non-Leukemia	560	95	106	109	117	140	120	153
Leukemia	15	71	151	178	371	794	1,287	1,616
Multiple Myeloma	2	100	65	235	-0-	700	-0-	1,100
Malignant Lymphomal	12	58	92	98	33	233	133	92
Esophagus	22	50	109	107	73	264	-0-	155
Stomach	224	92	102	96	92	104	119	146
Colon	20	85	90	95	110	180	80	340
Rectum	20	95	110	119	80	110	-0-	-0-
Pancreas	19	111	89	50	100	115	87	58
Other Digestive Organs	69	103	107	174	120	178	181	146
Lung	49	110	120	129	229	178	255	184
Breast	14	90	121	136	321	100	-0-	400
Uterus	36	131	139	86	125	122	92	156
Urinary Tract	12	92	100	138	183	233	258	92

Stomach. The high incidence of stomach cancer in the control population was not increased by low dose irradiation (Table 7-5). Thus, it appears to be doubtful that the "endemic cancer factor" of Japan and radiation are synergistic.

Colon, Rectum, and Pancreas. Radiation-induced colon cancer mortality data revealed hormesis with a threshold dose at about one hundred rads (Table 7-5). The data for rectum and pancreatic cancer show no correlation with radiation dose.

Other Digestive Organs. The dose-response curve for cancer mortality from other digestive organs show a gradual increased incidence correlated with dose (Table 7-5). However, details from only the Nagasaki cohort indicated no increase with dose for gall bladder or salivary gland cancers.[104] The dose-response data from liver cancer mortality showed a threshold at about two hundred rads.

Lung. The Nagasaki data suggested that radiation hormesis may be exhibited in lung cancer.[105] While no threshold was exhibited in data from the combined cities (Table 7-5), hormesis was exhibited in non-smokers.[106] The incidence of lung cancer per 1,000 persons was 16.6 in persons exposed to zero to nine rads (a category unfortunately substituted for zero rads) and only 7.5 in those exposed to less than one hundred rads.

Breast. Although radiation-induced breast cancer mortality is generally accepted for humans, both hormesis and the Hellström effect were found in the data from the combined Japanese cities (Table 7-5). When the zero and the one to nine rads were combined the combined results decreased the possibility of observing hormesis.[107]

Uterus and Prostate. Low doses of ionizing radiation may increase uterine cancer mortality; however, most higher doses do not appear to increase cancer mortality (Table 7-5). The Nagasaki data for prostate cancer mortality indicated hormesis at the lowest doses with no increased incidence with increased radiation.[108]

Urinary Tract. Radiation-induced urinary tract cancer mortality suggests hormesis at the lowest dose with a threshold above 22 rads and increased incidence at high doses (Table 7-5). The high doses at Nagasaki (primarily gamma radiation) gave no increased incidence while those at Hiroshima resulted in increased cancer mortality.

Thyroid. The incidence of radiation-induced thyroid cancer mortality in the Nagasaki study showed an increase with low doses and no correlation with higher doses.[109]

Summary of Japan Data. The data from Hiroshima and Nagasaki provide the best information regarding a dose-response relationship for radiation-induced cancer mortality. Probably little significant information will accrue after forty years. Several trends are evident. The hormesis model fits most cancer categories. The non-threshold model fits only the smoker lung cancer and "other digestive organs"; it is certainly a poor model for most of the data. Excepting leukemia, there was no statistically valid correlations of radiation dose with cancer mortality. The frequently noted decreased cancer mortality at 345 rads suggests the power of decreasing the activity of suppressor T cells, the Hellström effect. This could account for the low cancer incidences found for esophagus, colon, rectum, breast, uterus, pancreas, and multiple myeloma. Although these correlations are not statistically valid, the data tend to fit the hormesis model which was established with animal data.

Cancer mortality rates in 46,000 U.S. Army observers of atomic blasts indicate hormesis when the data are presented in a dose-response curve (Figure 7-14).

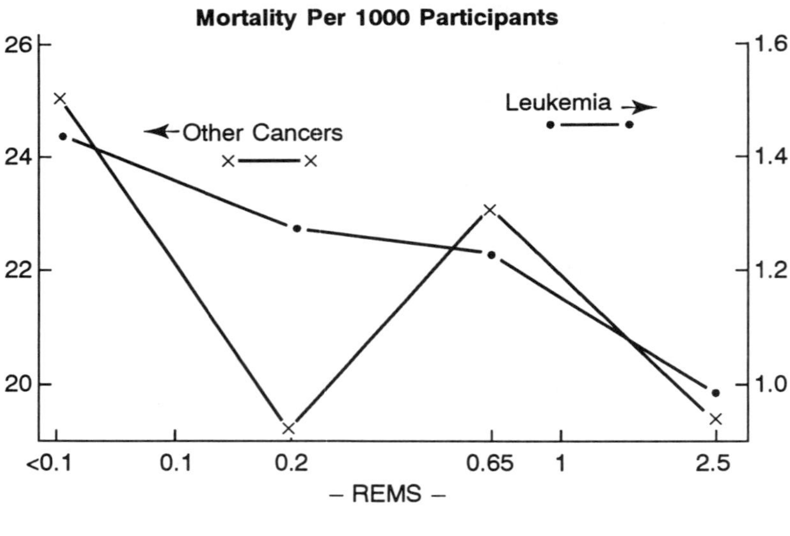

Figure 7-14: Cancer Mortality in Atomic Blast Observers. Mortality rates for leukemia and other cancers in 46,000 United States observers of atomic test blasts from 1953 to 1957.[110]

Radionuclide Exposures

Radium. Radium is a bone-seeking radionuclide which deposits in regions of new bone growth and later redistributes during the dynamic bone remodeling processes. Cells adjacent to such bone tissue receive the greatest doses of high LET radiation; these provoke primarily osteosarcomas with some fibrosarcomas and a mix of other sarcomas (the double names indicate that the tumor is malignant, *i.e.*, a cancer). Rowland *et al.* reviewed the dose-response relationships for radium-induced bone cancers.[111] Only one bone sarcoma was found in 413 known cases of non-medical intake of radium other than the radium dial painters. None of the 273 male dial painters developed bone cancer; perhaps males are more resistant than females. Of the 3055 female dial painters who started before 1950, 63 developed bone cancer. The dose-response curve (Figure 7-15a) indicates a threshold at about one hundred uCi radium. This cohort would be prime material for a hormetic study of the incidence of other tumors, infections, fecundity, and lifespan.

When the incidence of bone sarcomas, plus head carcinomas, were plotted against the log of the dose (Figure 7-15b), the results clearly indicated a threshold at about 300 rads over a period of fifty years. Evans commented: "A linear non-threshold model would have predicted in this group some fifteen radiogenic tumors between zero and one hundred rads. The probability of observing none, if a linear non-threshold model were correct, is one in five million. Clearly any non-threshold model used in much 'prudent' radiation protection work is strongly rejected by good data in the case of radium 226 and radium 228 in man."[112]

Iodine. Quimby and Werner investigated the possibility that iodine 131 used in treating hyperthyroidism might be carcinogenic and concluded that it was not.[113] While this opened the issue, it did not settle it. Boise and Land reviewed the evidence that thyroid nodules and cancers develop after excess radiation.[114] The 1,378 children of Utah who ingested iodine 131 in milk following the fallout from atomic bomb tests in the early 1950s showed no thyroid cancers and eighteen nodules (1.3

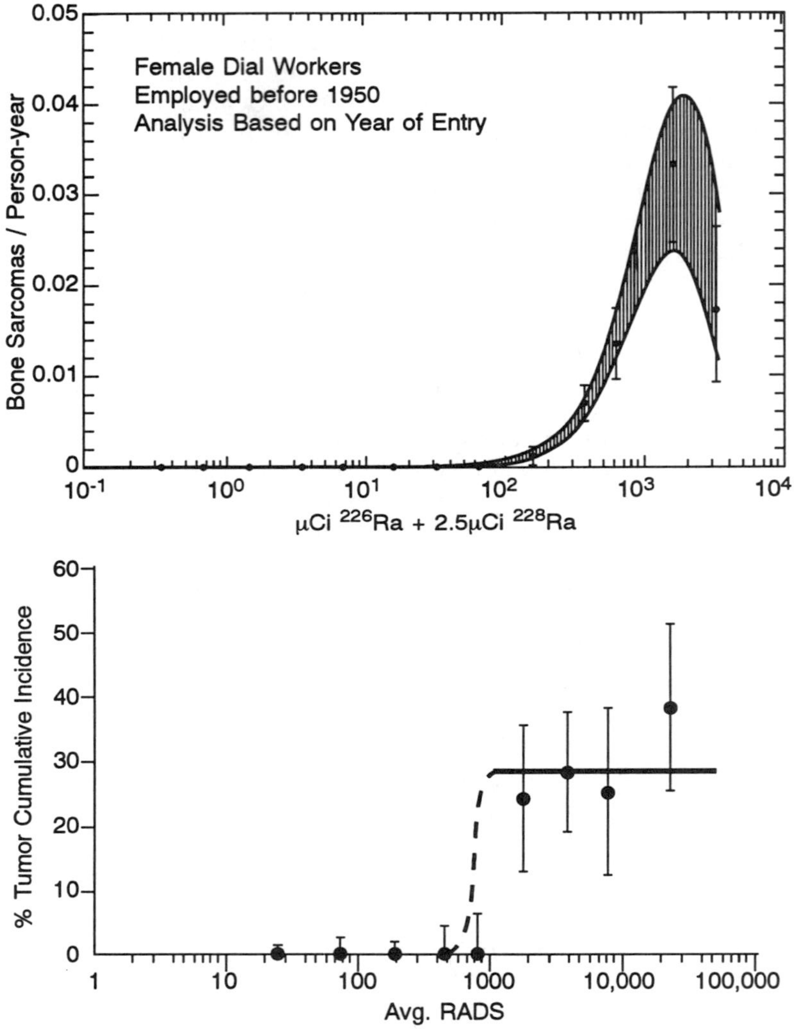

Figure 7-15: Bone Cancer in Radium Dial Painters.

15a - Radiation induced bone cancers, ordinant, in the radium dial painters after 33 to 45 years, plotted against the amount of radium intake, the abscissa.[115]

15b - Bone and head cancer incidence in radium dial painters compared with the total radiation (rads) for a period of 50 years.[116]

percent) while a cohort of 1,313 unexposed children who moved into the area after the fallout showed two thyroid cancers and nineteen nodules (1.4 percent).[117]

Saenger *et al.* reported that 22,000 hyperthyroid patients treated with iodine 131 showed a decreased leukemia rate (0.09 percent) when compared with 14,000 treated with surgery (0.15 percent).[118] The same group reported that iodine 131 treatment induced no excess cancers; the number of cancers which developed during the twenty years after treatment, involving about 600,000 person-years, was fewer when iodine 131 was used than when thyroidectomy or antithyroid drugs were used (Figure 7-16).[119] No new thyroid cancers developed in the iodine 131-treated patients while new cancers were found in the thyroidectomized patients after a ten-year period. Chapman summarized the viewpoint of this group: radioiodine is the preferred treatment and it costs less.[120]

The data from the United States are supported by results from Sweden. Holm and Holm *et al.* found that the diagnostic and therapeutic administration of iodine 131 to 10,133 patients did not induce thyroid cancers in excess of those found in carefully matched patients from the same hospitals.[122] The

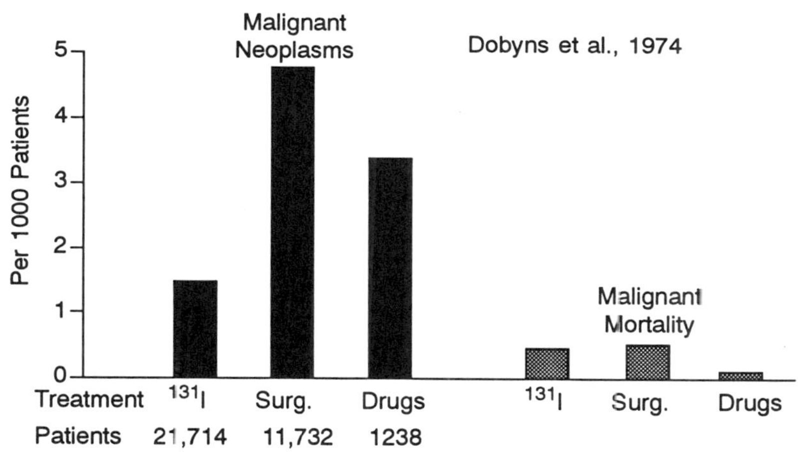

Figure 7-16: Thyroid Cancer in Treated Hyperthyroid Patients. Comparison of the numbers of cancers found in hyperthyroid patients following treatment by surgery, iodine 131, or drug therapy.[121]

administration of an average of 60 uCi dose of iodine 131 resulted in about one-tenth the number of cancers estimated by the linear model during the follow-up period of eighteen years. Holm suggested that radioiodine therapy may reduce the probability of developing thyroid cancers in adults.

Sanders and Kathren note that only one bone sarcoma originated at skeletal doses of less than one thousand rads in the two thousand German patients who were injected with radium 224 for tuberculosis or ankylosing spondylitis between 1944 and 1951.[123] They note that radiotherapy with greater than one thousand rads *occasionally* is correlated with increased bone tumors in adults.

Plutonium. Voelz, *et al.* studied 224 white males who had internal depositions of 10 nCi or more of plutonium 239, including plutonium 238, for an average of 26 years, beginning in 1955+8 years.[124] They report one hundred percent follow up involving 6,000 person-years with an average exposure of 19 nCi (<1 to 180 rem). Most exposures were by inhalation; the lungs then retain half of the body content for thirty years. For the eighteen men still living, the calculated cumulative tissue dose and individual dose rates (rads/rads per year) are: lung, 43,738/1.01; bone surface, 7,897/1.69; and liver, 527/0.112. The average subject's age was 64 in 1980 when the study was updated; at that time only 43 had died from all causes, a low rate attributable to worker selection and the "healthy worker effect." Mortality from cancer of the digestive tract and lymphoid tissues were close to the national averages (Table 7-6). No bone or liver cancers occurred. Mortality from lung cancer was definitely less than expected; less conservative models predicted 65 to 192 lung cancers for the exposures noted, conservative models would predict more. Excluded from the above were seventeen female workers who had an average of 9 nCi plutonium deposition and no cancer mortality.[125]

Uranium and Transuranic Elements. The main long-lived radionuclides to which nuclear workers are exposed are uranium, plutonium and other transuranic elements. Webster and Boice and Land critically reviewed the controversial literature in this area.[126] When age adjusted data was evaluated, the previously

Table 7-6: Thirty-Three Year Mortality in Plutonium Subjects

Cause of Death	*Observed*	*Expected [(a)]*
All	44	77.1
External	8	7.6
Respiratory	2	4.4
Circulatory	18	40.0
Cancer Total	8	14.9
Digestive	5	4.67
Respiratory	1	5.04
Bladder	1	0.43
Lymphopoietic	1	1.5

[(a)] U.S. population rates

reported excess lung cancer mortality disappeared. They both concluded that it is highly questionable whether the radiation exposure of nuclear workers results in any substantial increased risk for cancer. Gilbert and Marks reported that leukemia mortality in nuclear workers is only about half that expected nationally.[127] No increased mortality was found in Hanford workers for leukemia, lymphoma, or other solid tumors other than pancreatic tumors.[128] The last may be a "small sample artifact."[129] Darby and Reissland confirmed the negative associations between radiation levels and cancer induction in the Hanford workers.[130] They seriously criticized some of the previous studies which had indicated an excess cancer risk for this cohort.

The Canadian study eliminated the "healthy worker effect" by using different groups of workers which had the same preliminary examinations and health facilities. Preliminary results indicated that 18,500 past and present employees of Atomic Energy of Canada, Ltd. have less cancer mortality than the general population of Canada.[131] The possibility for the healthy worker effect also was negated when nuclear workers were compared with other workers in the same organization and with the same medical care. A study of 25,000 persons in three cohorts from Ontario Hydro indicated that total deaths and cancer mortality were lower in the nuclear workers than in the other two groups, thermal energy workers and other workers in the company. The nuclear workers also had less cancer mortality than was found in carefully matched controls in Ontario.

No excess cancer was found in 25,000 workers at the Portsmouth naval shipyard for nuclear powered vessels at Kittery, Maine.[132] The average lifetime occupational radiation dose for 7,615 of these workers was 2.8 rem (the range was 0.1 to 91 rem).

The China Study

During the past decade the High Background Radiation Research Group began an excellent "controlled ecology" study of two areas in China in order to determine whether there is a threshold for ionizing radiation for a variety of parameters.[133] Both areas are at the same altitude, average about 32 meters, with homogeneous radiation fields. The 73,000 inhabitants of the areas with monazite in the soil receive 330 mrem/year. This is the high radiation cohort. The low radiation cohort comprises 77,000 inhabitants in carefully chosen, very similar control villages and receives an average of 104 mrem/year. The difference in radiation levels appears to be attributable to radionuclides in the soil, which is reflected in food, water, air, and body tissues (Table 7-7). No differences were found in thirteen trace elements of the soil, e.g., zinc, copper, cobalt, manganese, and arsenic.[134]

Both areas have stable, high density populations of peasants of the Han race with little exposure to industrial carcinogens. Pesticides used were 125 g/mo and 84 g/mo for the control and high radiation groups, respectively. About two g/y per person of antibiotics and negligible amounts of other medicants were used in each group. Medical exposures were 5.2 and 6.5 exposures per person per year in the control and high radiation groups, respectively. Families living along the borders of the two areas are excluded from the study. There are relatively few visitors and the peasants travel very little, particularly the high radiation cohort. Food, clothing, and housing are largely from local sources. Families have lived here for two to fifteen generations, six or more for 67 percent of the control cohort and 91 percent of the high radiation cohort.

A medical survey indicated that the total chromosomal breakage and incidence of Downs Syndrome may be higher

Table 7-7: Radiation Sources in the China Study

Soil	*Unit*	*Control*	*High Radiation*
U	mg/kg	1.7 ± 0.7	7.7 ± 1.7
Th	mg/kg	7.9 ± 3.1	60.4 ± 28
Ra-226	10^{-9}Ci/kg	0.79 ± 0.25	3.91 ± 1.5
K	mg/kg	1.49 ± 0.4	2.56 ± 1.3
Air			
Indoors at 1m:	mR/mo	9.4	28.7
Outdoors at 1m:	mR/mo	5.9	14.5
Breath	mR/mo	9.3	25.0
Cosmic	mR/mo	2.3	2.3
Food & Water Intake			
Ra-226	pCi/d	4.81	14.81
Ra-228	pCi/d	7.84	30.03
Th-228	pCi/d	1.06	3.47
K-40	pCi/d	1.23	2.02
Body			
Total body Ra-226 (Calculated from Breath Rn)	n Ci	0.25	1.26
Total body Th-228	p Ci	32	100
Teeth Ra-226	f Ci/g	15.4 ± 3.5	52.2 ± 3.5
Teeth Ra-228	f Ci/g	7.7	40
Bone Marrow from Ra-228 and Daughters	mR/y	0.23	0.70
Bone, Surface Ra-226	mR/y	1.00	3.84
Bone Marrow Ra-226 (Est. from breath Rn)	mR/y	0.68	3.43
Bone Surface Ra-226 (Est. from breath Rn)	mR/y	3.83	13.82
Whole body UNSCEAR Calculated	m rem/y	67.5	182.4
Whole body, TLD dosimeters 10,000 persons/group	m R/y	104 ± 24	330 ± 55
Gonad dose	m R/y	81	209

($p>0.05$) and there is a statistically valid increase ($p<0.05$) in the two hit chromosomal breaks in the high radiation group (Table 7-8). However, "Chromosomal damage in an individual exposed to radiation has not been associated with subsequent illness."[135] Lymphocyte differention rates were greater in the inhabitants of the high radiation area of China than in the control cohort.[136]

A comparison of total cancer mortality, lung cancer mortality, and liver cancer showed the control cohort to have higher rates ($p<0.05$) than the high radiation cohort.[137] This difference persisted without statistical significance when the standard cancer mortality rates were $49.60/10^5$ and $51.00/10^5$ for the high and low radiation areas, respectively.[138] However, cancer mortality rates for females remained significantly lower ($p<0.05$). The leukemia mortality rate in the high radiation cohort was only half that of the control cohort.

Epidemiology

Frigerio *et al.* began a study of human cancer mortality with the presumption that background radiation must be carcinogenic.[139] They were forced to conclude, as did Oakley that it was not.[140] The risk of cancer is inversely correlated with background radiation. When age specific deaths were considered, the pattern of response was the same. This study has been repeated with regions, states and countries: the results are similar in all studies for 56 types of cancer mortality, including the classic example of radiation-induced cancer, leukemia (Figure 7-17).[141] None of the other correlations tested were as high as that of radiation and cancer.

Cohen found the negative correlation between age adjusted cancer mortality rates and background radiation levels in the 48 contiguous states to be highly significant ($p<0.001$).[142] Hickey *et al.* reported: "Bivariate correlation coefficients between radiation and mortality rates were significant for cancer of the lung and respiratory organs, cancer of the buccal cavity pharynx, cancer of the digestive organs and peritoneum, total cancer, and diseases of the heart."[143]

Eckhoff *et al.* states "—the leukemia mortality rate actually appears to decrease with increasing altitude"[145] for more than

Table 7-8: Medical Survey of High Radiation and Control Groups

Item	*Unit*	*Area Control*	*High Rads*
Genetics			
Children Surveyed	No.	13,987	13,425
Genetic Malformation or Disease	%	2.25	2.23
Total Chromosomal Breaks/Cell	X10[-3]	2.7	3.2
*Dicentrics (@ hit)/cell	X10[-4]	0.48	4.2
Total Hereditary Disease Rate	ratio	1	1
Down's Syndrome in Abnormals	%	0.31	1.04
Icthyosis Congenita Morbidity	%	11.1	9.5
Eye Malformations	%	0.5	0.7
Cancer			
Person/Years Surveyed	No.	563,507	531,972
*Cancer Mortality rate	%	49.86	41.09
*Lung Mortality	ratio	1	<1
Nasophargyngeal	ratio	1	1
*Liver	ratio	1	<1
Leukemia	ratio	1	1/2
Reproduction			
Spontaneous	%	10.7	7.8
Multiparity in Parturitions	%	3.1	3.8
Neonatal Mortality	%	36.0	27.3
*Infertility	%	46.4	15.3
Immunity			
Serum Immunoglobulins	ratio	1	1
T-Cells	ratio	1	<1
Lymphocytes PHA Stimulation	ratio	1	<1
T-Cell PHA Stimulation	ratio	1	>1
Blastogenic transformation of lymphocytes	ratio	1	>1
Life Span			
Person-Years	No.	141,827	148,527
Deaths/2 yr	%	0.649	0.651
Life expectancy-female	yr	75.8	69.3
Life expectancy-male	yr	67.2	66.5
* Life expectancy, over 40	ratio	1	>1

*Statistical difference at p = 0.05.

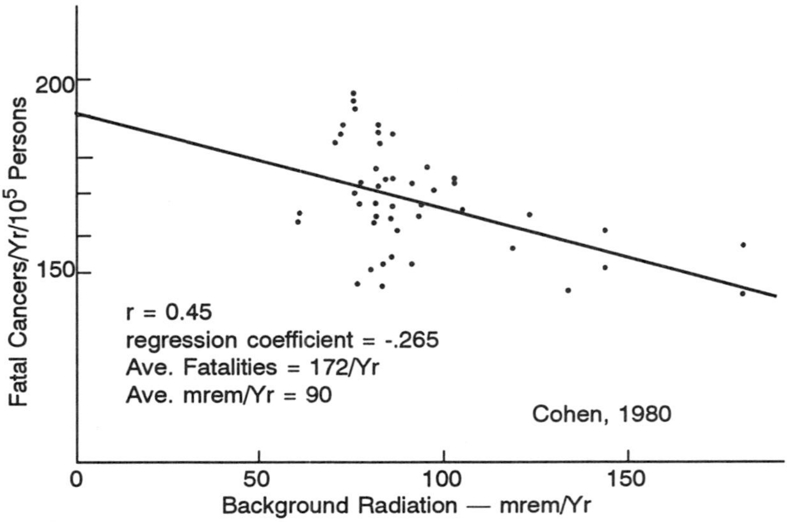

Figure 7-17: Cancer Epidemiology in the United States. There is a significant negative correlation (p.<0.03) between terrestrial radiation and leukemia mortality in the contiguous 48 United States.[144]

five thousand geographic areas of the United States. Yallow noted that Colorado residents show one of the lowest cancer death rates while receiving more radiation from cosmic rays, rocks, and mine tailings than is received by the Hanford workers.[146] Utah had the lowest mortality rate in spite of the fallout deposited from the Nevada atomic blasts in the 1950s. Yallow further suggests that up to fifteen rem radiation may be protective against malignancies, based upon finding of only fourteen cancers in the Utah cohort where 24 were expected.

Figerio *et al.* suggested that knowledge of the decreased cancer rate with increased background radiation levels is more pertinent to the life of everyday people than information about radiation exposures of victims of the atom bomb, uranium miners, or spondylitic patients.[147] Differences in background radiation would be found to be greater if consideration had been given to the type of housing and the concentrations of radon, which is generally elevated in homes with increased energy efficiency.[148]

Epidemiologic evidence clearly shows that low-level ionizing radiation is *not* associated with increased risk of cancer mortality. Low levels of ionizing radiation are clearly correlated with decreased cancer mortality risk. This agrees with information from accidental association with low doses of radiation in workers, medical patients, and atom bomb victims. The human data and the animal experiments give patterns of response which clearly fit the hormesis model for radiation-induced cancer.

Chernobyl

The Chernobyl nuclear power plant accident was a unique disaster in mankind's peaceful use of atomic power. If the intensity of study follows that of other exposed populations, these unfortunate individuals will contribute much information about the dose-response effects of different levels of ionizing radiation upon cancer. Hopefully, the deficiencies of the Japanese bomb study will be remedied by: (1) careful monitoring of fallout for individuals and different cohorts; (2) study of cohorts exposed to 0.1–0.9 rads; and (3) obtain previous irradiation data for each individual. The first must include information about the level of radionuclides present before Chernobyl. Consternation about the high levels of strontium 90 and cesium 137 in the food chain of Laplanders is unreasonable without noting this population has had a very high intake of these radionuclides for many generations.[149] The last appears to be an important factor in reducing radiation sickness and mortality.

Three decades from now, 95 percent of the cancer induction and mortality of Chernobyl will be known. Can the results be predicted with any degree of accuracy? They should be if the information gleaned from the past is utilized. The local population is similar in size to that of the Japanese bomb victims. The character of the radiation and the effects of fall-out also are not too different. A "Rosetta Stone" for radiation-induced cancer can be produced by compiling all the above information into one chart, deducing the frequency of each type of cancer for different doses and weighing the human values where information is adequate. The general information obtained can be used to deduce the patterns of cancer induction and mortality in any population, such as Chernobyl.

The Rosetta Stone for radiation-induced cancer involves a characterization of the dose-response curve from 64 animal and human cancer mortality studies. The composite curves for leukemia and the average from 23 other cancer types and organs are presented (Figure 7-18). The remarkable similarity of this curve to that from both animal data (Figure 7-9) and Japanese victims (Figures 7-12, 7-13) gives credence to the different parts of each curve. This provides a guide for anticipated cancer mortality for Chernobyl, or any other comparable exposed population. Predictably, populations receiving less

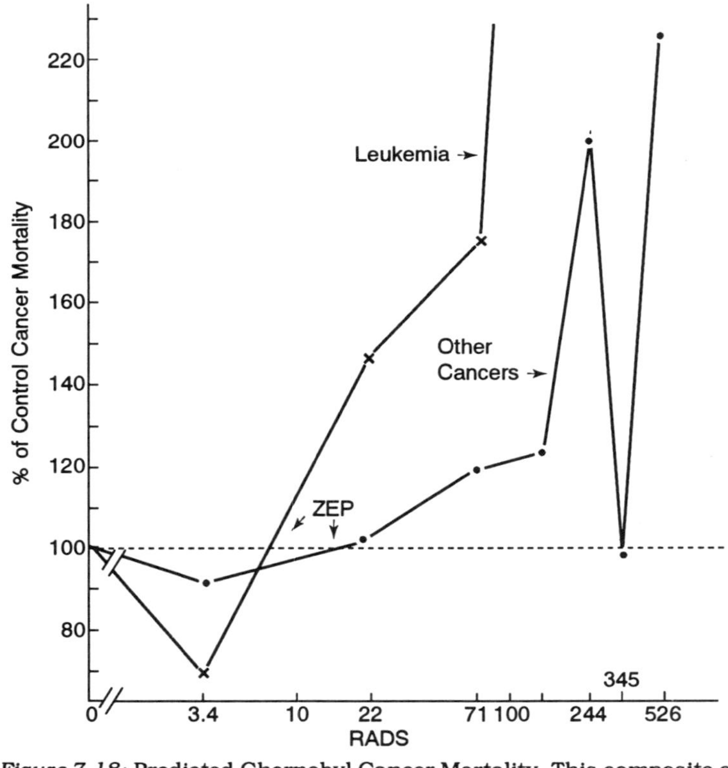

Figure 7-18: Predicted Chernobyl Cancer Mortality. This composite of leukemia and other cancer mortalities provides a realistic basis for estimating fatalities from acute human exposures.

than twenty rems whole body exposure may expect to have less cancer mortality than comparable unexposed populations. The exceptions to the patterns displayed were uterus and smokers' lung cancers which showed increased incidence with low doses. The pattern of coincidence will be clear by the turn of the century and the picture will be complete by the year 2030.

Hormesis in Lifespan

Since low levels of ionizing radiation appear to increase immune competence and reduce susceptibility to infection and cancer, the average lifespan should be increased. This is indeed the case for many invertebrate and vertebrate experiments where good objective data can be obtained and confirmed. Luckey cites 62 reports with invertebrates and 92 with vertebrates in which hormesis was noted for the average lifespan of lightly irradiated animals when compared with controls.[150] In some experiments the mortality of the control animals was so bad that they were not used in evaluation of the results.[151] Note that extensive use of closed colonies and specific pathogen-free animals, combined with a reluctance of radiobiologists to challenge with specific pathogens, would provide quite different conditions than were generally used before about 1960. Two exceptions where all the data is available are presented.

Animals

While radiobiologists discuss possibilities to perform a "megamouse" experiment to differentiate between threshold and nonthreshold models, they have ignored the "megafish" experiment reported in the last decade.[152] Not counting years of preliminary experiments to establish reliable methodology, this group marked and released over a million salmon into streams for their ocean sojourns of one to five years before returning to spawn after all the vigor of natural life. One study of whole body chronic cobalt 60 irradiation during egg incubation and salmonid development involved 315,000 controls and 314,000 irradiated fish or their siblings. About 35 rads of gamma radiation were administered to each at a rate of 0.5 rad/day. Of the identified fish that returned to spawn, 2,876 (0.913 percent) were controls and 3,562 (1.13 percent) were irradiated stock, 24 percent more than the controls. The natural environment allowed in this experiment distinguishes it from the usual laboratory experiments.

Spalding, *et al.* recently completed a study of lifespan in about 4,000 male mice exposed to daily cobalt 60 irradiation until the desired dose was attained.[153] The consistency of results is highlighted with mice started at two months of age (Figure 7-19). Irrespective of dose rate, each set showed better survival than controls ($p<0.05$) when fifty percent of the controls had died. This confirms other results which showed that low doses of radiation increased average lifespan. Individual animals do not survive to an unusual age; mid-life morbidity and mortality is lessened in the irradiated animals which gives a longer average lifespan for the group.

Humans

Although the risk of excess radiation to early radiologists was great,[154] in 1938 G. Pfahler noted that the response depended upon dosage and suggested that small doses of radiation might be beneficial: ". . . if you wish to live a long life, see to it that you spend part of it in an x-ray department."[155] Based upon lifespan and somatic and genetic effects, Henry suggested that chronic exposure to low levels of ionizing radiation was not harmful.[156] Hildreth noted that while the 204 radiologists born between 1860 and 1900 had a life expectancy of 43 years, 88 had lived past their 78th birthday.[157] Smith and Doll found that radiologists who began practice after 1920 had cancer mortality only 88 percent of that of other British physicians; radiologists who began after 1935 had only eighty percent that of cancer mortality in other British physicians.[158]

Mortality in the United States from chronic diseases correlates indirectly with background ionizing radiation according to many investigators who could not find valid correlations to other environmental, ethnic, social, or economic factors.[159] One possible explanation is that high background levels of ionizing radiation increase health and longevity. When compared with the average, the high death rate in the southeast coast of the United States is an anomaly in human population studies. Mortality rates correlate well with chronic diseases, but no

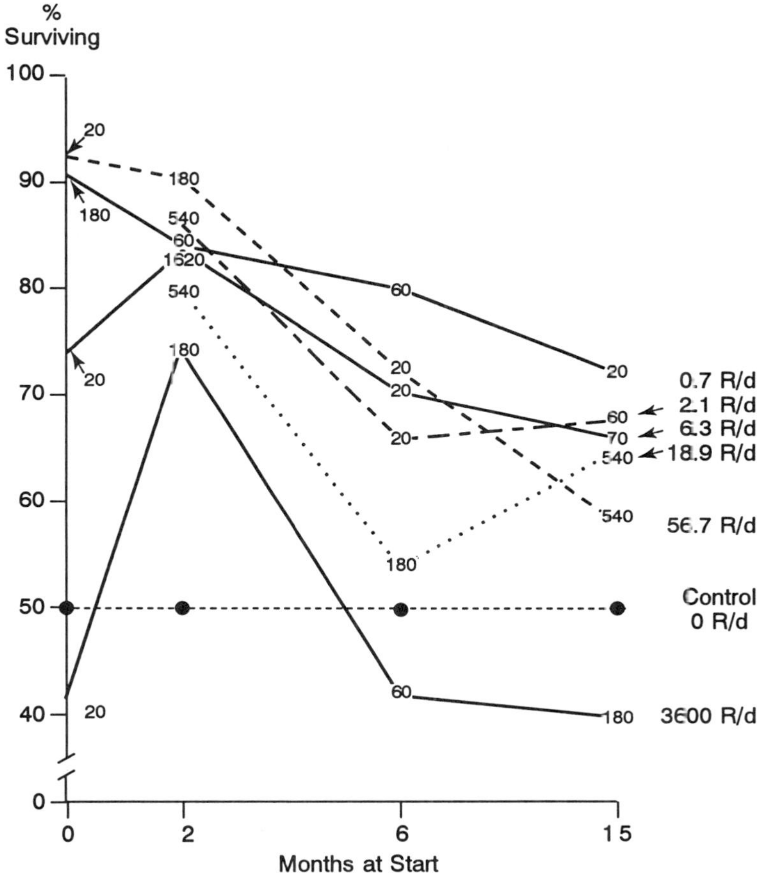

Figure 7-19: Radiation Hormesis in Mouse Lifespan. Comparison of the survival (ordinate) of chronically exposed mice at the time when one half of the control mice had died.[160] The abscissa gives the age of the mice at first exposure. The rate of gamma ray exposure for each cohort is listed on the right; the total rads administered to each group is indicated within each curve. Excepting the two closest to controls, all values were significantly different from the controls $(p<0.05)$.

factors can be identified to explain "the enigma of the southeast."[161] Cosmic radiation is lower in coastal areas than in mid-continent and the terrestrial ionizing radiation of the southeast coastal area averages 23 mrem/year.[162] This is about one-half that for the average United States and one fourth that of the Colorado[163] plateau where mortality from cancer and chronic diseases are lower than the average for the country.[164] This enigma appears to be explained by the epidemiologic correlation of low background ionizing radiation levels with death rates (Figure 7-20). The inverse correlation was very high.[165]

While spontaneous abortions and neonatal mortality appeared to be less in the high radiation group than in the control group in the China study, these differences have not yet reached statistical significance.[166] Infertility was significantly less ($p<0.05$) in the high radiation group than in the controls. The later study indicated that the high radiation cohort had more active immune function than the control cohort. Most comparisons of lifespan showed no differences; however, the lifespan of males over forty years old in the high radiation was increased ($p<0.05$) over that of controls.

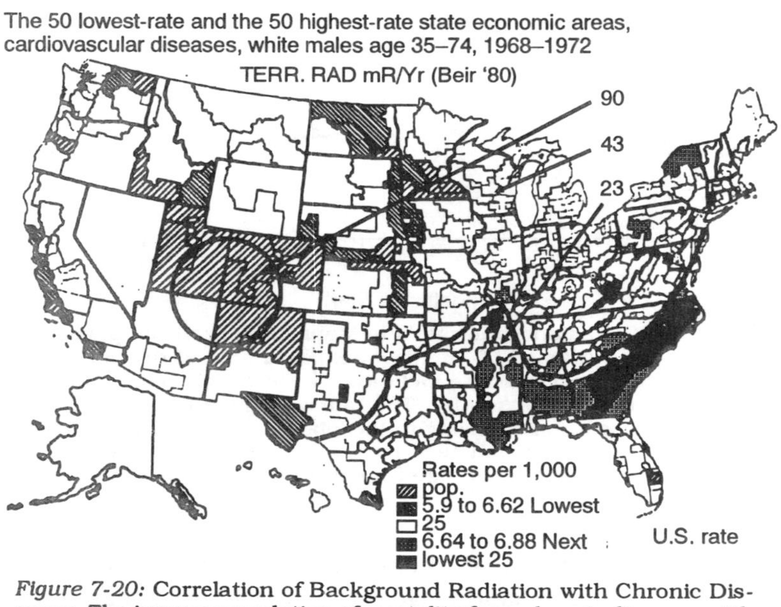

Figure 7-20: Correlation of Background Radiation with Chronic Diseases. The inverse correlation of mortality from chronic diseases with background ionizing radiation was significant ($p<0.001$).[167]

Information from Kerala, the most densely populated state of India, supplements the data from China and allows interpretation of the "paradox of Kerala."[168] The people of Kerala have the highest literacy rate and the best health status in India; yet their expenditure for health care was not appreciably above average. They have the lowest food intake and the least adequate diets; three-fourths of the families are deficient in calories and one-half are deficient in protein.[169] Yet, Kerala has fewer frank deficiencies in young children and lower gross mortality than any other state in India. This unusual health status is exemplified by the exceptional agile and exhaustive dancing in the ancient, ritualistic theater movements derived from kalarippayat, a possible precursor of kung-fu and karate.[170] A possible explanation of this paradox is the unusually high background radiation derived from the high radium and thorium content of some of their soil, water, and food. The average terrestrial radiation for the United States is 43 mrad/year; some of the Kerala population receives 1,000 to 3,000 mrad/year, a level comparable to the optimum in some of the animal experiments.

This brief summary showing increased average lifespan in animals, including humans, is well supported by the evidence of increased immune competence following low level irradiation. The decreased mid-life morbidity and mortality of irradiated cohorts compared with unirradiated controls results from faster wound healing and increased resistance to high doses of ionizing radiation, to infection and to cancer induction. The beneficial effects of ionizing radiation appear to include less infertility, fewer neonatal deaths, faster growth rates, greater visual, hearing and mental acuity, and improved general health during life. The average age of a given population is increased without individuals living to unusual ages. The mechanism of this apparent higher plateau of health must be important.

Radiation Hormesis as Nurture by an Essential Agent

Nurture is the action of the environment upon an organism. The literature suggests that low doses of ionizing radiation evoke reactions which appear to be beneficial to most organisms. This raises a tremendously important question. Is ion- izing radiation stimulatory as any toxic material might be, or is it an agent which is essential for vital physiologic functions? There are remarkably few well-defined lines between agents which are essential, stimulatory, or toxic.[171] The continuum of nurture, hormology, and toxicology is well illustrated by those metals which were first considered to be toxicants, then were recognized stimulants, and now are being examined as possible essential nutrients: selenium, tin, arsenic, silicon, strontium, and even lead.[172] The cumulative evidence for hormesis with ionizing radiation has focused a vital question. Is ionizing radiation essential for life?

One experimental approach to this question is to determine the effects of subambient levels of ionizing radiation, as illustrated left of the ordinant in Figure 7-21. If all quantities of the agent, ionizing radiation, were harmful, the data should approximate the straight line-S curve. Decreasing the amount of an agent showing a threshold should have no effect; this is the threshold-T curve. Stimulatory agents would show the hormesis peak and then the horizontal curve as background amounts were removed, the hormesis curve. Mercury or antibiotics would fit such a curve.[173] Only an essential agent, such as a vitamin or oxygen, would approximate the hormesis-H curve. Removal of an essential agent should disturb vital physiologic functions with development of a deficiency syndrome. This has been done for ionizing radiation in protozoa.[174]

The laboratory was in an underground vault to reduce cosmic radiation, inside a room with thick steel and lead walls to reduce earth radiation, and had carbon filters to reduce radon and other airborne radionuclides. The protozoa were grown in a special copper-cadmium clad lead box inside an air incubator. The culture medium was composed of reagent grade chemicals with natural potassium, which is radioactive, being replaced with the nonradioactive isotope, ^{39}KC1. Increased

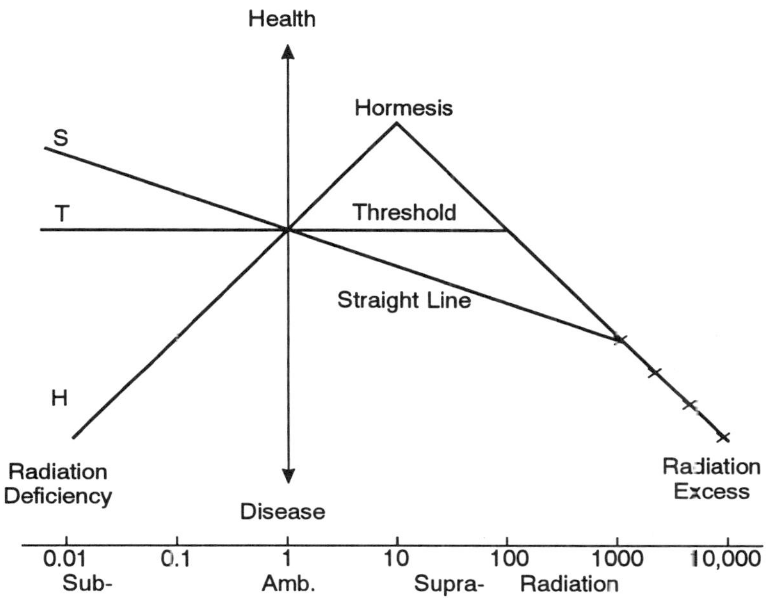

Figure 7-21: Models for Complete Dose-Response Curves. The postulated extension of each curve into subambient radiation levels suggests mechanism possibilities.

levels of ionizing radiation were obtained with the addition of ^{40}KCl for endogenous irradiation or with an external radiation source of $^{137}CsNO^{3}$.

The reproduction rate of the protozoan, *T. pyriformis*, in subambient radiation levels was statistically lower ($p<0.01$) than that of the control near ambient radiation levels, about 0.5mR/d (Figure 7-22). All cultures irradiated at levels greater than ambient replicated faster than the control culture. These results confirm those of Planel and co-workers with bacteria-fed protozoa and an alga.[175]

The question of stimulant or essential agent is partially answered. The microbic data clearly indicate that ionizing radiation is essential for fast growth rates, a most essential physiologic function in natural habitats where the battle for

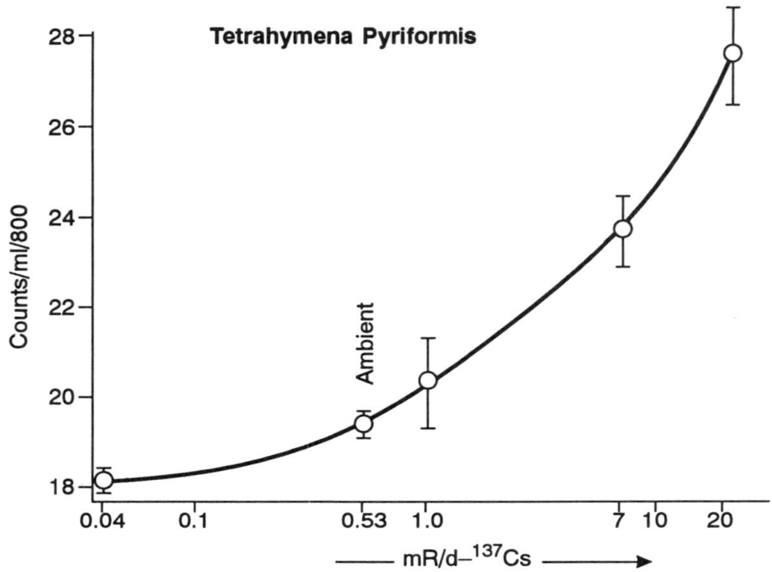

Figure 7-22: Effect of Radiation upon Protozoan Growth Rates. The growth (replication) rate of pure cultures of *T. pyriformis* in a chemically defined medium was directly related to the amount of ionizing radiation.

food supplies is won by those microbes which can outgrow their competitors. The tentative conclusions are astounding. Ionizing radiation is essential for life and the amount provided by usual background levels is less than optimum.

Discussion and Conclusion

Consistent and extensive literature of the past ninety years shows that microbes, plants, and animals are stimulated by low levels of ionizing radiation. This includes very important physiologic parameters such as growth, maturation, reproduction, mental acuity, immune competence, and reactions to stress. The resulting benefits include decreased infertility, improved resistance to infection, cancer, and radiation sickness, less mid-life mortality and consequently longer average lifespan. Although mechanisms are not well-defined, activation of the immune system may be more important than DNA repair.

The new evidence is that plant and animal microbes perform suboptimally when the radiation flux is lowered below ambient levels. A parabola (Figure 7-23) is the model which incorporates harmful effects of excess ionizing radiation, stimulation by low doses of the same agent, and evidence of a deficiency when the quantity of that agent is reduced below ambient levels.

A distinctive feature of this curve is its placement on the abscissa. This curve could represent vitamin A, vitamin C, iron, selenium, calcium, or any other essential agent where a given population normally received an inadequate supply of that agent. For such an agent considerations are made to supplement the deficient individuals. The usual scientific interpretation of the above information would conclude that ionizing radiation is an essential agent which is present in suboptimal

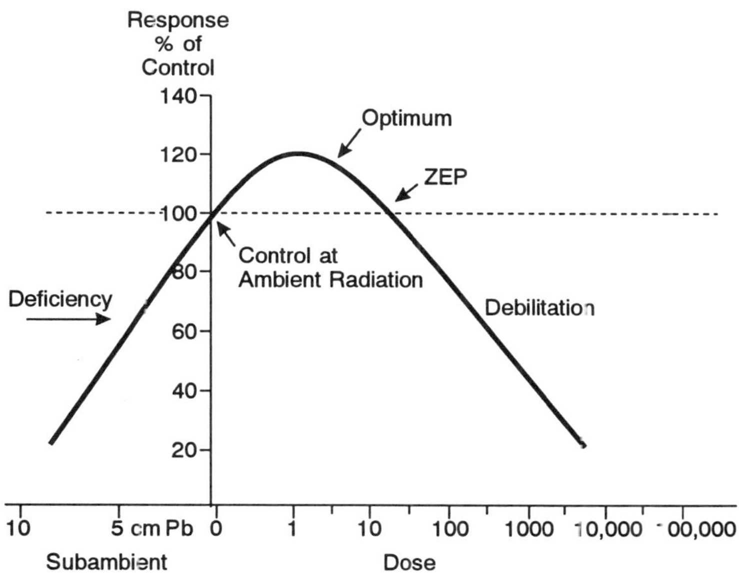

Figure 7-23: Generalized Complete Dose-Response Curve for the Effects of Ionizing Radiation upon Organisms.

quantities for most populations. This revolutionary concept is a reasonable interpretation of information from protozoa, invertebrates, and vertebrates, including humans. The increased physiologic function found with ionizing radiation is not due to low doses of a toxicant, as mercury or antibiotic; rather, it is due to insufficient amounts of an essential agent. The data suggest that in many populations more ionizing radiation is needed for optimum health.

One advantage of the model is the presence of a definitive zero equivalent point (ZEP). For any given parameter being studied, a few points above and below ZEP can establish an acceptable definition for a harmful dose, i.e., any dose greater than ZEP. However, matters become more complex when it is realized that there are many parameters, many different populations and many conditions which would give different ZEP values as well as the more important concept of different optimum levels for ionizing radiation. Pragmatic solutions to the quagmire of recommended allowances for optimum levels of ionizing radiation for farm animals and humans may follow the pattern set for other essential agents, committee evaluations, and consensus.

Extensive data from microbes, plants, invertebrates, and laboratory mammals give confirmation to the evidence from Japanese bomb survivors, accidentally exposed humans, the China study, and epidemiologic data from the United States. The consistency of results with low levels of ionizing radiation provides convincing evidence that reasonably increased levels of ionizing radiation may provide a new plateau of health.

The optimum acute dose for whole body radiation protection from cancer is about five rads of x- or gamma rays (Figure 7-17). Studies which did not show hormesis in radiation-induced cancer either used exposures greater than 25 rem or did not utilize whole body exposures. When spleens of mice were protected, decreased tumors were found ($p<0.05$) with 900 rads.[176] The optimum level of chronic ionizing radiation appears to be ten to one hundred times the background for most populations.

Acceptance of ionizing radiation as either a salutary or an essential agent which is usually present in suboptimal amounts is vital for future allocation of research funds, experimental design, government standards and regulations, and changed societal perception. This new concept encourages enlightened discussion of protection versus utilization of ionizing radiation as a natural resource.

Acknowledgment

The author acknowledges, with thanks, the suggestions of Pauline Luckey in the preparation of this manuscript.

Notes

1. C.M. Southam and J. Ehrlich, "Effects of Extract of Western Red-Cedar Heartwood on Certain Wood Decaying Fungi in Culture," *Phytopathology* 33:517 (1943).
2. BEIR Committee, *The Effects on Populations of Exposure to Low Levels of Ionizing Radiation*, Washington, D.C.: National Academy Press, 1980.
3. T.D. Luckey, "Ionizing Radiation Promotes Protozoan Reproduction," *Radiation Research* 108:215 (1986).
4. T.D. Luckey, *Hormesis with Ionizing Radiation*, Boca Raton: CRC Press, 1980; T.D. Luckey, "Physiological Effects from Low Levels of Ionizing Radiation," *Health Physics* 43:771 (1982); T.D. Luckey, "Beneficial Physiologic Effects of Ionizing Radiation," *Strahlenschutz in Forschung und Praxis* 24:184 (1985); *Radiation Hormesis*, Boca Raton: CRC Press, 1991.
5. T.D. Luckey, "Ionizing Radiation Hormesis of Non-Specific Immunity," *Microecology and Therapy* 11:113 (1981).
6. T.D. Luckey, "Ionizing Radiation Promotes Protozoan Reproduction," *loc. cit.*
7. H.F. Henry, "Is All Nuclear Radiation Harmful?," *Journal of the American Medical Association* 176:671 (1961).
8. W. Shrader, "Experiments with X-rays upon Germs," *The Electrical Engineer* 22:170 (1896).
9. G.F. Atkinson, "Report upon Some Preliminary Experiments with the Roentgen Rays on Plants," *Science* 7:7 (1898).
10. A. Venenziani, "On the Physiologic Action of Radium on *Opalina ranarum,*" *Zantrablatt fuer Physiologie* 18:130 (1904); R.E Zuelzer, "The Effect of Radium Rays on Protozoa," *Archiv fuer Protistenkunde* 5:358 (1905); C. Richet, "Influence of Radium Rays on Lactate Fermentation," *Archive Institute Physiologie* 3:130 (1905).
11. C.S. Gager, "Effects of Rays of Radium on Plants," *Memorial New York Botanical Garden* 4:1 (1908).

12. E. Kayser, "Influence of the Salts of Uranium on Alcoholic Ferments," *Comptes Rendus des Seances de l'Academy des Science*, Series D, 155:246 (1912).
13. H. Agulhon and R. Sazerac, "Activation of Bacterial Oxidation Processes by Uranium Salts," *Comptes Rendus des Seances de l'Academy des Science* 155:1186 (1912).
14. T.D. Luckey, *Hormesis with Ionizing Radiation*, *loc. cit.*; T.D. Luckey.
15. N.M. Berezina, A.M. Kuzin, and D.A. Kaushanskii, "Introduction into Agricultural Practice of Presowing Gamma Irradiation of Seeds Using the Production Model Gamma Installation KOLOS," *Atomic Energy, USSR* 37:43 (1975); A.M. Kuzin, *Stimulating Action of Ionizing Radiation Residual Chemicals on Biologic Processes*, Moscow: Atomizdat, 1977; J. Simon and S.H. Bhattacharya, *The Present Status and Future Prospects of Radiation Stimulation in Crop Plants*, Budapest, Budai Janos, 1977; T.D. Luckey, *Hormesis with Ionizing Radiation*, *loc. cit.*
16. I. Fendrik and H. Glubrecht, "The Effect of Small Radiation Doses on the Yield of Strawberries," *Gartenbauwissenschaft* 37:155 (1972).
17. H.B. Seuberling, "X-ray Induced Stimulation of Cell Division of *Euglena gracilis*," *Stimulation Newsletter* 1:4 (1970).
18. I. Fendrik and H. Glubrecht, *loc. cit.*
19. A. Feldman, "Study of Radiostimulation. III. Influence of Temperature on the Radiation Induced Growth Stimulation in *Lemna minor L.*," *Radiation Botany* 11:59 (1971).
20. A. Feldman, *ibid.*
21. L. Zelles and D. Ernst, "Modification of the Effect of UV Irradiation on Pine Pollen by Isolated Cell Constituents," *Radiation and Environmental Biophysics* 11:271 (1978).
22. A.M. Kuzin, *loc. cit.*
23. J. Simon and S.H. Bhattacharya, *loc. cit.*
24. C. Melville, "An Apparent Beneficial Effect of Gamma Radiation on the Flour Mite," *Nature* (London) 181:1403 (1958).
25. T.D. Luckey, *Hormesis with Ionizing Radiation*, *loc. cit.*; T.D. Luckey, "Physiological Effects from Low Levels of Ionizing Radiation," *loc. cit.*
26. C. Melville, "An Apparent Beneficial Effect of Gamma Radiation on the Flour Mite," *loc. cit.*
27. B.L. Strehler, "Origin and Comparison of the Effects of Time and High Energy Radiations on Living Systems," *Quarterly Review of Biology* 14:117 (1959); B.L. Strehler, "Studies on the Comparative Physiology of Aging. III. Effects of X Radiation Dosage on Age Specific Mortality Rates of *Drosophila melanogaster* and *Campanularia flexuosa*," *Journal of Gerontology* 19:83 (1964).
28. E. Lorenz, "Some Biologic Effects of Long-Continued Irradiation," *American Journal of Roentgenology* 63:176 (1950); E. Lorenz, L.O. Jacobson, W.E. Heston, M. Shimkin, A.B. Eschenbrenner, M.K. Deringer, J. Doniger, and R. Schweistal, "Effects of Long-Continued

Total Body Gamma Irradiation on Mice, Guinea Pigs and Rabbits. III. Effect on Lifespan, Weight, Blood Picture and Carcinogenesis, and Role of Intensity of Radiation," in R.E. Zirkle, editor, *Biological Effects of External X and Gamma Radiation*, Vol. 1, New York: McGraw-Hill, 24.

29. J.J. Morris, unpublished report (see T.D. Luckey, *Hormesis with Ionizing Radiation, op. cit.*, 60); T.D. Luckey, *Hormesis with Ionizing Radiation, op. cit.*; T.D. Luckey, "Physiological Effects from Low Levels of Ionizing Radiation," *loc. cit.*
30. E. Lorenz, "Some Biologic Effects of Long-Continued Irradiation," *op. cit.*
31. J.J. Morris, *loc. cit.*
32. T.D. Luckey, *Hormesis with Ionizing Radiation, loc. cit.*
33. A.A. McDowell, "Transfer by Normal and Chronic Whole-Body Irradiated Monkeys of a Single Learned Discrimination Along a Peripheral Cue Gradient," *Journal of Genetic Psychology* 97:41 (1960).
34. T.D. Luckey, *Hormesis with Ionizing Radiation, op. cit.*, Tables 4, 5, and 7.
35. S. Muramatsu, T. Sugahara, T. Tsuchiya, and Y. Okazawa, "Effects of Chronic Low Dose Irradiation for Three Successive Generations on the Breeding Behavior of Mice," *International Journal of Radiation Biology* 8:523 (1964).
36. J.W. Gowan and J. Stadler, "A Quantitative Study of Lifetime Sickness and Mortality and Progeny Effects Resulting from Penetrating Radiation," *Progress Report*, (11–1)–107, Washington, D.C.: U.S. Atomic Energy Commission, 80; D.F. Riordan, "Effect of Continuous Low Level Gamma Irradiation on Successive Generations of *Dahlbominus fuscipennis* (Zett.) (*Hymenoptera: Eulophidae*)," *Canadian Journal of Zoology* 42:685 (1964); K. Nishio, T. Megumit, and M. Yonezawa, "Effect of ^{137}Cs and ^{90}Sr Administered Continuously and Through Generations of Mice," *Annual Report of the Radiation Center of Osaka Prefecture* 8:123 (1967); N.R. French and H.W. Kaaz, "The Intrinsic Rate of Increase of Irradiated Peromyscus in the Laboratory," *Ecology* 49:1172 (1968); N.R. French, B.G. Maza, H.O. Hill, A.P. Schwaden, and H.W. Kaaz, "A Population Study of Irradiated Desert Rodents," *Ecological Monographs* 44:45 (1974).
37. I.I. Kaplan, *Clinical Radiation Therapy*, New York: P.B. Hoeber, 1949.
38. J.F. Spalding, M.R. Brooks, and G.L. Tietjen, "Comparative Litter Reproduction Characteristics of Mouse Populations for 82 Generations of X-Irradiated Male Progenitors," *Proceedings of the Society for Experimental Biology and Medicine* 166:237 (1981).
39. S. Jablon, "Follow-up Studies on A-Bomb Survivors: Implications for Radiological Protection," Health Physics Society 15th Mid-Year Symposium, Albuquerque.
40. S. Muramatsu *et al.*, *loc. cit.*
41. T.D. Luckey, *Hormesis with Ionizing Radiation, loc. cit.*; T.D. Luckey,

"Physiological Effects from Low Levels of Ionizing Radiation," *loc. cit.*
42. O. Schurch and E. Tschudi, "Experimental Study of Wound Healing by Electric Coagulant and Radium," *Strahlentherapie* 32:143 (1929); S. Fukase, "The Influence of X-rays on Wound Healing," *Strahlentherapie* 36:102 (1930); L. Freund, "The Therapeutic Action of X-Rays on Fresh Wounds," *Deutsche Zeitschrift fuer Chirurgie* 222:416 (1930); H. Buhtz, "Study of the Pathologic Physiology of Inflammation. X. On the Influence of X-rays on Wound Inflammation," *Frankfurt Zeitschrift fur Pathologie* 44:57 (1933); M. Lupo and G. Pisani, "X-ray Stimulation of Cicatrization of Experimental Skin Injuries in Rabbits," *Ricerea Scientifica* 21:956 (1951).
43. S. Grigorescu, "The Antiblastic Action of Low Doses of Ionizing Radiations," in S.M. Milcu, M. Nicul, and M. Racovaeunu, editors, *Low Level Radiation in Medicine, Biology and Agriculture*, Bucharest: Academy of Science, 1968, 147–148.
44. M.A. Omarov, "Effect of Multiple Total Body Radiation on Healing of Bone Fractures," *Medical Radiology* 18:64 (1973).
45. J. Maisin, E. van Duyse, A. Dunjic, J. van der Merckt, A. Wambersie, and D. Werbrouck, "Acquired Radio-Resistance, Radio-Selection and Radio-Adaptation," in A.A. Buzzati-Traverso, editor, *Immediate and Low Level Effects of Ionizing Radiations*, London: Taylor and Francis, 1960, 183–194; K. Nishio, "Effects of ^{90}Sr and ^{137}Cs Administered Continuously to Mice. I. Sex Ratio and Radioresistance," *Stimulation Newsletter* 1:13 (1970).
46. T.D. Luckey, *Hormesis with Ionizing Radiation*, *op. cit.*, Tables 4, 5, and 7.
47. Anonymous, "Character of Roentgen Rays," *The Electrical Engineer* 22:176 (1896).
48. H. Gerhartz, "Diphtheria and X-rays," *Berlin klinik Wochenschrift* 46:1800 (1909); J.P. Bisgard, H.B. Hunt, and R.H. Dickinson, "Effect of X-ray Irradiation upon Bacterial Toxemia in Rabbits," *Radiology* 43:330 (1944).
49. L.A. Terzian, "The Effect of X Irradiation on the Immunity of Mosquitoes to Malarial Infection," *Journal of Immunology* 71:202 (1953); A. Adler, "Influence of Some Environmental Factors on the Growth and Development of Silkworms Treated with Gamma Radiation in Their Embryo Phase," *Zootechnica. Med. Vet.* 12:45 (1969).
50. V.K. Russ, "Consensus of the Effect of X-rays on Bacteria," *Hygiene* 56:341 (1909); G.A. Sacher, "On the Statistical Nature of Mortality, with Especial Reference to Chronic Radiation Mortality," *Radiology* 67:250 (1956); G.A. Sacher, "Survival of Mice under Duration of Life Exposure to X-rays at Various Rates," Washington, D.C.: Office of Technical Publications, U.S.D.C., 1956, 435–463.
51. S.Z. Liu, "Some Advances in Radiation Immunology," *Radiation Protection* 5:129 (1986); S.Z. Liu, P.X. Xiao, S.Y. Ma, G.Z. Xu, C.H. Tian, H.Y. Yu, and L.M. Zhang, "Studies on Cell Mediated Immunity of the

Inhabitants in an Area of High Natural Radioactivity in Guangdong Province," *Chinese Journal of Radiological Medicine and Protection* 2(2):64 (1982); S.Z. Liu, G.Z. Xu, X.Y. Li, F.C. Xia, H.Y. Yu, J. Qi, F.L. Wang, and S.K. Wang, "A Re-Study of Immune Functions of the Inhabitants in High Background Radiation Area in Guangdong," *Chinese Journal of Radiological Medicine and Protection* 5:124 (1985); S. Liu, W. Liu, and J.B. Sun, "Radiation Hormesis—Its Expression in the Immune System," *Health Physics* 52:579 (1987).

52. E.T. Bloom, E.L. Korn, and T. Makinodan, "Delayed Effects of Low Dose Radiation of Cellular Immunity in Atomic Bomb Survivors Residing in the United States," *Health Physics* 52:585 (1987). See also *Radiation Research* 96:399 (1986).
53. S. Liu *et al.*, "Radiation Hormesis—Its Expression in the Immune System," *op. cit.*
54. I. Hellström and K.E. Hellström, "Antitumor Effect of Whole-Body X-Irradiation: Possible Role of X-ray Sensitive T Suppressor Cell Population," *Transplantation Proceedings* 11:1073 (1979); R.E. Anderson, G.M. Troup, S. Tokuda, and N.L. Warner, "Radiation-Induced Augmentation of Tumor Rejection in Mice," *Proceedings of the Federation of American Societies for Experimental Biology* 39:331 (1980).
55. T.D. Luckey, "Hormesis with High LET Radiation Induced Cancer," *Zeitschreift fuer Physikalische Medizin, Balneologie, Medizale Klimatologie* 13:11 (1984).
56. BEIR Committee, *loc. cit.*
57. T.D. Luckey, *Hormesis with Ionizing Radiation, op. cit.*, Tables 4, 5 and 7.
58. H.T. Mulholland and R.B. Conolly, "DNA Damage by ^{60}Co Radiation and MNNG in Cultures of Newborn Rat Cutaneous Epidermal Keratinocytes," *Federation Proceedings Abstracts* 431:593 (1984); G. Olivieiri, J. Bobycote, and S. Wolff, "Adaptive Response of Human Lymphocytes to Low Concentrations of Radioactive Thymidine," *Science* 223:594 (1984).
59. J.F. Spalding *et al., loc. cit.*
60. J.V. Neel, C. Satoh, and H.B. Hamilton, "Search for Mutations Affecting Protein Structure in Children of Atomic Bomb Survivors, a Preliminary Report," *Proceedings of the National Academy of Sciences, USA* 77:4221 (1980); W.J. Schull, M. Otakee, and J.V. Neel, "Genetic Effects of the Atomic Bomb: Reappraisal," *Science* 213:1220 (1981); S. Jablon, *loc. cit.*
61. I. Hellström and K.E. Hellström, *loc. cit.*
62. R.J. North, "Cyclophosphamide-Facilitated Adoptive Immuno-Therapy of an Established Tumor Depends on Elimination of Tumor-Induced Suppressor T-Cells," *Journal of Experimental Medicine* 155:1063 (1982).
63. R.D. Anderson *et al., loc. cit.*
64. S.Z. Liu, *loc. cit.*

65. E. Lorenz, *loc. cit.*
66. T.D. Luckey, "Introduction to Comparative Animal Nutrition," in M. Rechigl, editor, *Comparative Animal Nutrition*, Vol. 1, Basel: S. Karger, 1976, 1–22; T.D. Luckey, "Comparative Nutrition," in M. Rechigl, editor, *CRC Handbook Series in Nutrition and Food*, Cleveland, CRC Press, 1977, 1–18; T.D. Luckey and B. Venugopal, *Metal Toxicity in Mammals. 1. Physiologic and Chemical Basis for Metal Toxicity*, New York: Plenum Press, 1977.
67. D. Grahn, "Biologic Effects of Protracted Low Dose Radiation Exposure of Man and Animals," in R.J.M. Fry, D. Grahn, M.L. Griem, and J.H. Rust, editors, *Late Effects of Radiation*, London: Taylor and Francis, 1970, 101–136.
68. J.B. Murphy and J.J. Morton, "The Effect of Roentgen Rays on the Rate of Growth of Spontaneous Tumors in Mice," *Journal of Experimental Medicine* 22:800 (1915).
69. H. Lisco, H.S. Ducoff, and R. Baserga, "The Influence of Total Body X-Irradiation on the Response of Mice to Methylcholanthrene," *Bulletin of the Johns Hopkins Hospital* 103:101 (1958).
70. H. Lisco *et al.*, *loc. cit.*
71. K. Nishio *et al.*, *loc. cit.*
72. J. Furth and M.C. Boon, "Enhancement of Leukemogenic Action of Methylcholanthrene by Pre-Irradiation with X-rays," *Science* 98:138 (1943); F.H. Figge, "Cosmic Radiation and Cancer," *Science* 105:323 (1947); S.G. Ong, "Cosmic Radiation and Cancer. I. Influence of Cosmis Radiation at Sea Level on Induced Cancer," *Scienctia Sinica* 13:61 (1963).
73. A.C. Upton, M.L. Randolph, and G.S. Conklin, "Late Effects of Fast Neutrons and Gamma Rays in Mice as Influenced by the Dose Rate of Irradiation: Induction of Neoplasia," *Radiation Research* 41:467 (1970).
74. R.L. Ullrich, M.C. Jernigan, G.E. Cosgrove, L.C. Scatterfield, N.D. Bowles, and J.B. Storer, "The Influence of Dose and Dose Rate on the Incidence of Neoplastic Disease in RFM Mice after Neutron Irradiation," *Radiation Research* 68:115 (1976); R.L. Ullrich, M.C. Jernigan, and J.B. Storer, "Neutron Carcinogenesis. Dose and Dose-Rate Effects in BALB/C Mice," *Radiation Research* 72:487 (1977); R.L. Ullrich and J.B. Storer, "Influence of Irradiation on the Development of Neoplastic Disease in Mice. I. Reticular Tissue Tumors. II. Solid Tumors. III. Dose-Rate Effects," *Radiation Research* 80:303, 317, 325 (1979); J.B. Storer, L.J. Serrano, E.B. Darden, M.C. Jernigan, and R.L. Ullrich, "Life Shortening in RFM and BALB/C Mice as a Function of Radiation Quality, Dose, and Dose Rate," *Radiation Research* 78:122 (1979).
75. A.M. Brues, G.A. Sacher, M.P. Finkel, and H. Lisco, "Comparative

Carcinogenic Effects by X Radiation and P^{32}," *Cancer Research* 9:545 (1949).

76. D. Grahn, R.J.M. Fry, and R.A. Lea, "Analysis of Survival and Cause of Death Statistics for Mice under Single and Duration-of-Life Gamma Irradiation," in W. Vishniac, editor, *Life Sciences and Space Research,* Vol. 10, Berlin: Academic Verlag, 1972, 175–190.
77. D. Grahn, G.A. Sacher, R.J.M. Fry, J.M. Rust, E.M. Cooke, and E. Staffeldt, "Survival of Mice under Daily Gamma Irradiation: Progress Report for Incidence of Some Tumors," *Annual Report, Biological and Medical Research Division: ANL-7535,* Argonne: Argonne National Laboratory, 68:74 (1968).
78. D.J. Mewissen and J.H. Rust, "Reticuloendothelial Neoplasms in C57 Black Mice after Fast Neutron Irradiation at Low Doses," *U.S. AEC Conference 740930,* Oak Ridge: U.S. AEC, 1976.
79. D. Nolibe, R. Masse, and J. LaFuma, "The Effect of Neonatal Thymectomy on Lung Cancers Induced by Plutonium Diozide," *Radiation Research* 87:99 (1981).
80. D. Grahn, G.A. Sacher, *et al.,*"Survival of Mice under Daily Gamma Irradiation," *loc. cit.*
81. D.F. Cahill, J.F. Wright, J.H. Godbold, J.M. Ward, J.W. Laskey, and E.A. Tompkins, "Neoplastic and Lifespan Effects of Chronic Exposure to Tritium. II. Rats Exposed *in utero,*" *Journal of the National Cancer Institute* 53:371 (1975).
82. A.C. Upton, T.T. Odell, Jr., and E.P. Sniffen, "Influence of Age at Time of Irradiation on Induction of Leukemia and Ovarian Tumors in RF Mice," *Proceedings of the Society of Experimental Biology and Medicine* 104769, 1960; U. Reincke, E. Stutz, and G. Wagner, "Tumoren einmaliger Rontgenbestrahlung Weisser Ratten in verschiedenem Lebensalter," *Zeitschrift fuer Krebsforschung* 66:165 (1964); W. Friedberg, G.D. Hanneman, D.N. Faulkner, B.R. Neas, G.E. Cosgrove, Jr., and E.G. Darden, Jr., "Fast-Neutron Irradiation of Mouse Embryos in the Pronuclear Zygote Stage: Mortality Curves and Neoplastic Diseases in 30-Day Postnatal Survivors," *Proceedings of the Society for Experimental Biology and Medicine* 151:808 (1976).
83. E.W. Webster, "On the Question of Cancer Induction by Small X-ray Doses," *American Journal of Roentgenology* 137:647 (1981).
84. *Ibid.*; J.D. Boice, Jr. and C.E. Land, "Ionizing Radiation," in D. Schottenfeld and J.F. Fraumeri, editors, *Cancer Epidemiology and Prevention,* Philadelphia: W.B. Saunders, 1982, 231–253.
85. W.W. Sutow, D.J. Fernbach, and T.J. Vietti, *Clinical Pediatric Oncology,* St. Louis: C.V. Mosby Co., 1984.
86. E.W. Webster, *loc. cit.*
87. BEIR Committee, *loc. cit.*
88. F.W. Gunz and H.R. Atkinson, "Medical Radiation Leukemia; a Retrospective Survey," *British Medical Journal* 1:389 (1964); A. Linos, J.E.

Gray, A.I. Orvis, R.A. Kyle, W.M. O'Fallon, and L.T. Kurland, "Low-Dose Radiation and Leukemia," *New England Journal of Medicine* 302:1101, 303:114, 814 (1980).

89. G.W. Beebe, H. Kato, and C.E. Land, "Studies of the Mortality of A. Bomb Survivors. 6. Mortality and Radiation Dose, 1950–1974," *Radiation Research* 75:138 (1978); C.L. Sanders and R.L. Kathren, *Ionizing Radiation: Tumorigenic and Tumoricidal Effects*, Columbus: Battelle Press, 1983.
90. H. Kato and W.J. Schull, "Studies of the Mortality of A-Bomb Survivors. 7. Mortality, 1950–1978. Part 1. Cancer Mortality," *Radiation Research* 90:395 (1982); T. Wakabayashi, H. Kato, T. Ikeda, and W.J. Schull, "Studies of the Mortality of A-Bomb Survivors. Report 7. Part III. Incidence of Cancer in 1959–1978, Based on the Tumor Registry, Nagasaka," *Radiation Research* 93:112 (1983).
91. N.A. Fujita, A.A. Awa, D.A. Pierce, H. Kato, and Y. Shimizu, "Reevaluation of Biological Effects of Atomic Bomb Radiation by Changes of Estimation Dose," in *Biological Effects of Low-Level Radiation*, Vienna: International Atomic Energy Agency, 1983, 55–60.
92. T.D. Luckey, *Hormesis with Ionizing Radiation*, *loc. cit.*
93. I. Schmitz-Feuerhake and P. Carbonell, "Evaluation of Low Level Effects in the Japanese A-Bomb Survivors after Current Dose Revisions and Estimation of Fallout Contribution," in *Biological Effects of Low-Level Radiation*, Vienna: International Atomic Energy Agency, 1983, 45–53.
94. C.D. Robinette, S. Jablon, and T.L. Preston, "Studies of Participants in Nuclear Tests," *Final Report DOE/EV/01577 1.*, National Research Council, Washington, D.C., 1–47.
95. C.L. Sanders and R.L. Kathren, *loc. cit.*
96. H. Kato and W.J. Schull, *loc. cit.*
97. *Ibid.*
98. K.E. Hellström and I. Hellström, "Antigen Specific Suppressor ('Blocking') Factors in Tumor Immunity," in A. Nowotry, editor, *Pathological Membranes*, New York: Plenum Press, 1985.
99. E.T. Bloom *et al.*, *loc. cit.*
100. C.E. Land, "Estimating Cancer Risks from Low Doses of Ionizing Radiation," *Science* 209:1197 (1980); T. Pagnementa, personal communication, 1983.
101. T. Pagnementa, *loc. cit.*
102. H. Kato and W.J. Schull, *loc. cit.*
103. T. Wakabayashi *et al.*, *loc. cit.*
104. *Ibid.*
105. *Ibid.*
106. H. Kato and W.J. Schull, *loc. cit.*
107. *Ibid.*
108. T. Wakabayashi *et al.*, *loc. cit.*

109. *Ibid.*
110. C.D. Robinette *et al.*, *loc. cit.*
111. R.E. Rowland, A.F. Stehney, and H.F. Lucas, "Dose-Response Relationships for Radium-Induced Bone Sarcomas," *Health Physics* 44: suppl. 1:15 (1983).
112. R.D. Evans, "Inception of Standards for Internal Emitters, Radon and Radium," *Health Physics* 41:437 (1981).
113. E.H. Quimby and S.C. Werner, "Late Radiation Effects in Roentgen Therapy for Hypothyroidism: Their Possible Bearing on the Use of Radioactive Iodine," *Journal of the American Medical Association* 140:1046 (1949).
114. J.D. Boice, Jr. and C.E. Land, *loc. cit.*
115. R.E. Rowland *et al.*, *loc. cit.*
116. R.D. Evans, *loc. cit.*
117. M.L. Rallison, B.M. Dobyns, and F.R. Keating, "Thyroid Disease in Children," *American Journal of Medicine* 56:457 (1974); E.W. Webster, *loc. cit.*
118. E.L. Saenger, G.E. Thoma, and E.A. Thompkins, "Incidence of Leukemia Following Treatment of Hyperthyroidism; Preliminary Report of the Cooperative Thyrotoxicosis Therapy Follow-up Study," *Journal of the American Medical Society* 205:855 (1968).
119. B.M. Dobyns, G.E. Sheline, J.B. Workman, E.A. Thompkins, W.M. McConahey, and D.V. Becker, "Malignant and Benign Neoplasms of the Thyroid in Patients Treated for Hyperthyroidism; a Report of the Cooperative Thyrotoxicosis Therapy Follow-up Study," *Journal of Clinical Endocrinology and Metabolism* 38:976 (1974).
120. E.M. Chapman, "History of the Discovery and Early Use of Radioactive Iodine," *Journal of the American Medical Association* 250:2042 (1983).
121. B.M. Dobyns *et al.*, *loc. cit.*
122. L.E. Holm, "Thyroid Treatment and Its Possible Influence on Occurence of Malignant Tumors after Diagnostic ^{131}I," *Acta Radiologica Oncology* 19:455 (1980); L.E. Holm, G. Lundell, and G. Walter, "Incidence of Malignant Thyroid Tumors in Humans after Exposure to Diagnostic Doses of Iodine-131. I. Retrospective Cohort Study," *Journal of the National Cancer Institute* 64:1055 (1980). See also *New England Journal of Medicine* 303:188, 217 (1980).
123. C.L. Sanders and R.L. Kathren, *loc. cit.*
124. G.L. Voelz, J.F. McInroy, G.S. Wilkinson, G.L. Tiejen, and J.W. Healy, "Mortality Study of Los Alamos Workers with Higher Exposure to Plutonium," *Proceedings of the Midyear Topical Meeting*, Albuquerque, 1983.
125. *Ibid.*
126. E.W. Webster, *loc. cit.*; J.D. Boice, Jr. and C.E. Land, *loc. cit.*
127. E.S. Gilbert and S. Marks, "An Analysis of the Mortality of Workers in

a Nuclear Facility," *Radiation Research* 79:122 (1979). See also 83:740.

128. G.B. Hutchinson, B. MacMahon, S. Jablon, and C.E. Land, "Review of Reports by Mancuso, Stewart and Kneale of Radiation Exposure of Hanford Workers," Health Physics Society Midyear Symposium, Albuquerque, 1979.
129. E.W. Webster, *loc. cit.*
130. Darby & Reissland (1981) "Low Level Ionizing Radiation and Cancer," *J. Roy. Stat. Soc.*, Series A: Gen. part 3, 144, 298.
131. J.D. Abbat, T.R. Hamilton, and J.L. Weeks, "Epidemiological Studies in Three Corporations Covering the Canadian Nuclear Fuel Cycle," in *Biological Effects of Low Level Radiation*, Vienna: International Atomic Energy Agency, 1983, 351–361.
132. NIOSH, "NIOSH," *Chemical and Engineering News* 60:22 (1980); E.W. Webster, *loc. cit.*
133. HBRRG (High Background Radiation Research Group), "Aspects of Environmental Radiation and Dosimetry Concerning the High Background Radiation Area in China," *Journal of Radiation Research* (Tokyo) 22:88 (1981). See also *Science* 209:877 (1981); W. He, R. Feng, M. Liang, M. Chen, L. Chen, R. Li, and G. Zhang, "A Summary of Third Stage Investigations on Cancer Mortality in High Background Radiation Area," *Chinese Journal of Radiological Medicine and Protection* 5:109 (1985).
134. HBRRG, *loc. cit.*
135. E.W. Webster, *loc. cit.*
136. S.Z. Liu *et al.*, "Studies on Cell Mediated Immunity of the Inhabitants in an Area of High Natural Radioactivity in Guangdong Province," *loc. cit.*; S.Z. Lui, P.X. Xiao, S.Y. Ma, G.Z. Xu, C.H. Tian, H.Y. Yu, and L.M. Zhang, "Studies on the Immune Status of Inhabitants in an Area of High Natural Radioactivity in China," *Proceedings of the 7th International Congress of Radiation Research* C1–20, 1983; S.Z. Liu *et al.*, "A Restudy of Immune Functions of the Inhabitants in High Background Radiation Area in Guangdong," *loc. cit.*; S. Liu *et al.*, "Radiation Hormesis—Its Expression in the Immune System," *loc. cit.*
137. S. Zhai, X. Lin, T. Pan, W. He, R. Feng, M. Chen, S. Li, L. Chen, R. Li, and H. Yie, "An Interim Report on the Survey of Mortality from Malignant Tumors in High Background Area of Guangdong," *Chinese Journal of Radiological Medicine and Protection* 2(2):48 (1982).
138. W. He *et al.*, *loc. cit.*
139. N.A. Frigerio, K.F. Eckerman, and R.S. Stowe, *Carcinogenic Hazard from Low-Level, Low-Rate Radiation*, Part 1, Rep. ANL/ES–26, Argonne: Argonne National Laboratory, 1973; N.A. Frigerio and R.S. Stowe, "Carcinogenia and Genetic Hazard from Background Radiation," Vol. 2, *Biological and Environmental Effects of Low Level Radiation*, Vienna: International Atomic Energy Agency, 1976, 385–393.

140. D.T. Oakley, "Natural Radiation Exposure in the U.S.," *Thesis (USEPA Report ORD/SID–72/1)*, Boston: Harvard School of Public Health, 1972.
141. N.D. Eckhoff, J.K. Shultis, R.W. Clack, and E.R. Ramer, "Correlation of Leukemia Mortality Rates with Altitude in the United States," *Health Physics* 27:377 (1974); T.J. Mason and R.W. Miller, "Cosmic Radiation at High Altitudes and U.S. Cancer Mortality," *Radiation Research* 60:302 (1974); A.P. Jacobson, F.A. Plato, and N.A. Frigerio, "The Role of Natural Radiations in Human Leukemogenesis," *American Journal of Public Health* 66:31 (1976); B.S. Sanders, "Low-Level Radiation and Cancer Deaths," *Health Physics* 34:521 (1978); J.J. Cohen, "Natural Background as an Indicator of Radiation Induced Cancer," *Proceedings of the 5th International Radiation Protection Association*, Jerusalem, March 1980; R.J. Hickey, E.J. Bowers, D.E. Spence, B.S. Zemmel, A.B. Clelland, and R.C. Clelland, "Low Level Ionizing Radiation and Human Mortality: Multi-Regional Epidemiological Studies," *Health Physics* 40:625 (1981); R.J. Hickey, A.B. Clelland, R.C. Clelland, B.S. Zemmel, and E.J. Bowers, "Multivariate Analysis of U.S. State Data," *American Journal of Physical Anthropology* 54:233 (1981); R.S. Yallow, "Reappraisal of Potential Risks Associated with Low-Level Radiation," *Comments on Molecular and Cellular Biophysics* 1:149 (1981); H.T. Sauer, M.R. Dicks, and R.F. Stearley, "Low-Level Radiation and the Risk of Dying," American Public Health Meeting, Montreal, 1982.
142. J.J. Cohen, *loc. cit.*
143. R.J. Hickey *et al.*, "Multivariate Analysis of U.S. State Data," *loc. cit.*
144. J.J. Cohen, *loc. cit.*
145. N.D. Eckhoff *et al.*, *loc. cit.*
146. R.S. Yallow, *loc. cit.*
147. N.A. Frigerio *et al.*, *Carcinogenic Hazard from Low-Level, Low-Rate Radiation, loc. cit.*
148. F. Steinhausler, I. Uzunov, and E. Phol, "The Main Inconsequences in the Present Radiological Protection Concept for the General Population," *Health Physics* 49:1229 (1985).
149. T.D. Luckey, *Hormesis with Ionizing Radiation, loc. cit.*
150. *Ibid.*; T.D. Luckey, "Physiological Effects from Low Levels of Radiation," *loc. cit.*; T.D. Luckey, "Beneficial Physiologic Effects of Ionizing Radiation," *loc. cit.*
151. E. Lorenz *et al*, *loc. cit.*; D. Grahn *et al.*, "Analysis of Survival and Cause of Death Statistics for Mice under Single and Duration-of-Life Gamma Irradiation," *loc. cit.*
152. K. Bonham and L.R. Donaldson, "Low Level Chronic Irradiation of Salmon Eggs and Alevins," in *Disposal of Radioactive Wastes into Seas, Oceans and Surface Waters*, Vienna: International Atomic Energy Agency, 1966, 869–884; L.R. Donaldson and K. Bonham, "Effects of

Chronic Exposure of Chinook Salmon Eggs and Alevins to Gamma Irradiation," *Transactions of the American Fisheries Society* 99:112 (1970). See also 93:333 (1964); W.K. Hershberger, K. Bonham, and L.R. Donaldson, "Chronic Exposure of Chinook Salmon Eggs and Alevins to Gamma Irradiation: Effects on Their Return to Freshwater as Adults," *Transactions of the American Fisheries Society* 107:622 (1978).

153. J.F. Spalding, R.G. Thomas, and G.L. Tietjen, "Lifespan of C57 Mice as Influenced by Radiation Dose, Dose Rate, and Age at Exposure," *Report No. UC-48; LA-9528*, Los Alamos: Los Alamos National Laboratory, 1982.
154. P. Brown, *American Martyrs*, Springfield: C.C. Thomas, 1936.
155. R. Hildreth, *From X-ray Martyrs to Low Level Radiation*, Kalamazoo: Industrial Graphics Services, 1981.
156. H.F. Henry, *loc. cit.*
157. R. Hildreth, *loc. cit.*
158. P.G. Smith and R. Doll, "Mortality from Cancer and All Causes among British Radiologists," *British Journal of Radiology* 54:187 (1981).
159. N.A. Frigerio and R.S. Stowe, *loc. cit.*; N.D. Eckhoff *et al.*, *loc. cit.*; B.S. Sanders, *loc. cit.*; J.J. Cohen, *loc. cit.*; R.J. Hickey *et al.*, "Low Level Ionizing Radiation and Human Mortality," *loc. cit.*; R.J. Hickey *et al.*, "Multivariate Analysis of U.S. State Data," *loc. cit.*; HBRRG, *loc. cit.*; H.T. Sauer, "Geographic Patterns in the Risk of Dying and Associated Factors, Ages 35–74 Years," *DHHS Publication PHS 80-1402* (CNTIS No. HRP-0903230), 1980; H.T. Sauer *et al.*, *loc. cit.*
160. J.F. Spalding *et al.*, *loc. cit.*
161. H.T. Sauer, *loc. cit.*
162. BEIR Committee, *loc. cit.*
163. *Ibid.*
164. H.T. Sauer, *loc. cit.*
165. H.T. Sauer *et al.*, *loc. cit.*
166. HBRRG, *loc. cit.*
167. H.T. Sauer *et al.*, *loc. cit.*
168. S.J.R. Kemala, "Child Health and Nutrition," *Nutrition Notes* 17:14 (1981).
169. P.G.K. Panikar, "Resources Not the Constraint on Health Improvement: A Case Study," *Economics and Political Weekly* 14:1803 (1979).
170. J.H. Mazo and D. Mehta, "Kerala," *GEO* 3:40 (1981).
171. T.D. Luckey, "Hormology with Inorganic Compounds," *Environmental Quality and Safety*, Supplement 1:81 (1975); T.D. Luckey, "Introduction to Comparative Animal Nutrition," *loc. cit.*; T.D. Luckey and B. Venugopal, *loc. cit.*
172. T.D. Luckey, "Hormology with Inorganic Compounds," *loc. cit.*; T.D. Luckey, "Comparative Nutrition," *loc. cit.*
173. T.D. Luckey, "Antibiotics in Nutrition," in H.S. Goldberg, editor,

Antibiotics, Their Chemistry and Non-Medical Uses, Princeton: D. Van Nostrand Co., 1959, 174–341.

174. T.D. Luckey, "Ionizing Radiation Promotes Protozoan Reproduction," *loc. cit.*
175. H. Planel, J.P. Soleilhavoup, R. Tixador, F. Croute, and G. Richoilley, "*Paramecium aurelia* as a Cellular Model Used for Studies of the Biological Effects of Natural Ionizing Radiation or Chronic Low-Level Irradiation," in *Methodology for Assessing Impacts of Radioactivity on Aquatic Systems*, Technical Report Series No. 190, Vienna: International Atomic Energy Agency, 1979, 335–346; A. Conter, D. Dupouy, and H. Planel, "Demonstration of a Biological Effect of Natural Ionizing Radiations," *International Journal of Radiation Biology* 43:421 (1983).
176. W.G. Heston, E. Lorenz, and M.K. Deringer, "Occurrence of Pulmonary Tumors in Strain A Mice Following Total Body X-Radiation and Injection of Nitrogen Mustard," *Cancer Research* 13:573 (1953).

SEVEN A

DISCUSSION OF LUCKEY'S PAPER

Shu Zheng Liu

The nature of health effects of low-level radiation has been a problem of considerable controversy. Ionizing radiation has been confirmed to be carcinogenic within a certain dose range, and this has for many years been of concern to the public as well as to governmental organizations and expert bodies. Radiation carcinogenesis is considered to be a stochastic effect. It has been taken for granted by many that the risk of cancer due to low-level ionizing radiation can simply be extrapolated from data obtained with large doses, using some mathematical models, whether linear, quadratic, or linear-quadratic. Some radiation biologists believe that we should adopt the linear model which states that the risk is directly proportional to the radiation dose received, right down to zero dose. Actually, there have been reports in the literature opposing such predictions. More than two decades ago, H.F. Henry presented the concept that low-level radiation may be beneficial, stating that a significant and growing amount of experimental information indicated that the overall effects of chronic exposure at low levels might not be harmful (Henry). In recent years attention has been paid by many investigators to the possibly entirely different effects of low-level ionizing radiation from those of high doses. In Professor T.D. Luckey's monograph "Hormesis with Ionizing Radiation" (Luckey), an extensive survey of literature was made. This has aroused vivid discussion since its publication (Hickey; Liu, 1985a; Conference on Radiation Hormesis at Oakland, CA., 14–16 August 1985). In his paper, "Hormesis and Nurture with Ionizing Radiation," presented to the present Conference, more than fifty new references since 1980 were cited giving more support to the premise.

Professor Luckey has collected data on bacteria, plants, invertebrates, and vertebrates, including mammals, to illustrate the stimulating effect of low-level radiation on physiological functions, including immune competence. Special emphasis was given in this paper to radiation hormesis expressed in the immune system. The immune status of the population and the individuals may be closely related to lifespan and to cancer induction and mortality. Though the theory of immune surveillance against the growth of transformed cells into malignant tumor has been doubted by some authors (Möller and Möller; Klein and Klein; Spärck), the consensus still holds that the non-specific and specific components of host immune systems constitute one of the most important defense mechanisms against the establishment and growth of tumor induced by various environmental agents. As it has been repeatedly disclosed in humans as well as in experimental animals in our laboratory, low-dose radiation has a stimulatory effect on some parameters of the immune system. It was found that the morphologic transformation rate of peripheral blood lymphocytes in response to PHA (phytohemagglutinin) stimulation was increased in inhabitants in an area of high natural radioactivity in Guangdong, China (Liu *et al.*, 1982; 1983; 1985), though the number of T (thymus-dependent) lymphocytes might be somewhat lowered (Table 7A-1).

Animal experiments have confirmed these observations. Single whole body x-irradiation with 25 to 75 mGy (2.5–7.5 rad) caused a definite increase in the PFC (plaque forming cell) reaction of the splenocytes in C57BL/6 mice (Figure 7A-1) (Liu and Liu, 1986). Continuous γ-irradiation with a dose rate of 5.4 mGy/6h/d also led to enhancement of PFC reaction of the spleen at a cumulative dose of 65 mGy (in two weeks) (Figure 7A-2) (Liu and Liu, 1987). The NK (natural killer)

Table 7A-1: Morphological Transformation of Peripheral Blood Lymphocytes in Response to PHA in Inhabitants of High Natural Radioactivity (HNR) Area

Age	*Morphologic Transformation Rate*		%
	HNR Area	*Control Area*	*P*
16–55	75.0 ± 0.90 (148)*	70.6 ± 0.99 (148)	<0.01
16–25	77.2 ± 1.12 (82)	71.3 ± 1.28 (84)	<0.01
45–55	72.2 ± 1.40 (66)	69.8 ± 1.57 (64)	>0 05

* Mean ± S.D. (n)

activity of the splenocytes in this same strain of mice showed marked increases at single whole body doses of 75 and 500 mGy when the effector to target cell ratio was 200:1 (Figure 7A-3), using YAC-1 cells labeled with deoxyuridine (^{125} I-UdR) as the target cells. This increased NK activity also was noticed at single whole body x-irradiation with doses of 25 to 75 mGy when effector to target cell ratio was 50:1 and 100:1 (Figure 7A-4) (Fan and Liu, 1989).

The enhancement of immunologic functions under such small doses could not be explained by a change of ratio of helper to suppressor T lymphocytes. In our laboratory, acid naphthylacetate esterase (ANAE) cytochemical staining was used as a cytoplasmic marker to distinguish T and B (bursa-dependent) lymphocytes and the subsets of T lymphocytes. The spot-granular pattern of ANAE stain was taken as a marker of the helper T cells and the scattered granular pattern for the suppressor T cells. Under single and continuous irradiation in doses causing stimulation of PFC reaction and NK activity, there was no statistically significant change in the percentage of these subsets (Liu and Liu, 1990) (Figure 7A-5, 7A-6). Recent studies using more sophisticated methods in identifying the T subsets, such as surface antigens and surface receptors, also supported the above conclusion (Liu, 1989a; Liu, 1989b).

At the same time we looked at the reactivity of thymocytes and splenocytes to interleukin 1 and interleukin 2, respectively, under different radiation doses. It was found that the reactivity

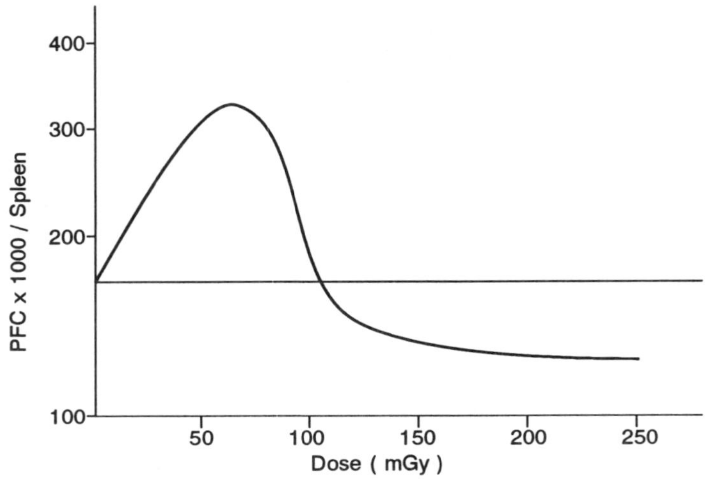

Figure 7A-1: PFC reaction of mouse spleen following single small doses of whole body X-irradiation.

$Y=10^{(2.221-0.001233X+0.02008X^2-0.002043X^3)}$

Dose range of X=0-100 mGy

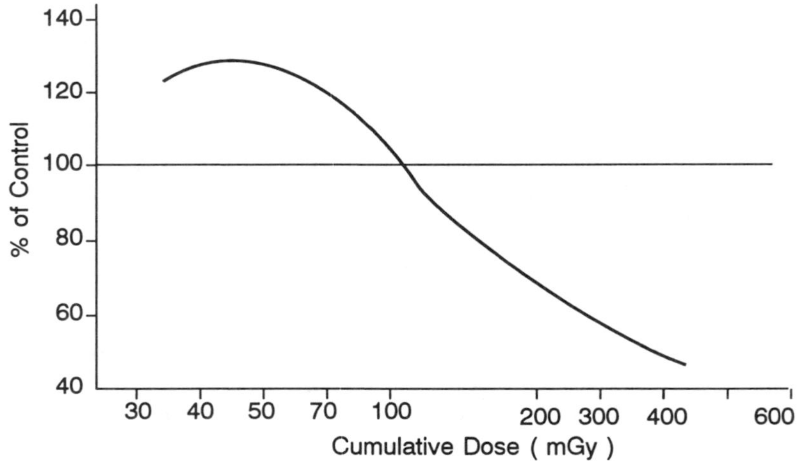

Figure 7A-2: PFC reaction of mouse spleen after continuous γ-irradiation.

$Y = -38.078+263.603X-126.873X^2+16.866X^3$

X = logarithm of dose; $R^2=0.9996$

Y = effects

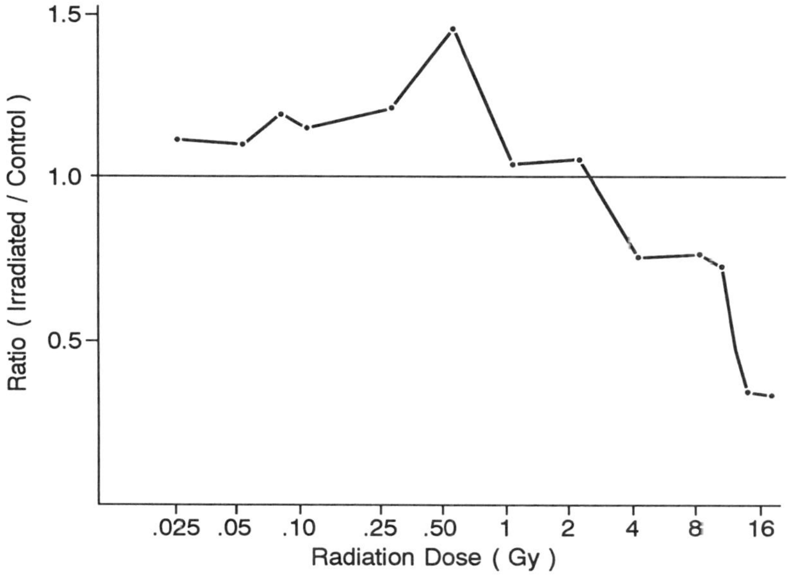

Figure 7A-3: Natural killer cell activity after single whole-body X-irradiation.

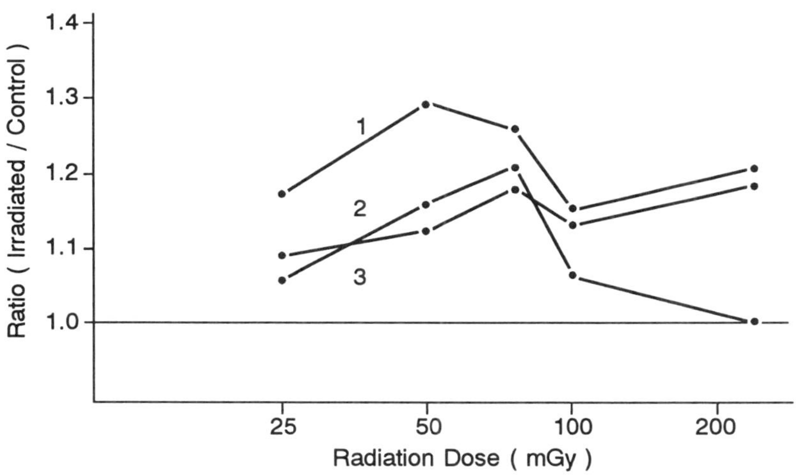

Figure 7A-4: Natural killer cell activity after whole-body X-irradiation with different effector to target ratios.

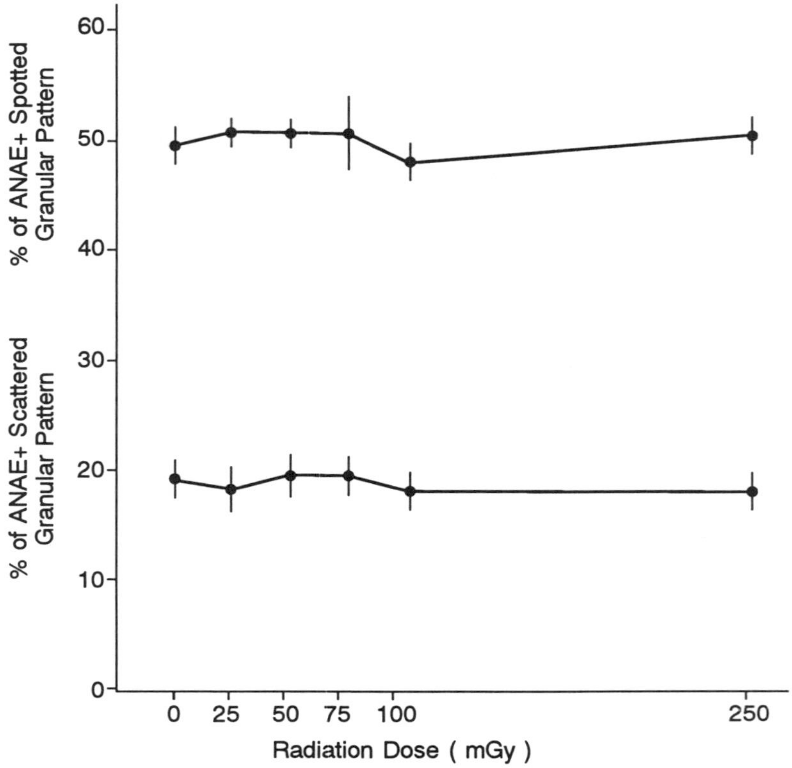

Figure 7A-5: Granular pattern of ANAE+ lymphocytes in mouse peripheral blood after small doses of whole-body X-irradiation.

of the thymus to interleukin 1 and that of the spleen to interleukin 2 tended to increase under the above mentioned dose range (Figures 7A-7, 7A-8, 7A-9). These experimental data point to the possibility of an increase in release of immunoenhancing lymphokines under low-level radiation resulting in an amplification of the immunologic processes (Liu, 1989a). Recent work in our laboratory also showed a tendency of the splenocytes from mice irradiated with low doses to produce more interleukin 2 in response to ConA (concanavalin A) stimulation. Therefore, under such small doses as 25 to 75 mGy,

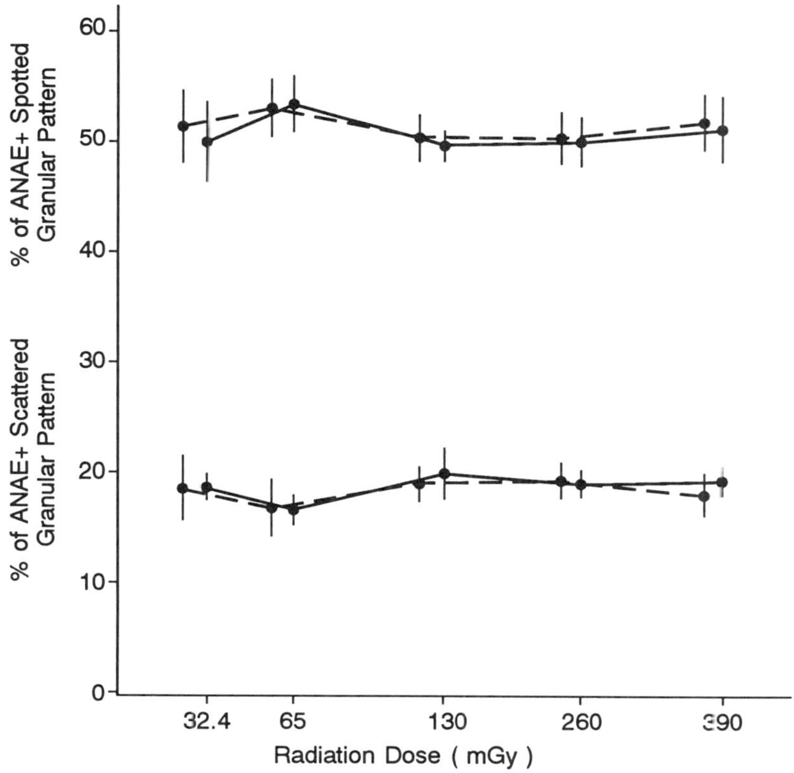

Figure 7A-6: Granular pattern of ANAE+ lymphocytes in mouse peripheral blood after continuous γ-irradiation.

---- Sham-exposed; — Exposed

the enhancement of immunologic functions could not be explained as the result of preferential damage to or killing of the suppressor T cells.

In his paper, Professor Luckey mentioned the so-called Hellström effect (Hellström and Hellström). It was demonstrated that a single whole body irradiation with 4.5 Gy would increase the tumor suppressing effect of the host previously immunized with the specific tumor antigen due to damage of the relatively radiosensitive suppressor T cells. However, in our experiments,

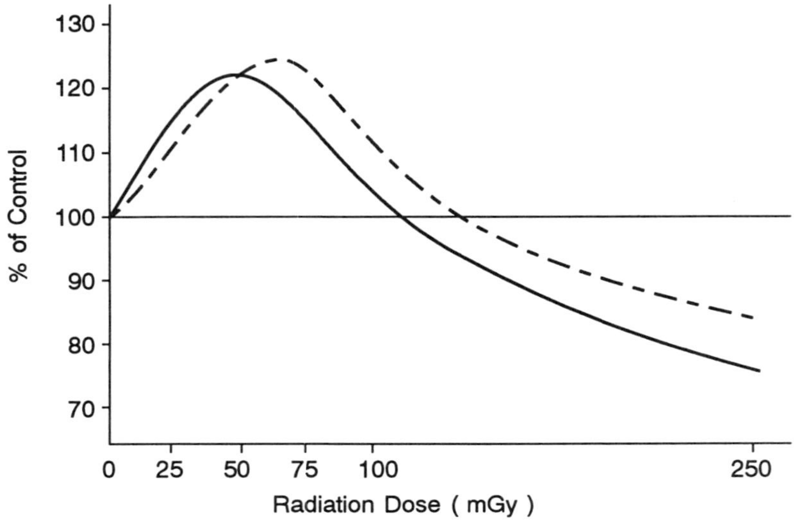

Figure 7A-7: Reaction of thymus to IL-1 following single small doses of whole-body X-irradiation.

CPM/thymus; _._._ Cell number/thymus

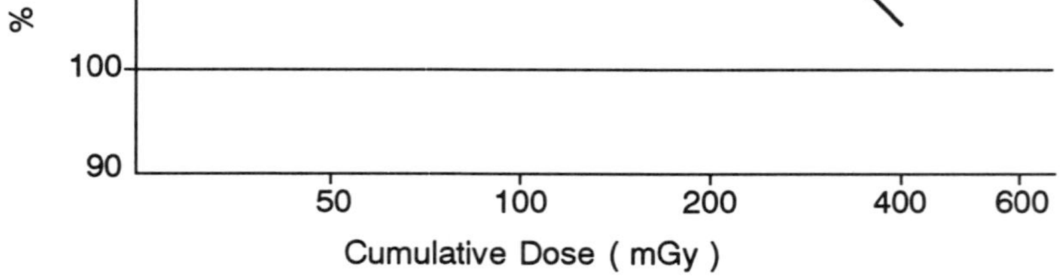

Figure 7A-8: Reactivity of thymocytes to IL-1 after continuous γ-irradiation in mice.

$Y = 75.561 + 42.527X - 9.282X^2$

X: logarithm of dose; Y: effects

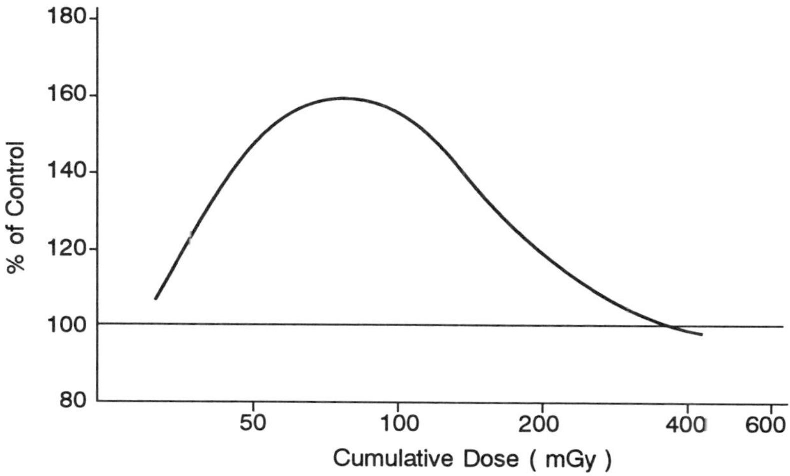

Figure 7A-9: Reactivity of splenocytes to IL-2 after continuous γ-irradiation in mice.
Y = -11.550+145.172X–32.193X²
X: logarithm of dose; Y: effects

4 Gy whole body x-irradiation actually profoundly suppressed the PFC reaction (Figure 7A-10) (Ju *et al.*) as well as the NK activity (Fan and Liu, 1989) (Figure 7A-3). It should be pointed out that the Hellström experiment was done under specific circumstances. The mice had been immunized with tumor antigen before the whole body irradiation with 4–4.5 Gy. That means all the T subsets, including the suppressor, helper, and cytotoxic T lymphocytes, had already been activated before irradiation. It is well-known that activation of lymphocytes, either with antigens or with mitogens, would markedly increase their radio-resistance, though the relative radio-sensitivity of the suppressor T cells might still be higher than that of the helper and cytotoxic T cells. Under such circumstances, a single acute dose of 4 Gy would result in a preferential damage to the suppressors manifesting in an enhancement of immune reaction against the tumor cells. However, without prior activation through immunization with the tumor antigen, a medium dose such as 4–4.5 Gy would kill the suppressor as well as the

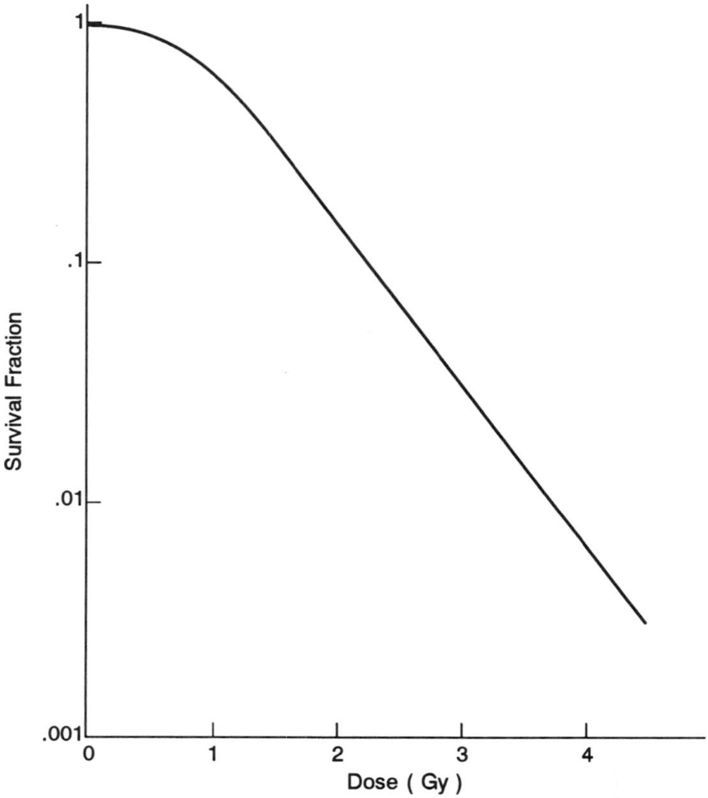

Figure 7A-10: Dose-response curve of PFC in mouse spleen.

helper and the killer T cells, resulting in an overall depression of the immune competence. There have been many reports on the enhancement of immune reactions by radiation. Both in vivo and in vitro experiments showed enhancement of immune reaction by medium radiation doses only when antigen stimulation preceded irradiation, and immuno-enhancement occurred only with smaller doses (e.g., within 0.25 Gy) when irradiation preceded antigen stimulation (Anderson *et al.*; Liu, 1985a). Therefore, the experiment of Hellström and Hellström cannot be taken as a general phenomenon.

More than half of the space of Professor Luckey's paper is devoted to hormesis in radiation-induced cancer. Both experimental and epidemiological data were presented to illustrate the phenomenon of hormesis for a wide variety of radiation-induced cancers. It is clear that the linear no-threshold model does not fit all cases. Many radiation-induced cancers have an actual dose threshold, such as radium-induced osteosarcoma. Hormesis has been observed in some instances. Some remarks should be made on the "China study," which was cited in Professor Luckey's paper. According to the 1982 report (Zhai *et al.*), with 531,972 man years for the area of high natural radioactivity, and 563,507 man years for the control area, the standardized mortality rates for all cancers were $41.19/10^5$ and $49.86/10^5$ for the two areas respectively ($p<0.05$). For specific cancers it was found that the mortality rates for liver and lung cancer were significantly lower in the area of high natural radioactivity ($p<0.05$). However, in the 1985 report (He *et al.*) with 764,696 man years for the area of high natural radioactivity and 777,482 man years for the control area, the standardized mortality rates for all cancers were $44.60/10^5$ and $51.00/10^5$ respectively ($p>0.05$), being lower in the area of high natural radioactivity, though having not reached a significant level statistically. On the other hand, the difference in mortality rates for all cancers in the female populations of the two areas was statistically significant, being $32.30/10^5$ for the area of high natural radioactivity and $41.90/10^5$ for the control area ($p<0.05$). In conclusion, we can say that with the survey of three-quarters of one million man years, at least no increase of cancer mortality was found in an area with natural radioactivity three times as high as the control. This study has provided a rather stable human population under continuous low-level irradiation for generations, which undoubtedly will yield valuable information when long enough follow-up observations are made.

Another human study is the Japan atomic bomb survivors which provides data of remote effects following acute exposure. The data were given in Table 7-5 and Figure 7-12 in Professor Luckey's paper (Kato and Schull) in which the 10–90 mGy (1–9 rad) group demonstrated a lower than control mortality rate for all cancers, especially for leukemia, malignant lymphoma, and some other cancers. However, the authors considered the difference to be of no statistical significance.

The dose-response curves given in Figure 7-8 and Figure 7-9 in Professor Luckey's paper deserve special consideration. The V-shaped curve in Figure 7-8c may not be a universal phenomenon. As it has been pointed out in a previous paragraph, the so-called Hellström effect which occurs only under specific circumstances should be scrutinized more closely. Actually most animal and human data with acute radiation of 4–4.5 Gy did not show a marked drop of cancer mortality rate (Upton; Kato and Schull). I think the preferential killing of host suppressor T cells in the dose range around 4–4.5 Gy could not be considered as an inevitable constituent part of the dose-response curve for acute exposures, and it is even more doubtful for chronic irradiation. Therefore, for human cancer induction or mortality the hormesis phenomenon may only be manifest in the region of low doses confined to 10 to 100 times the natural background. And further epidemiological data are needed even for such a generalization.

Another aspect worth mentioning is the rate of chromosomal aberrations occurring under low-level radiation. Most studies, including examination of the inhabitants living in the area of high natural radioactivity in China (Chen *et al.*) and the atomic bomb survivors in Japan (Awa and Nakano), as well as other surveys (Wang *et al.*), disclosed a definite increase of chromosomal aberrations in the peripheral blood lymphocytes. Although "chromosomal damage in an individual exposed to radiation has not been associated with subsequent illness" (Webster), in any case it stands for definite structural changes of genetic material. Some relate these damages with possible cell transformations. It has been confirmed that under conditions of low-level radiation the repair synthesis of DNA might be stimulated, expressed in the form of an increase in unscheduled DNA synthesis (UDS) (Tuschl *et al.*; Liu *et al.*, 1985; 1987) (Table 7A-2). This may be the result of activation of repair enzymes under low-level radiation. The whole picture might be visualized as follows: With such adaptive mechanisms the host may still not be able to properly repair all damages of the genetic material caused by radiation and some transformed

Table 7A-2: DNA Repair Synthesis (UDS) of Peripheral Blood Lymphocytes in Males (DPMuv/DPMc)

Age	*HNR Area*	*Control Area*	*P*
16–25	2.3 ± 0.4 (10)*	1.9 ± 0.6 (13)	< 0.05
45–55	2.0 ± 0.4 (10)	1.6 ± 0.4 (12)	< 0.05

* Mean ± SD(n)

cells may appear. If the immune competence of the host is deranged by relatively large doses of radiation, the transformed cells may have a chance to develop into an established tumor. On the other hand, if the immune competence of the host is enhanced by low-level radiation, such newly transformed cells may be eradicated or kept dormant by the defense mechanisms, preventing the establishment of a clinical cancer. Elucidation of the whole cascade of events of such mechanisms will be of great importance in consolidating the thesis of hormesis with ionizing radiation and promoting its practical application in health sciences.

The experimental data presented in Professor Luckey's paper have clearly demonstrated the "essentiality" of ionizing radiation for proper growth and reproduction of protozoa. If this is proven to be true also for mammals and humans, it could mean that ionizing radiation should be considered as an indispensable element for optimal health. At present, such a tentative conclusion can only be drawn by inference.

In conclusion, Professor Luckey's paper has clearly illustrated the phenomenon of radiation hormesis in its broad sense, by an extensive review of the existing literature. It also has re-emphasized the concept that low-level radiation reasonably above background may not be harmful to human health. The theory of radiation hormesis undoubtedly will promote research in radiation biology and radiation epidemiology. Further confirmation of this new concept with forthcoming radiobiologic and epidemiologic data would influence government standards and regulations and give great impetus to the theory and practice of radiological sciences.

References

Anderson, R.E., I. Lefkovits, and G.M. Troup. 1980. "Radiation-Induced Augmentation of the Immune Response." *Contemporary Topics in Immunobiology*. Volume #11 (Ed. N.L. Warner). New York: Plenum Press, 11:245.

Awa, A. and M. Nakano. 1985. "Radiation Induced Chromosome Aberrations in A-Bomb Survivors, Hiroshima and Nagasaki." *Chinese Journal of Radiological Medicine and Protection* 5:78.

Chen, D., C. Zhang, S. Yao, C. Gao, W. Shen, J. Gao, T. Jiang, X. Go, and C. Li. 1985. "Further Investigation on Chromosome Aberrations in Lymphocytes of Inhabitants in High Background Radiation Area in Yangjian." *Chinese Journal of Radiological Medicine and Protection* 5:116.

Fan, X.H. and S.Z. Liu. 1989. "Radiation Effect on Splenic NK Cells in Mouse." *Journal of Norman Bethune University of Medical Sciences* 15:551.

He, W., R. Feng, M. Liang, M. Chen, L. Chen, R. Li, and G. Zhang. 1985. "A Summary of Third Stage Investigations on Cancer Mortality in High Background Radiation Area." *Chinese Journal of Radiological Medicine and Protection* 5:109.

Hellström, I. and K.E. Hellström. 1979. "Antitumor Effect of Whole-Body X-Irradiation: Possible Role of an X-Ray-Sensitive T Suppressor cell Population." *Transplantation Proceedings* 11:1073.

Henry, H.G. 1961. "Is All Nuclear Radiation Harmful?" *Journal of American Medical Association* 176:671.

Hickey, R.J., E.J. Bowers, and R.C. Clelland. 1983. "Radiation Hormesis, Public Health, and Public Policy. A Commentary." *Health Physics* 44:207.

Ju, G.Z., X.H. Fan, and S.Z. Liu. 1985. "Radiation Effects on Antibody-Forming Cells in Mouse Spleen." *Chinese Journal of Radiological Medicine and Protection* 5:331.

Kato, H. and W.J. Schull. 1982. "Studies of the Mortality of A-Bomb Survivors. 7. Mortality, 1950–1978. Part 1. Cancer Mortality." *Radiation Research* 90:395.

Klein G. and E. Klein. 1977. "Immune Surveillance Against Virus-Induced Tumors and Nonrejectability of Spontaneous Tumors: Contrasting Consequences of Host Versus Tumor Evolution." *Proceedings of National Academy of Science (USA)* 74:2121.

Liu, S.Z. 1985a. "Some Advances in Radiation Immunology." *Radiation Protection* 5:129.

__________. 1985b. *Radiation Immunology.* Beijing: Peoples' Health Publishing House.

__________. 1989a. "Radiation Hormesis: A New Concept in Radiological Science." *Chinese Medical Journal* 102:750.

__________. 1989b. "Advances in Studies of Immunological Effect of Low Dose Radiation." *Foreign Medicine: Radiation Medicine and Nuclear Medicine* 13:203.

__________, P.X. Xiao, S.Y. Ma, G.Z. Xu, C.H. Tian, H.Y. Yu, and L.M. Zhang. 1982. "Studies on Cell-Mediated Immunity of the Inhabitants in an Area of High Natural Radioactivity in Guangdong Province." *Chinese Journal of Radiological Medicine and Protection* 2(2):64.

__________, P.X. Xiao, S.Y. Ma, G.Z. Xu, C.H. Tian, H.Y. Yu, and L.M. Zhang. 1983. "Studies on the Immune Status of Inhabitants in an Area of High Natural Radioactivity in China." *Proceedings of the 7th International Congress of Radiation Research.* C1-20.

__________, G.Z. Xu, X.Y. Li, F.C. Xia, H.Y. Yu, J. Qi, F.L. Wang, and S.K. Wang. 1985. "A Restudy of Immune Functions of the Inhabitants in High Background Radiation Area in Guangdong." *Chinese Journal of Radiological Medicine and Protection* 5:124.

__________, W.H. Liu, and J.B. Sun. 1987. "Radiation Hormesis—Its Expression in the Immune System." *Health Physics* 52:579–583.

Liu, W.H. and S.Z. Liu. 1986. "Stimulatory Effect of Low Dose Whole-Body Irradiation on Splenic PFC Reaction in Mice." *Chinese Journal of Radiological Medicine and Protection* 6:77.

__________. 1987. "Effect of Continuous γ-irradiation of Antibody Forming Cells of Mouse Spleen and a Preliminary Study of Its Mechanism." *Chinese Journal of Radiological Medicine and Protection* 7:336.

__________. 1990. "Effect of Low Dose Radiation on Subpopulations of Lymphocytes and T Cell Subsets in Mice." *Journal of Norman Bethune University of Medical Sciences* 16:1.

Luckey, T.D. 1980. *Hormesis with Ionizing Radiation.* Boca Raton: CRC Press.

Möller, G. and E. Möller. 1975. "Considerations of Some Current Concepts in Cancer Research." *Journal of the National Cancer Institute* 55:755.

Spärck, J.V. 1981. "The Immune System Is a Prerequisite of Cancer: The Hypothesis of Neoplastic Determinants." *The Immune System. 1. Past and Future.* (Eds. C.M. Steinberg and I. Lefkovits) Basel: S. Karger, 339.

Tuschl, H., R. Kovac, and H. Altmann. 1983. "UDS and SCE in Lymphocytes of Persons Occupationally Exposed to Low Levels of Ionizing Radiation." *Health Physics* 45:1.

Upton, A.C. 1961. "The Dose-Response Relation in Radiation Induced Cancer." *Cancer Research* 21:717.

Wang, Z., C. Huang, R. Zhang, S. Shao, and B. Yang. 1984. "Analysis of Chromosome Aberrations in Medical Diagnostic X-Ray Workers in China." *Chinese Journal of Radiological Medicine and Protection* 4:46.

Webster, E.W. 1981. "On the Question of Cancer Induction by Small X-Ray Doses." *American Journal of Roentgenology* 137:647.

Zhai, S., X. Lin, T. Pan, W. He, R. Feng, M. Chen, S. Li, L. Chen, R. Li, and H. Yie. 1982. "An Interim Report on the Survey of Mortality from Malignant Tumors in High Background Area of Guangdong." *Chinese Journal of Radiological Medicine and Protection* 2(2):48.

EIGHT

NUCLEAR WASTE DISPOSAL

THE NATURE OF THE PROBLEM

Jerry J. Cohen

Introduction

In ancient times, myths (beliefs not necessarily based upon fact) became embedded in the folklore of a culture over long periods of time by passing from generation to generation. Often such myths were embellished and amplified with each passage. Laws and rules governing society, such as the witchcraft laws in colonial America, were predicated on such beliefs since they came to be regarded as fundamental truths. Today, in the age of mass communication, myths can become established far more quickly. The advent of science during the last few centuries may have had a mitigating effect on adherence to mythology, particularly in modern societies, but this is by no means always the case. The folklore regarding nuclear waste presents a particular case in point where beliefs, not supported by science and logic, have played a major role in the development of our policies, rules, and laws.

The two major beliefs implicit in deriving our nuclear waste policies are:

(1) Nuclear waste is extremely toxic. Its disposal poses a danger of immense magnitude. The requirements for its complete isolation from the biosphere suggest that escape of the most minute quantity from isolation would pose an extremely serious threat to the environment and the health and safety of anyone who might become exposed to it.

(2) Because of the long half-lives of many component radionuclides, the disposal of nuclear waste presents mankind with a threat of unprecedented duration. For example, nuclear waste disposal has been characterized as a 250,000-year problem (ten times the half-life of plutonium-239). Such a problem is considered unique since human institutions are simply not capable of coping with problems that persist for unimaginably long time periods.

The Nuclear Waste Policy Act (NWPA) of 1982 embodied both of these concerns in defining high-level waste (HLW) as "highly radioactive material resulting from the reprocessing of spent nuclear fuel" and "other highly radioactive material that . . . requires permanent isolation." An understanding of the general character of concern related to nuclear waste can be gained from sampling views on the subject expressed in the literature. For example, Nader and Abbotts state:

> Storage of nuclear waste is much more than a problem of technology. Safe storage requires stable geological formations, a guarantee which is beyond the promise of technology. Safe storage also requires the development of stable human institutions to exist for thousands of years to prevent the waste from leaking and contaminating the biosphere. It should be remembered that Neanderthal man appeared "only" about 75,000 years ago.

A long period of isolation from the biosphere also is emphasized by the American Physical Society in its report on nuclear fuel cycles and waste management. As summarized in *Physics Today,* this report points out:

> Nuclear waste contains fission products and long-lived transuranic elements. Hundreds of thousands of years may be required for adequate isolation from the biosphere. The long time scales have raised institutional as well as technical questions about the viability of nuclear waste management.

The Clergy and Laity Concerned (CALC) similarly stated:

> There is no known way to dispose of radioactive wastes safely. Because they last so long (plutonium 239 alone remains dangerous for 250,000 years—longer than the existence of all civilization), the dangers don't diminish, they persist and grow.

Finally, Alvin Weinberg, former director of the Oak Ridge National Laboratory, described the problem in the following terms:

> We nuclear people have made a Faustian Bargain with society. On the one hand we offer . . . an inexhaustible source of energy . . . But the price we demand of society for this magical energy source is both a vigilance and a longevity of our social institutions that we are quite unaccustomed to.

The above quotations are a small sampling of a ponderous amount of literature supporting the perception that nuclear waste management presents us with a problem of unprecedented dimensions. A comprehensive summary of viewpoints on the nuclear waste problem may be found in a review document (Cohen *et al.*, 1983) compiled by the Lawrence Livermore National Laboratory (LLNL). From this review, it is apparent that the negative views toward nuclear waste management are long on rhetoric and short on quantification. Although it is clear that such views are deeply felt, they are generally not based on scientific rationale.

It can be conservatively estimated that to date, hundreds of millions of dollars have been spent in this country alone, in an attempt to develop an acceptable solution to the nuclear waste disposal problem. Current plans call for the expenditure of billions more. Yet, in the minds of the vast majority of the public, the problem is perceived as intractable and evokes great apprehension. The suggestion for the siting of a nuclear waste repository at any specific location stimulates fierce opposition. The so-called NIMBY (not in my backyard) syndrome has become a fixture in our society. National efforts to develop the first nuclear waste repository seem to be all but paralyzed (Marshall).

In this atmosphere, any suggestion that the high level of concern and apprehension related to nuclear waste might be technically and logically unfounded would appear to border on heresy. Yet, that is precisely the point that this paper will attempt to make. Stated succinctly, the potential hazard of nuclear waste is no greater than that of many other commonly accepted industrial activities in today's world and the concern related to its longevity (long half-life) is absurd when compared to current levels of concern related to use of stable toxic elements (e.g., lead, cadmium, mercury) which last forever. The next section of this paper will review technical data on the comparative hazards of nuclear waste which indicate that the problem is far from serious. The last section will explore reasons on how the disparity between actual risks of nuclear waste and their perception evolved.

Hazard Assessment

As previously discussed, the major concerns related to nuclear waste management can be expressed in terms of hazard and longevity. These concerns may be paraphrased as follows: First, nuclear waste is extremely toxic. The radioactive waste from a single nuclear reactor is enough to poison the entire population of the world several times over. It could cause malignancy and other diseases to exposed populations and genetic defects to their descendants. Second, because of the extremely long half-life of plutonium and some of the other components, its toxicity will persist for thousands and perhaps millions of years.

Both of these statements are true. However, when viewed in a different perspective, they lose their specter of severity. For example, a valid analogy to the first statement would be the observation that considering such items as cleaning compounds, pesticides, and other chemicals, there is enough toxic material in the average supermarket or hardware store to poison everyone in the community, if not the entire state. The problem has been one of confusing toxicity with hazard. The mere existence of a toxic substance does not constitute a hazard, unless that substance is readily available for dissemination and assimilation in the human body.

Consider, for example, that the lead used in manufacturing automobile batteries in this country each year is also sufficient, if properly distributed, to poison the entire world population. Although long half-lived radionuclides in radioactive waste may persist for centuries or millennia, lead, being a stable element, will exist forever. In addition, lead is also a carcinogen and a mutagen. Nevertheless, lead in automobile batteries is not generally considered to be a serious environmental threat, simply because of its low availability for human assimilation. The annual production of lead in this country, if administered by ingestion, would be sufficient to kill far more people than the annual amount of plutonium produced under the most ambitious nuclear power production program conceivable.

Similarly, one can calculate that the annual emission of lead from auto exhausts in this country, if properly distributed, also would be sufficient to kill a large fraction of the world population. However, even in this readily dispersed and available form, the effects have not been all that catastrophic. They have been of concern, but certainly not catastrophic, simply because the exhausted lead is naturally dispersed in such a way that only a minuscule fraction ever finds its way into human tissues. In comparison, the nuclear waste situation is neither unique nor unusual in its potential for hazard.

A difficulty in assessing the degree of hazard associated with nuclear waste (or any other potential danger) is in determining a meaningful quantitative measure of its severity. To measure how dangerous something is, what "yardstick" do we use? In the previous chapter, the number of lethal doses was used (i.e., by dividing the amount of material in question into equal quantities, and determining the maximum number of lethal doses to humans, what could result?). Using this same approach, Cohen (1976) summarized the potential hazard for various toxic materials produced annually in the USA and compared the number of lethal doses to that of the waste produced from the annual production of nuclear power. The results are shown in Table 8-1. From these data, it would appear that the potential hazard from nuclear waste is not remarkable. Although the "number of lethal doses" approach to hazard assessment may be technologically crude, it is certainly preferable to judging the magnitude of the problem by the number of articles on the subject in the news media or severity of rhetoric contained in them.

Table 8-1: Annual U.S.A. Production of Some Toxic Materials

Material		*Number of Lethal Doses*
• Via Inhalation		
Chlorine		4×10^{14}
Phosgene		2×10^{13}
Ammonia		6×10^{12}
Hydrogen Cyanide		6×10^{12}
Nuclear Waste	10 yr. old	2×10^{11}
	500 yr. old	5×10^{10}
• Via Ingestion		
Barium		9×10^{10}
Copper		8×10^{10}
Arsenic		1×10^{10}
Lead		4×10^{9}
Nuclear Waste	10 yr. old	3×10^{10}
	500 yr. old	1×10^{7}

Number of lethal doses represented by the annual USA production of some toxic materials. From Cohen (1976).

Perhaps a more technically descriptive (and accurate) method of expressing degree of hazard is the hazard index (HI). The HI is the quotient of the quantity of hazardous material in question divided by the concentration limit considered acceptable for public drinking water supplies. It can be expressed in units of volume and is indicative of the volume of water required to dilute a given quantity of toxic material to a concentration level considered safe for drinking. As the potential hazard increases, larger volumes of water will be required for dilution to safe levels. The index can, therefore, serve as a measure of relative toxicity. Such indices have been applied in several previous studies on nuclear waste (Haug; Hamstra; Clairborne; and Gera and Jacobs, 1971), all of which indicate the potential danger is not particularly severe.

Other useful insights may be derived from application of the hazard index. For example, the consequence of nuclear waste disposal has been characterized as "so dangerous that it threatens the very existence of life on this planet" (Abalone Alliance). In this regard, it should be noted that underground burial of nuclear waste would not be the first introduction of toxic materials into the geologic structure of the earth. Indeed, nature already has incorporated considerable quantities of toxic minerals in the earth's crust. Table 8-2 lists the estimated crustal abundance of naturally occurring toxic elements along with the hazard index for each. Table 8-3 provides analogous data for waste resulting from nuclear power production. From these data, assuming the accumulation of nuclear waste from 100 million megawatt-years of power production (1,000 power reactors each rated at 1,000 MWe capacity and all operating for 100 years) were buried deep in the earth's crust, the net hazard index of the earth's crust would increase by a factor of 1/100 of 1 percent for 10 year-old waste and only 1/10,000,000 of 1 percent after 1,000 years of decay. This would not appear to be a major perturbation on the earth's content of toxic materials. However, it could be argued that this assessment is inappropriate since toxic minerals have a widespread distribution while

Table 8-2: Crustal Abundance of Toxic Elements

Toxic Element	*Crustal Abundance (GM)*	*Hazard* Index (m^3)*
Arsenic	1.0×10^{20}	2.0×10^{21}
Barium	8.6×10^{21}	8.6×10^{21}
Cadmium	3.6×10^{18}	3.6×10^{20}
Chromium	4.0×10^{21}	8.0×10^{22}
Lead	3.2×10^{20}	6.4×10^{21}
Mercury	1.0×10^{19}	5.0×10^{21}
Selinium	1.8×10^{18}	1.8×10^{20}
Uranium	4.0×10^{19}	3.5×10^{17}
		Total 10^{23} m^3

Crustal abundance in grams of some toxic elements and the corresponding hazard index (HI), or volumes of water in cubic meters required to dilute them to a level safe for drinking.

* Hazard Index: Volume of water required to dilute indicated quantity of material to level of safe drinking water standards.

Table 8-3: Hazard Index of Nuclear Waste Produced from 10^8 MWe-Yr* of Power Production

Age of Waste Waste (Yr)	*Hazard Index (m^3)*
100	7.0×10^{16}
10^3	5.0×10^{13}
10^4	1.0×10^{13}
10^5	4.0×10^{12}
10^6	1.0×10^{12}

Same as Table 8-2 for nuclear waste produced from 100 million MWe-years of power production.
* One thousand nuclear power reactors each operating for 100 years and each capable of generating 1000 MWe.

nuclear waste would be concentrated in only a few repositories. Nonetheless, such comparisons can be considered valid in light of the fact that toxic elements in the earth's crust are also found in concentrated form in mineral ore bodies. Take for example, Figure 8-1, from the environmental impact statement on the Management of Radioactive Waste (DOE/EIS-0046F) which shows how the relative hazard index of nuclear waste compares to that of certain toxic ore bodies. In this assessment, the relative hazard index for a typical (0.2 percent) uranium ore body (from which the nuclear fuel was extracted in the first place) is assigned a value of unity, and the values for other toxic ore bodies are proportioned to it. Even relatively fresh nuclear waste (when it is at its most toxic level) poses no greater hazard than a mercury ore body. After a few thousand years of decay, the relative hazard of the waste becomes equivalent to that of a uranium ore body from which the nuclear fuel was originally extracted. A more comprehensive survey on the relative hazards of nuclear waste and toxic minerals may be found in Tonnessen and Cohen.

Another insight that may be gained from the analysis in Figure 8-1 is that after a few thousand years, the potential hazard from the nuclear waste becomes *less* than that of the original uranium ore body from which it was extracted. Other

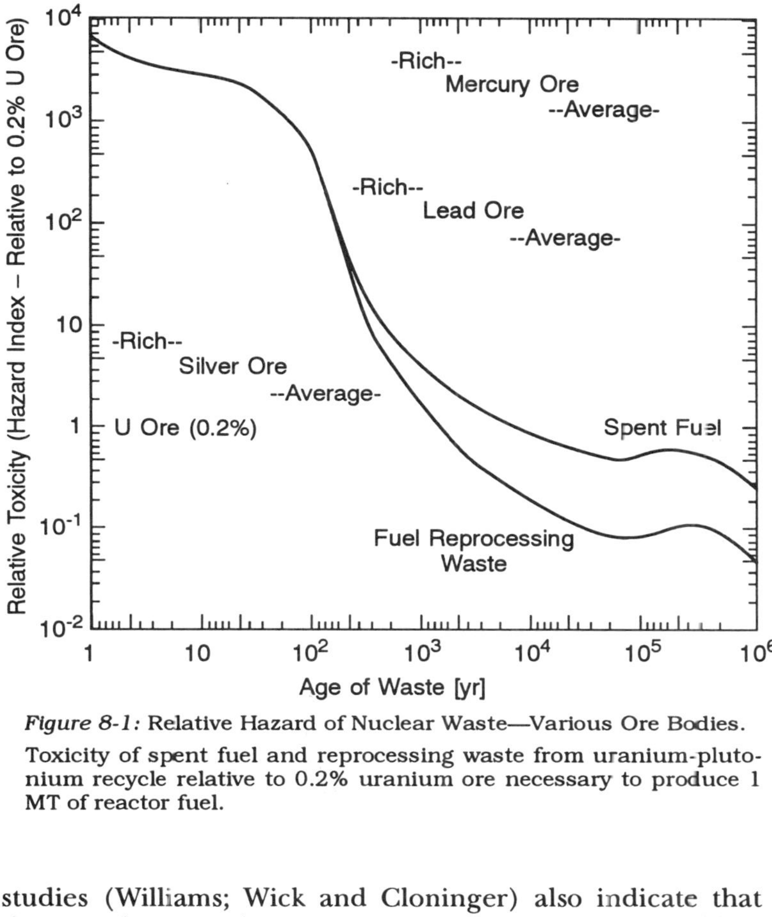

Figure 8-1: Relative Hazard of Nuclear Waste—Various Ore Bodies. Toxicity of spent fuel and reprocessing waste from uranium-plutonium recycle relative to 0.2% uranium ore necessary to produce 1 MT of reactor fuel.

studies (Williams; Wick and Cloninger) also indicate that a deep geologic nuclear waste repository poses comparable (or less) hazard than do typical uranium ore deposits. Indeed, considering extended time periods (beyond a few millennia), nuclear fission power production has been characterized as a means of "cleansing the earth of radioactivity" (Cohen, 1977). To gain further insight, consider the analysis presented in Figure 8-2. This graph follows the hazard index of one kilogram of natural uranium under two possible scenarios. In the first scenario (solid line), it is assumed that every uranium atom in

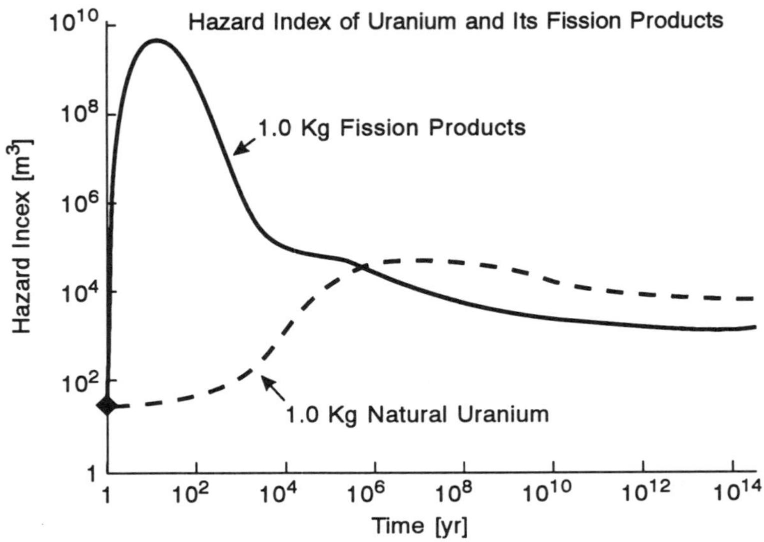

Figure 8-2: Hazard Index of Uranium and Its Fission Products. Hazard index (HI) of one kilogram of natural uranium and its natural decay products as a function of time if left undisturbed in the ore body (dashed line) and if completely fissioned in a reactor (solid line). Note that stable end products of the undisturbed ore are more toxic than those of the fission product decay.

the kilogram of material is fissioned. The initial hazard index of the fission products becomes very high since extremely radiotoxic nuclides such as strontium 90 and cesium 137 are included. With time, these fission products decay and the hazard index decreases accordingly as shown. Eventually, there is ingrowth of the stable end products of radioactive decay, including such elements as bismuth and arsenic. Note that the material never becomes totally innocuous since the stable end products are also toxic.

In the second scenario (dotted line), it is assumed that the kilogram of natural uranium remains intact and undisturbed in its ore body. In time, with the radioactive decay of the uranium isotopes, we observe the effects of the ingrowth of radium, a radionuclide whose unit toxicity is significantly greater than the parent uranium. In about one million years, a condition of secular equilibrium is reached where one curie of radium exists for each curie of uranium remaining. Eventually, after a few billion years, the entire system decays to stable lead, which itself is very toxic and significantly more toxic than the stable end products of fission product decay. The crossover of the two curves at about one million years indicates that there is some validity to Cohen's statement about the cleansing of the earth. It all depends upon the period of time over which one chooses to view the problem.

Again, it might be argued that the hazard index approach is simplistic and inappropriate since it considers only the inherent toxicity of the material but not its relative environmental mobility, its biological availability, or its potential for eventual assimilation in human tissue. To rectify this deficiency, Smith and his associates at LLNL developed the Geotoxicity Hazard Index (GHI) which considers all of the above factors, in addition to the environmental persistence of the hazardous material (Smith *et al.*). The GHI provides an index related to the overall hazard from either naturally occurring or man-made toxic materials buried underground. Figure 8-3 shows the results of applying the GHI to one cubic meter of various materials, including ten- and one hundred-year-old nuclear waste. This assessment also indicates that nuclear waste presents no unique or unusual hazard. Toxic elements in mineral form are at least as likely to be leached and transported as vitrified nuclear waste. For example, Goldberg estimates that over thirty million kg of lead and ten million kg of arsenic are leached from their natural mineral state in the 48 contiguous States annually finding their way to surface waters and eventually discharging into the ocean.

The accumulation of toxic elements in the world's ocean waters over geologic ages has resulted in a considerable pool of natural "pollution." Table 8-4 presents the oceanic burden of certain toxic elements, along with the associated toxicity index. Considering only these data, it can be calculated that a uniform distribution of the dissolved one thousand-year-old nuclear waste from the production of one hundred million

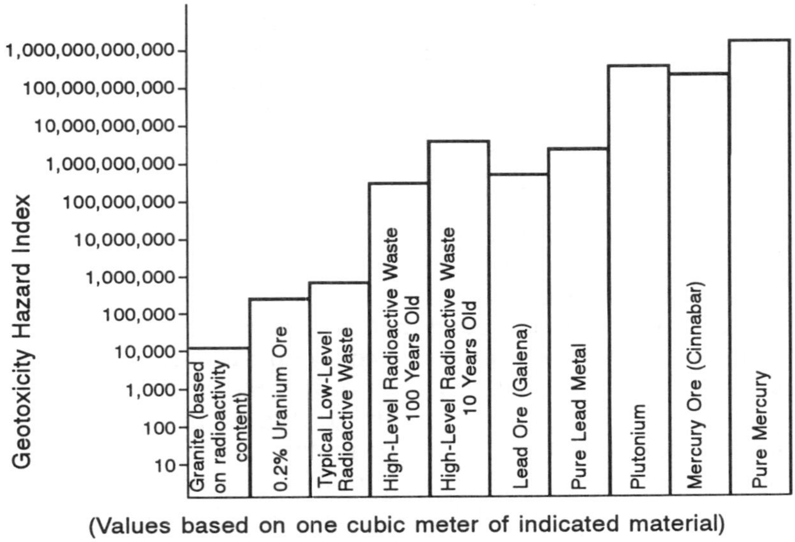

Figure 8-3: Relative Hazard of Various Materials Buried Underground. (Based on Ingestion Toxicity). Relative hazard index (HI) for one cubic meter of various materials buried underground based on the ingestion toxicity.

megawatt-years of electricity into oceanic waters (certainly a "worst case") would increase the average oceanic toxicity by 1/100 of 1 percent. It is truly unfortunate that extreme environmental concern (sometimes bordering on hysteria), related to oceanic disposal of radioactive waste, has precluded serious consideration of this alternative. Systematic study (Cohen, 1983) has shown that oceanic disposal would likely provide the safest method for disposing nuclear waste, In addition, it would be the least expensive and most readily implemented.

The discussion to this point has focused on the relative hazard of nuclear waste and has attempted to show that this hazard poses neither a unique nor an unusual threat. The second major concern, related to the long half-life components of nuclear waste, also can be shown to be untenable for two basic reasons: First, it would seem inconsistent, if not illogical,

Table 8-4: Toxic Elements in the World's Oceans

Toxic Element	*Oceanic Burden (GM)*	*Hazard Index (M^3)*
Arsenic	8×10^{15}	1.6×10^{17}
Barium	4×10^{16}	4.0×10^{16}
Lead	3×10^{15}	6.0×10^{16}
Mercury	3×10^{14}	5.0×10^{16}
Selenium	3×10^{13}	3.0×10^{17}
		6.1×10^{17}

Total burdens of toxic metals in the ocean in grams and the corresponding hazard indices (HI) in cubic meters of water required to dilute to standard for drinking water.

to worry about plutonium with its 24,000 year half-life and not devote equal or far greater concern to the stable toxic elements such as lead, cadmium and mercury, which will persist forever. One possible explanation for this disparity is an apparent mystique surrounding plutonium, that apparently can engender special fears. For example, in an early draft of its policy statement on nuclear energy, the National Council of Churches labeled plutonium "an intrinsically evil element." However, in light of scientific analysis, plutonium does not seem to live up to the sinister image portrayed in the public media and reflected in related policies and regulations. For example, via ingestion (the predominant pathway of concern associated with geologic disposal of nuclear waste), plutonium, per unit of radioactivity, is two thousand times less toxic than radium, which is ubiquitous in nature. Of course, plutonium also can be considered ubiquitous since over ten tons of this material were disseminated throughout the world during the period of atmospheric nuclear weapons testing. No discernibly adverse biological effects have been observed as a result of them. This is understandable considering, for example, that the oceanic inventory of natural radium is about eighty million tons, which is equivalent in ingestion radiotoxicity to one hundred billion tons of plutonium. For the sake of perspective, it also should be noted that about two hundred fifty tons of plutonium would be produced by one hundred million megawatt years of nuclear power generation.

Second, a basic radiation physics principle indicates that special concern related to long half-life radionuclides is unwarranted. The specific activity of any radioactive isotope is inversely proportional to its half-life. Simply stated, the longer its half-life, the less radioactive it is! A recent study (Cohen *et al.*, 1986) noting this problem proposed a semantic solution. It suggested that only radionuclides with a half-life of less than one million years be called radioactive. Those with half-lives greater than one million, but less than one trillion years could be termed radiopassive and those with half-lives exceeding a trillion years, would be radioquiescent. Such a classification system might tend to counter perceptions such as that expressed by W.D. Rowe, former head of the EPA Office of Radiation Programs, who stated: "We feel that some of the long half-lived materials such as plutonium can indeed be very dangerous because of their half-lives." It is somewhat of an enigma why half-life, per se, is equated with danger; yet, this concept seems to be well-embedded in current mythology.

A manifestation of this phenomenon can be found in the extensive research related to evaluation of iodine 129, a major component of nuclear waste with a half-life of fifteen million years (Kocher; McKay; Oztunali and Petschk). This research was likely motivated by concern over the extremely long half-life of the radionuclide. Yet, it has been calculated (Cohen *et al.*, 1986) that if every atom of iodine in the human body were iodine 129 (an absurdly conservative assumption), the resultant whole-body equivalent dose would be less than 600 mrem/yr, close to the allowable limit for public exposure to radiation. This should not be surprising, considering that iodine 129 is barely radioactive in the first place.

From the material discussed in this section, it would appear that the perceived hazard related to nuclear waste is well out of proportion to actual levels of danger likely to be posed. Possibly more interesting than assessing the hazards of nuclear wastes is a study of how the wide disparity between actual and perceived hazard came to exist. This topic is dealt with in the next section.

Perceptions and Policies

Previous discussion supports the concept that problems related to nuclear waste management and disposal are not nearly as serious or intractable as is commonly believed. Yet, this concept itself would appear to be incredible. One might reasonably ask: If nuclear waste were in fact a relatively minor problem, why then after decades of research costing hundreds of millions of dollars, have we no accepted solution? Why is a multi-billion-dollar program pursuant to the NWPA necessary to develop a nuclear waste repository? Could the decision-makers be unaware of the data and insights presented in the previous section? Certainly that information was available in the open literature. Could it be that they are aware of this material but have chosen to ignore it in the formulation of policies? Surely it would be reasonable to conclude that if, in fact, nuclear waste disposal posed no serious technological problem, current plans and policies constitute a gross "overkill." Could it be true that, as S. Fred Singer concluded: "Nearly everyone agrees privately that safe disposal of spent fuel or other high-level radioactive material is not a technical problem, but a political one?" Questions such as these are, in my opinion, by far the most interesting and the most important to consider in gaining an understanding of the nuclear waste issue. If one accepts the view apparently held by the majority of scientists working in the nuclear waste field that public apprehension regarding the problem is grossly exaggerated, then it is reasonable to ask how this condition came to exist. How did the myth evolve?

The causes for the current climate of apprehension toward nuclear waste are certainly complex and would likely involve political, social, and psychological as well as technical factors. I have pondered the situation at some length and have thought it would be fascinating to perform an anthropological study on the various groups involved with the nuclear waste problem similar to the studies done on primitive and advancing societies. Unfortunately, I do not have the expertise to do such a study. Nonetheless, from several years of personal experience and association with others working in nuclear waste research, I have formed certain impressions.

To explore the evolution of nuclear waste mythology, it can be useful to look at the role that various groups have played in influencing public opinions and policies. Probably the most vocal and outspoken of these groups are the so-called environmentalists, represented by such organizations as Friends of the Earth and the Abalone Alliance. Frequently, members of the technical community have accused these groups of exaggerating the nuclear waste problem beyond all reasonable bounds and needlessly inflaming public fears in their zeal to preserve what they envision to be a pristine environment. Others have described them as "coercive utopians" (Metzger) and advocates of halting all technical progress for which the development of nuclear power provides them with a rallying point. Although the environmental groups are seen as a potent force in today's society, I, for one, do not believe they have been nearly as effective in influencing either public opinion or government policies as my colleagues in the technical community give them credit for. The vast majority of the general public has the sense to see these "environmentalists" for what they are and takes this into account.

The news media also have been accused of playing a nefarious role in needlessly inflaming public fears with frequent horror stories on the consequences of nuclear waste disposal. To become a hero to the media, it seems that all one needs to do is view with alarm any real or imagined environmental problem. It matters little whether the frightening story has any basis in fact or logic. Fear sells. It is certainly the job, if not the duty, of the news media to bring any real or potential health and safety threat to the public's attention. They do not have capability or responsibility to evaluate the technical validity of every news item that comes to their attention. They do have the responsibility to remain financially solvent by selling newspapers or maintaining high viewer ratings. Since sensational or fearful news is more likely to attract attention, the media has a natural and understandable bias toward such stories. In reporting them, media is just "doing its thing" and, I believe, the majority of the public has the sense to take this fact into consideration.

In evaluating the role that the public itself has had in nuclear waste developments, I cannot agree with my colleagues in the technical community who disparage non-technical people for their failure to understand the problem. Regarding nuclear waste, the public has been rational. Their fears and apprehensions are understandable, given the information available to them. Discounting, for the moment, the media horror stories on the subject, suppose that one totally unfamiliar with the subject were to gain their entire knowledge from a review of government laws, regulations and policies related to nuclear waste. Suppose that in addition they were to review all of the government-sponsored research on the subject and evaluate the enormous budgets expended on that research. The only conclusion that one could arrive at by rational inference is that nuclear waste: (1) is hazardous beyond all comprehension, and (2) poses a threat of unprecedented duration. I can sympathize with the public's fear. Were I not technically conversant with the subject I would share that fear. It is unreasonable to expect the general public to have the same understanding of complex technical subjects as do scientists. They depend on guidance from the technical community to gain that understanding.

In viewing the role of the technical community who perform the nuclear waste research along with the political officials who provide the necessary taxpayer funds and the bureaucrats who administer those funds and write the policies, I am reminded of a quotation from the comic strip Pogo, "We have met the enemy, and he is us."

To solve the problem in this country we have managed to set up a strong, heavily funded nuclear waste establishment of scientists and bureaucrats. We have simultaneously given that establishment a strong disincentive to solve the problem. Previously in this paper, arguments were presented indicating that from a technological standpoint, the nuclear waste problem was relatively minor. Obviously, it is difficult, if not impossible, to obtain huge levels of funding to support research on relatively minor problems. Since self-perpetuation is a fundamental characteristic of bureaucracies, especially of those with giant budgets, there seems little incentive for the beneficiaries to discourage the perception that nuclear waste management is anything but a serious and complex problem that requires extensive research and development efforts. It is difficult to be optimistic that an acceptable solution to the waste management problem will be found as long as those with the responsibility for solving it would find it to their disadvantage to do so.

I am not implying that all of the individuals and organizations involved in nuclear waste research are sinister and would place their self-interest above the public welfare. Perhaps, in some cases that description might be appropriate; however, for the most part I believe those scientists and managers in nuclear waste are basically trying to do the best job they can within a system that appears to have been designed for failure.

What caused this sorry state of affairs? In considering this question, I am reminded of an idea that Leonard Sagan of the Electric Power Research Institute facetiously "proposed" several years ago. In addressing the "Environmental Crisis," Sagan, with tongue in cheek, suggested we recycle human waste by universally applying a medical procedure in which a connection is established between the rectum and the stomach—a single closed loop system to reduce human pollution and assure recycling. He then went on to conjecture on the reaction of various groups:

> Government regulatory agencies concerned that the primary loop might fail will demand a second loop—the so-called "double loop system."
>
> Conservationists will declare that the proposal has no merit at all since small releases will continue to occur. They will demand zero release.
>
> Lawyers will feel that the issue could only be resolved in court, particularly since the surgeons themselves are not free of taint.
>
> Industry will respond with full-page ads which point out that the small amounts to be released from the loop are far less than the background levels to which we have always been exposed, particularly in families which own cats and dogs.
>
> Public relations men will propose painting the effluent white.
>
> Economists will argue that we can accomplish the same thing by raising dime toilets to a quarter.

The point Sagan wanted to make was this: In discussions of the environmental crisis, each discipline tends "to see its own expertise as providing solutions superior to all others." Thus, instead of a cooperative effort by interested parties, there has developed an atmosphere of suspicion and hostility.

An early and significant milestone in the evolution of our waste management policies occurred in the mid-1950s with the report of a National Academy of Science panel recommending the geologic disposal of nuclear waste in salt deposits. Since that panel was composed predominantly of geologists, it was apparently pre-ordained that nuclear waste was to be a geological problem. It remains that way to this day. Singer (1985) appropriately describes the Nuclear Waste Policy Act of 1982 as being a "geologist's full employment act." He further asserts that this legislation, ". . . is known irreverently as the Nuclear WPA—and for good reasons. NWPA may be the largest public-works program ever foisted onto the American public by Congress. Unlike cross-country canals, flood control, and other water-management projects, there is not even a useful output here. Just one hundred billion dollars (or so, depending on inflation), spent over 25 years, with a bunch of people digging deep holes in the ground and another bunch filling them in."

After the geologists staked their claim on this valuable territory, the materials scientists came onto the scene proclaiming the need for development of every more stable waste forms to assure that there would be no dissolution of the buried deadly waste in groundwater. Later, the systems analysts boarded the "gravy train" and applied increasingly more complex calculational models to predict the radiological consequences of these geologically-isolated, chemically-stabilized, nuclear wastes in future millennia. The results of these calculations generally indicated those consequences would be minuscule; however, because of the vast uncertainties in their models and calculational assumptions, it was usually concluded that further reinements in the modeling process were required. Commenting on the nature of expertise being brought to bear on the nuclear waste problem, Ida Hoos observed, "If the only tool one has is a hammer, all problems tend to be treated as if they were nails."

Topping the melange of technical talents working the nuclear waste problem are battalions of government bureaucrats who administer the taxpayers' and ratepayers' financial contributions to the effort. Finally, orchestrating the whole show are the politicians applying their well-known wisdom and judgement to assure that the public is well-protected against the dire threat that nuclear waste poses. Do you ever wonder why, after three decades of intensive research and the expenditure of a significant part of our national wealth and resources on the nuclear waste problem, we seem no closer to an accepted solution than we were in the beginning? Perhaps the insights brought out in this paper will provide some clue as to the answer.

It somehow seems appropriate to close this discussion on the nature of the nuclear waste problem with an excerpt from Alice in Wonderland:

> "I was wondering what the mouse-trap was for," said Alice. "It isn't very likely there would be any mice on the horse's back."
>
> "Not very likely, perhaps," said the Knight; "but, if they *do* come, I don't choose to have them running all about."
>
> "You see," he went on after a pause, "it's as well to be provided for *everything*.
>
> That's the reason the horse has all those anklets around his feet."
>
> "But what are they for?" Alice asked in a tone of great curiosity.
>
> "To guard against the bites of sharks," the Knight replied. "It's an invention of my own . . . "
>
> Lewis Carroll
> "Through the Looking Glass"

I sometimes conjecture that an untold part of this tale was that much of the wealth of the realm was spent to study the properties of the anklet metal to assure it would be impenetrable to shark teeth. So successful was this effort, that not one horse was afflicted by shark bite throughout the entire kingdom, even those horses that didn't wear them. Unfortunately, half the horses in the kingdom starved to death because there were no research funds left to study methods for the prevention of oat blight.

References

Abalone Alliance. 1979. "The Leaky Nuclear Fuel Cycle." *Radioactive Times.* San Francisco: Waller Press.

Clairborne, H.C. 1975. "Effect of Actinide Removal on the Long-Term Hazard of High Level Waste." Report ORNL-TM-4724. Oak Ridge National Laboratory.

CALC (Clergy and Laity Concerned). 1980. *Threats to Human Survival.* (Pamphlet available from CALC, 198 Broadway, New York, N.Y. 10038.)

Cohen, B. 1976. "Environmental Impacts of High Level Radioactive Waste Disposal." *IEEE Transactions on Nuclear Science* N5-73(1):56

__________. 1977. "High Level Radioactive Waste from Light Water Reactors." *Review of Modern Physics* 49:1.

__________. 1983. "Ocean Dumping of High Level Waste—An Acceptable Solution We Can Guarantee." *Nuclear Technology* 47:163.

Cohen, J.J., C.F. Smith, and F.J. Ciminesi. 1983. "Nuclear Waste—A Perspectives Document." Report UCRL-15530. Livermore: Lawrence Livermore National Laboratory.

__________, C.F. Smith, and A.J. Toy. 1986. "Characterizing the Risk of Long-Lived Radionuclides in Waste." *Transactions of the American Nuclear Society* 52:37.

DOE/EIS-0046F. 1979. "Final Environmental Impact Statement on Management of Commercially Generated Radioactive Waste." Washington D.C.: U.S. Department of Energy.

Gera, F. and D.G. Jacobs. 1971. "Hazard Potential of Radioactive Wastes." In *Proceedings of an International Conference on Radioecology Applied to Protection of Man and His Environment. EUR-4800.* Rome.

__________ and D.G. Jacobs. 1972. "Considerations in Long-Term Management of High-Level Radioactive Wastes." Report ORNL-4762. Oak Ridge National Laboratory.

Goldberg, E.D. 1979. "Assimilative Capacity of U.S. Coastal Waters for Pollutants." Washington D.C.: U.S. Department of Commerce.

Hamstra, J. 1975. "Radiotoxic Hazard Measure for Buried Solidified Radioactive Wastes." *Nuclear Safety* 16(2):

Haug, H.O. 1977. "Some Aspects and Long-Term Problems of High-Level and Actinide-Contaminated Spent Fuel Reprocessing Wastes from the Nuclear Fuel Cycle." Report ORNL-TR-4302. Oak Ridge National Laboratory.

Hoos, I. 1978. "Statement in Proceedings of a Public Forum on Environmental Protection Criteria for Radioactive Waste." Report ORP/CSD-78-2. Washington D.C.: U.S. Environmental Protection Agency.

Kocher, D.C. 1979. "A Dynamic Model of the Global Iodine Cycle for the Estimation of Dose to the World's Population from Releases of Iodine-129 to the Environment." Report ORNL/NUREG-59. Oak Ridge National Laboratory.

Marshall, E. 1986. "Nuclear Waste Program Faces Political Burial." - *Science* 233:835.

McKay, H.A.C. 1980. "Background Considerations in the Immobilization of Volatile Radionuclides." Report IAEA-SM-245/8. Vienna: International Atomic Energy Agency.

Metzger, H. Peter, 1979. "The Coercive Utopians." Denver: Public Service Company of Colorado. (Originally presented at the National Meeting of the American College of Nuclear Medicine, 28 April 1978.)

Nader, R. and J. Abbotts. 1979. *The Menace of Atomic Energy*. New York: W.W. Norton and Co.

Oztunali, O.I. and K.D. Petschek. 1983. "Re-examination of Radiotoxicity of Iodine-129." *Waste Management* '83. Tucson: University of Arizona.

Physics Today. 1977. "The Nuclear Fuel Cycle: An Appraisal." *Physics Today* (October 1977):32–39.

Rowe, W.D. 1976. "What is EPA's Role in Radiation?" *EPA Journal* 2(2):14.

Singer, S.F. 1985. "Seeking Some Ways to Store Nuclear Waste." *Wall Street Journal* (March 29, 1985).

Smith, C.F., J.J. Cohen, and T.E. McKone. 1980. "A Hazard Index for Underground Toxic Material." Report UCRL-52889. Livermore: Lawrence Livermore National Laboratory.

Tonnessen, K.A. and J.J. Cohen. 1976. "Summary of Naturally Occurring Hazardous Materials in Deep Geologic Formations." Report UCRL-78746. Livermore: Lawrence Livermore National Laboratory.

Weinberg, A.M. 1972. "Social Institutions and Nuclear Energy." *Science* 177:27.

Wick, O.J. and M.O. Cloninger. 1980. "Comparisons of Potential Radiological Consequences from a Spent-Fuel Repository and Natural Uranium Deposits." Report PNL-3540. Richland: Battelle-Pacific Northwest Laboratory.

Williams, W.A. 1980. "Population Risks from Uranium Ore Bodies." Report EPA 520.3-80-009. Washington D.C.: U.S. Environmental Protection Agency.

EIGHT A

DISCUSSION OF COHEN'S PAPER

Alvin Weinberg: I agree with almost everything Jerry Cohen says in his paper.

I agree wholeheartedly that of the two technical issues that confront nuclear energy—one is the possibility of an accident and the other is the sequestering of the wastes. The sequestering of the wastes is essentially not a technical problem, in the sense that there is almost no plausible or even implausible, scenario that can impose a non-stochastic radiation dose on anybody.

There are a few quite implausible scenarios involving digging down into an aquifer that is very close to a high-level repository which conceivably could lead to a non-stochastic dose, but most people don't take these scenarios very seriously. I, therefore, agree with Jerry Cohen that wastes are not a technical problem—and yet they clearly are a problem.

Jerry Cohen concedes in his paper that waste disposal is a terrible problem even though it is not a technical problem. We're spending $500 million a year in the United States and we are farther from solving the waste problem then when we started. I disagree with Jerry when he says that it's all a put-on on the part of the people making money by being in the waste disposal business. The waste problem is one of perception.

Jerry did not bring out sufficiently in his paper, a point that was brought home to me in a conversation that I had just a month ago with Dr. Eklund, the former head of the International Atomic Energy Agency. He said to me: "Nuclear energy cannot survive in the western, democratic, open societies unless the attitude of people towards low-level radiation is changed, unless people recognize that 20 percent, 30 percent, even doubling the background, is not a very important thing." My main criticism of Jerry Cohen's paper is his failure to stress that the resolution of the waste issue depends on people recognizing that the consequences of exposures at levels which are of the order of the natural background are just not very important environmental questions.

Let me conclude with the following story: At the Oak Ridge National Laboratory we have had a good deal of success with a permanent method of sequestering waste, which is called hydrofracture. You drill a hole, about one thousand feet deep into horizontally-bedded strata. The strata are cracked by very high hydraulic pressure applied through a slot in a pipe encased in the drill hole. You then inject a grout (a liquid cement) containing the radioactivity to be sequestered. The radioactive grout spreads out, and it solidifies and stays there forever in the crack produced by the hydraulically-induced fracture. The Oak Ridge National Laboratory has been disposing of its low-level wastes in this way—about a million Curies over the past twenty years. About twenty of these hydro-fracture disposals have taken place during this time. The injections have been monitored by three or four monitoring wells in the vicinity. Earlier this year, radioactivity was discovered in one of those monitoring wells. The amount was not serious—it was of the order of a few Curies, but it was something that they had not expected. And you ask: Well, how could that have been when everything looked so perfectly clear-cut? You put the liquid cement down there, it's radioactive, it solidifies and it stays there for millions of years. Well, because in actual practice the situation is not as simple as that. For example, after you actually inject the grout you have to pour water down to clean out the thing. The grout is, therefore, diluted with a little water. The water, which is labile, doesn't have very much radioactivity, but people get very nervous.

The upshot of the matter is, the Department of Energy closed the operation down. Bang, just like that. You can argue until you are blue in the face that there is no possibility that anybody is going to be hurt by the hydro-fracture disposals, you can say that the few Curies that were measured are not very much—but you are then underestimating what the public's attitude toward any radioactivity is. And the public's attitude has nothing to do with people in the radioactive grout business wanting to hang onto their jobs!

I have, therefore, concluded that in an open democracy the only way that the waste disposal problem can be solved, even after people get over their unjustifiable concern about the effect of low-level radiation, is by massive payment—we are going to have to bribe people to take the waste. When I suggest bribing people I am often hissed at. All I can say is that a bribe that is large enough is a golden opportunity. (I understand that active measures are underway to legislate massive compensation—around one hundred million dollars a year—for a state that accepts the Monitored Retrievable Storage Facility, or the Permanent Waste Repository.)

Enders Robinson: When I went to college, no university student worthy of the name would treat astrology with anything other than contempt. It was the most disgusting, stupid superstition of 4,000 years. And now we face a society in which a large fraction of those with a university education will treat it, at the very least, semi-seriously.

Here we are talking about nuclear wastes. It is essentially a problem where, with even one thousand nuclear power plants, we are talking about a ton a year of high-level waste. Now if this isn't the most manageable, trivial kind of technological question. I don't care how dangerous it is. In comparison with the explosives that we deal with regularly, or with the toxic gases that we tote around and use for manufacture of wide ranges of chemicals essential to our use, I just cannot understand it. As I say, we are talking about a ton a year—*one ton each year*—of high-level waste. Don't tell me that this is some insuperable problem. The whole thing is a trivial matter.

Jerry Cohen: I had not heard of the termination of Oak Ridge's waste disposal by hydro-fracture. As you describe the situation, it was closed down by the DOE rather abruptly. I believe that in doing so, the DOE is, in effect, telling the public that this is really a dangerous practice. So it's natural that people would become concerned about it. After sinking millions of dollars in it, they abruptly stop the operation because of what they say is public concern. I maintain that's a self-fulfilling prophesy. In other words, such action stimulates concern.

A typical response of my colleagues researching the nuclear waste problem (when I kid them that I think some of their programs have no technical purpose) is: that they are aware that perhaps there isn't much technological merit, but they point out that I don't realize the level of concern of the public. Therefore, they have to complete whatever research they are doing in order to gain public confidence, regardless of the cost of that research. Now I subscribe that given that the attainment of public confidence is the stated objective of this effort, not only will such effort be ineffective, it will be downright counter-productive. If you want to scare the hell out of people, spend lots of money on research to solve a problem and do so very conspicuously. Such practice is particularly effective when they don't trust you in the first place. It's a very effective way of stimulating apprehension. As I see it, that's exactly what our government programs have been doing. If they were designed to stimulate public fear and concern then they were brilliantly conceived.

Jovan Jovanovich: I would like to make a couple of perhaps unconventional comments. Concerning the nuclear waste, I have taken the unconventional position; ideal technology doesn't exist. If we do want the electricity, we can get it either by producing fission products and transuranics (that you might call either resource or waste), or by producing carbon dioxide. So the question is, not whether I want fission products—of course, I don't want fission products. But I can't have it both ways. What do I want? Carbon dioxide or fission products? When the question is asked that way, it takes on a completely different dimension.

The second comment is: I'm really not convinced whether fission products and the transuranics are waste or resource. If we are talking about the economic system of the United States that is producing 15 percent of its electrical energy with nuclear electricity with no intention of processing the waste—or spent fuel, for fission products—then of course it is waste. But if we talk about active growth, which will use a lot more energy than today and which will have to run on nuclear power—because there is really nothing else—then we are talking about tens of thousands, maybe hundreds of thousands of nuclear reactors producing a lot of spent fuel which will have a lot of valuable materials in it—not only plutonium, which will be essential for breeders in the program, but it also will have some non-radioactive platinum and a group of other useful elements which are present in non-negligible quantities. Once you start processing the fuel, it's certainly possible to separate long-lived transuranics and put them back into the reactors and burn them—actually get rid of them in that way. That little bit of cesium 137 and strontium 90 and a few other elements that stay in may have to be kept somewhere for a few centuries but then the problem is solved. It's very unconventional, I agree, but that's the way it looks to me.

Max Singer: On this question of the rationality, of the fear and how to deal with it—I don't want to sound like I'm talking about a conspiracy—but, essentially, there is a way of thinking—a set of ideas, and habits; and there are people who have found out what's going on in the United States today and to some extent in the world. It has a political side and it has a side concerned with why we don't trust our government to carry some low-level wastes from here to there. There's a way of thinking, I call it low national morale—a lack of self-respect, first for the country, and to some extent for the species—the idea that rocks are equal to man, at least trees and streams. Whereas, a couple of generations ago, we believed that man was sacred and that the world was ours, to respect it for the long-term, but the measure of what was good was man. This whole pattern of thinking is what's being reflected, in a subtle way, in this rejection of nuclear energy of the government when it transports nuclear wastes. And that going along with these fears and finding justification for them—sympathy for them—without recognizing the pattern that they are coming from, is to forfeit the battle. Part of the evidence has to do with understanding political patterns; a part of the evidence has to do with technological inconsistency. The two things together are what is confirming. So that not recognizing this, being sympathetic with it and just going around it, I think, leads to the kind of pattern of governmental behavior and public behavior which builds up and doesn't deal with the problem.

That's why I think that what's called for is to have a positive philosophy or view of the world which you hold up against the false view of the world that is being propagated. It has to do with economic development and world poverty. It has to do with whether technology is basically a friend or an enemy. It has to do with whether the United States and the Soviet Union are morally equivalent. It has to do with a number of things. And I think that we have to have, and put before people, a realistic view of the world in all of its aspects rather than going along with the great mass of our intellectual media leadership which has got this other point of view which they are insidiously putting-in in all kinds of settings. In each case you have a question, do you sort of accept and quibble with it around the edges? or do you recognize what's behind it and confront it overall? It's important to understand this is an overall struggle of ideas, to recognize all of its manifestations and to have an overall approach to winning it because that's what this issue is concerned with, as well as many other issues.

Hugh Ellsaesser: Another aspect of this struggle of ideas which came to me while I was studying the problem of air pollution (Ellsaesser, 1971) is the problem of keeping the success goals of the individual aligned with the welfare of the species. The individual is going to do those things which he perceives as successful for him regardless of what it does for society. Somehow we have to align those success goals for the individual so that he works toward the welfare of the species and the planet rather than to their detriment, as sometimes happens now. I can think of two motivating factors which bother me, one of which is that we are getting people who find it profitable merely to complain. Ralph Nader comes to mind, but a better example is provided by the contingency fee lawyers who have restricted, or even eliminated, many beneficial voluntary efforts of the past by skyrocketing the cost of liability insurance.

Another thing that has happened that I am disturbed about is that scientific research today is funded virtually exclusively by the Federal Treasury. Now, if you want to get funds to research a particular problem, you have to convince Congress that it's a serious problem. You don't do things anymore because you are going to learn something. You do it because you're protecting the world from carbon dioxide, or from freons, or from ultra-violet radiation, or from something else.

NINE

POSSIBLE REPLACEMENTS FOR FOSSIL FUELS AND COMPARATIVE ENVIRONMENTAL ASPECTS

Jovan V. Jovanovich

Introduction

Shall the world always stay divided into rich and poor, into the affluent industrial and primitive agricultural? Will the societies that are presently undeveloped pass through their own industrial revolution and eventually catch up with high income countries of today? Can we imagine, sometime in the future, our globe with ten billion people and five billion cars?

In previous papers (Jovanovich, 1986a, 1986b; Jovanovich and Rotty), some statistical information was reviewed which showed that the whole world, developed and undeveloped, has been moving in the direction of global affluence. Will this process continue into the future? We, of course, do not know, but what we do know is that an essential prerequisite for the industrialization of the whole world is an abundant supply of energy, a supply that will not have adverse effects on the environment we live in and depend on.

How much energy will be needed when, and if, the world achieves this global affluence? Assuming that the energy consumption per person will stay at some "saturation" level, let us say at the same level as it is today in the United States and Canada, and that the world population will stabilize at ten billion people, the demand for energy will increase roughly tenfold over the present demand. (While it is possible to argue that the above assumed "saturation" level is reasonable, one can also present arguments that it is too low or too high.)

In Figure 9-1 are reproduced past and possible future commercial energy consumption curves (Jovanovich, 1986a). From 1925 to 1965, the data were taken from Darmstadter, and after 1965 from BP. In the 1920s, the industrially developed world, composed of western Europe, North America, Japan and Australia, consumed over 85 percent of all energy used in the world. These same countries, with some but not very significant exceptions, form today the Organization for Economic Cooperation and Development (OECD) and consume only 50 percent of the world energy. It should be noticed that the non-OECD world has been doubling its energy consumption every eleven years, increasing it by thirty times during the last fifty-five years. The lines extrapolating energy consumption into the future indicate that most of the increase in energy consumption is expected to come from the undeveloped part of the world and that this tenfold increase in energy demand may be reached as soon as half a century from now.

The object of this paper is to survey the existing and emerging energy technologies, the available energy resources and the impact such high consumption of energy would have on the environment. The time period under consideration will be approximately the next half-century.

The final conclusion reached is that there is no reason why the future industrialization of the world should be constrained by the lack of available energy supplies, or by the detrimental impact of energy technologies on the environment.

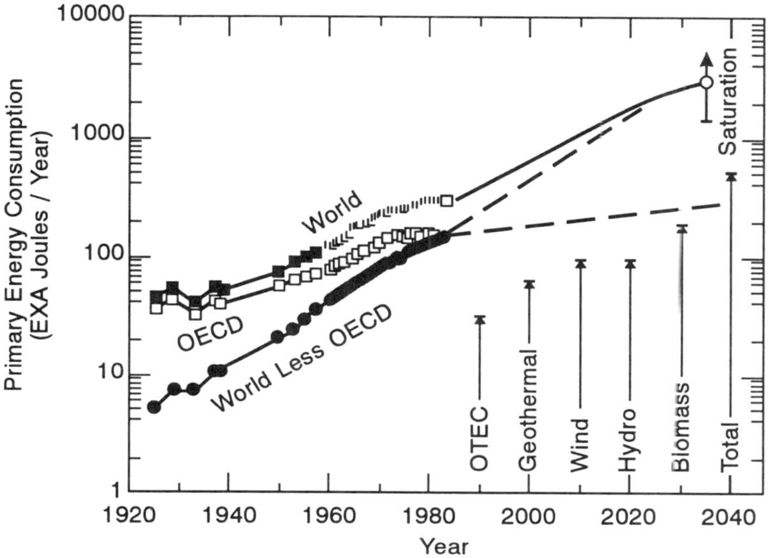

Figure 9-1: World consumption of commercial energy (top curve) is divided into two world regions, one industrially well-developed (OECD) and the other, the rest of the world. Possible future energy consumption for the whole world, OECD countries and all other non-OECD countries is crudely outlined. The short dashed line is a historical trend for the non-OECD countries. The heavy line is the sum of this line and the long dashed line, the low projection for OECD countries in the IIASA study (Hafele *et al.*). In the top right hand corner is indicated what the energy consumption would be a few decades from now, when, and if, the whole world becomes as affluent as the United States and Canada are today. In the bottom right hand corner the upper limits of technical potential for several renewable energy sources (Hafele *et al.*) are indicated with arrows and bars.

Energy Technologies and Resources

Where could energy needed by an affluent future world come from? An optimist will answer that new technologies using renewable and "clean" energy sources will provide all that energy, if only enough money is invested into properly selected research areas. It would be nice if this statement could be true, but it is not.

Existing Technologies

Practically, there are only five technologies that today supply the industrialized world's commercial energy. They are: coal, oil, gas, hydroelectricity, and nuclear electricity. The contribution of these five technologies to the world's use of energy, as well as the total world consumption of energy during the last six decades, is shown in Figure 9-2. The data before 1965 were taken from Darmstadter and after 1965 from British Petroleum.

Fossil Fuels. After wood, coal is the oldest fuel used by man. It was used in China for smelting copper at least 3,000 years ago. By the end of the seventeenth century 200,000 tons of coal a year were produced in England (*Encyclopedia International*). As illustrated in Figure 9-2, in the 1920s, well over 80 percent of energy used in the world was produced by burning coal. Today, coal supplies only 30 percent of the world's energy.

At the present stage of technological development, oil is indispensable as raw material for the production of liquid fuels so essential in transportation. Since 1973, oil is widely being replaced by gas for heat production and by coal and uranium for electricity generation. Figure 9-2 shows that the world oil consumption has been decreasing since 1979.

Hafele *et al.* estimate the fossil fuel resource base as follows: conventional oil has perhaps up to 400 TWyr of energy; heavy oil and tar sands another 400 TWyr; shale oil, 60 TWyr; natural gas, 350 TWyr; and coal 2,400 TWyr, the total of all being 3,610 TWyr. The present rate of energy consumption is about 9 TWyr/yr (1 TWyr = 31.5 EJ). Thus, at the present rate of consumption, fossil fuels would last for about four hundred years, and at the assumed "saturation" consumption about forty years. At "saturation," if conventional oil were the only fuel available it would last less than five years. From these very crude estimates it follows that even if energy consumption grew at rates as fast as outlined in Figure 9-1, that is, increasing by ten times within the next half a century, there would be enough fossil fuels, mainly coal, to power the world to the middle of the next century, but not much longer.

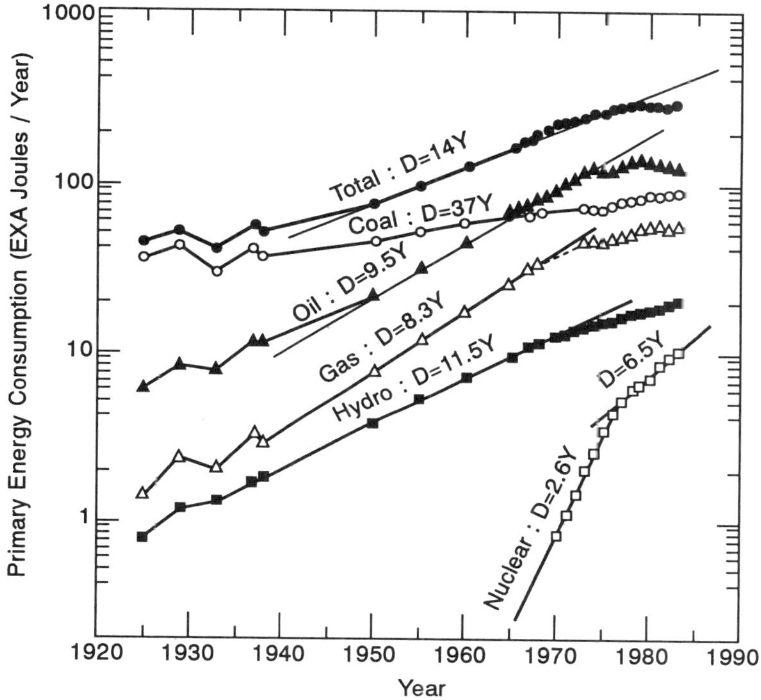

Figure 9-2: The growth of world consumption of industrially produced and commercially sold energy for the period between 1925 and 1983 is illustrated in units of exajoules (EJ) per year. (1 exajoule per year is equal to 31.71 gigawatts.) The data points represent the total energy consumption ([dark circle]) and consumption of five major energy sources, oil ([dark triangle]), coal ([circle]), gas ([triangle]), hydroelectricity ([dark square]) and nuclear electricity ([square]). Hydro- and nuclear electricity are shown as their heat equivalents (1KWh = 3.6 MJ). Where data follow a pure exponential, straight lines are drawn and times in years needed to double the amount (called D) are explicitly shown.

Hydroelectricity. At present, about 20 percent of world electricity and 5 percent of world energy is produced by hydro-power (cf. Figure 9-2). Hydroelectricity is a very cheap source of high-quality energy. Unfortunately, the world resources of hydro-power are quite limited. As Figure 9-1 shows, the technical potential of hydro resources is much smaller than the present world consumption of energy. The contribution of hydro to the energy needs of an affluent world will be essentially negligible.

Nuclear Fission. The use of nuclear fission for the production of electricity is a relatively new technology, practically used (cf. Figure 9-2) since the 1960s and presently supplying about 3 percent of the world's commercial energy.

The process of nuclear fission can be used in two types of reactors, the so-called "thermal" reactors and the "breeder" reactors. In thermal reactors, only the naturally occurring and very rare isotope, uranium 235, is burned. In breeders, uranium 238, an isotope 130 times more abundant, is also used. This is achieved by converting uranium 238 into plutonium 239, which is then used in much the same way as uranium 235.

At present, only thermal reactors are commercially available. The working prototypes of breeder reactors, capable of producing electricity in large quantities, also have been constructed, but are not used yet on a commercial basis. They produce about two times more expensive electricity.

The resources of uranium are estimated by Hafele *et al.* to be about twenty-five million tons, using only "high-grade" ores, ores with uranium concentrations higher than 0.1 percent. One hundred tons of natural uranium used in thermal reactors produce about three GWyr of thermal energy. As the present world consumption of commercial energy is nine TWyr (=9000 GWyr), it follows that the present world would need about three hundred thousand tons of uranium per year if it ran on uranium only. In that case, the estimated resources of high-grade uranium ores could power an affluent world only for about eight years.

If the world would use breeder reactors, which use uranium roughly one hundred times more efficiently than thermal reactors, the world's "high-grade" uranium supplies of twenty-five million tons would last at "saturation" for eight centuries, not just eight years.

Low-grade uranium ores with uranium content between a few parts per million (ppm) and a few hundred ppm are thousands of times more abundant than high-grade ores (Deffeyes and MacGregor). When thorium resources, which can be used in breeders but not in thermal reactors, are added, it is clear that fission resources used in breeders can power an affluent world at "saturation" level, not just for thousands, but for millions of years.

Unfortunately, even with this scheme there are difficulties.

If energy consumption in the undeveloped world continues to rise as indicated in Figure 9-1 and if the bulk of this increase should come from thermal and breeder reactors, a temporary difficulty with supplies of uranium and plutonium would arise. While the breeder reactors would be able to provide enough energy at the assumed "saturation" for many centuries, once the steady state is reached, breeder reactors will not be able to convert uranium 238 into plutonium fast enough to supply fuel for all new breeder reactors during the times of fast growth of energy consumption. The reasons are that the production of nuclear energy should be doubled every few years until "saturation" is reached. The supply of plutonium to be initially loaded into new reactors must grow at the same rate. Unfortunately, present breeders can double the amount of plutonium loaded into them only in fifteen to twenty years. Thus, the world supplies of energy could not be bootstrapped by breeders alone at a rate which would double the world energy consumption every fifteen years or so. This problem has been considered by the World Energy Conference (Foster *et al.*), and the conclusions were illustrated in a diagram which is reproduced in Figure 9-3.

The anticipated plutonium shortage can be resolved by developing accelerator breeders (Davies), a technology which is scientifically feasible but has not yet been developed (see the next section).

Emerging Technologies

Renewable Energy Sources. Renewable hydroelectricity is an existing technology and was discussed in the previous section.

Other potential renewable energy sources, wind electricity, biomass fuels, geothermal heat, OTEC (Ocean Thermal Energy Conversion) electricity, and tidal power could not provide more than a few percent of the total energy needed by an affluent world. The technical potential of these fuels (Hafele *et al.*) on the global scale is illustrated in Figure 9-1. The practical,

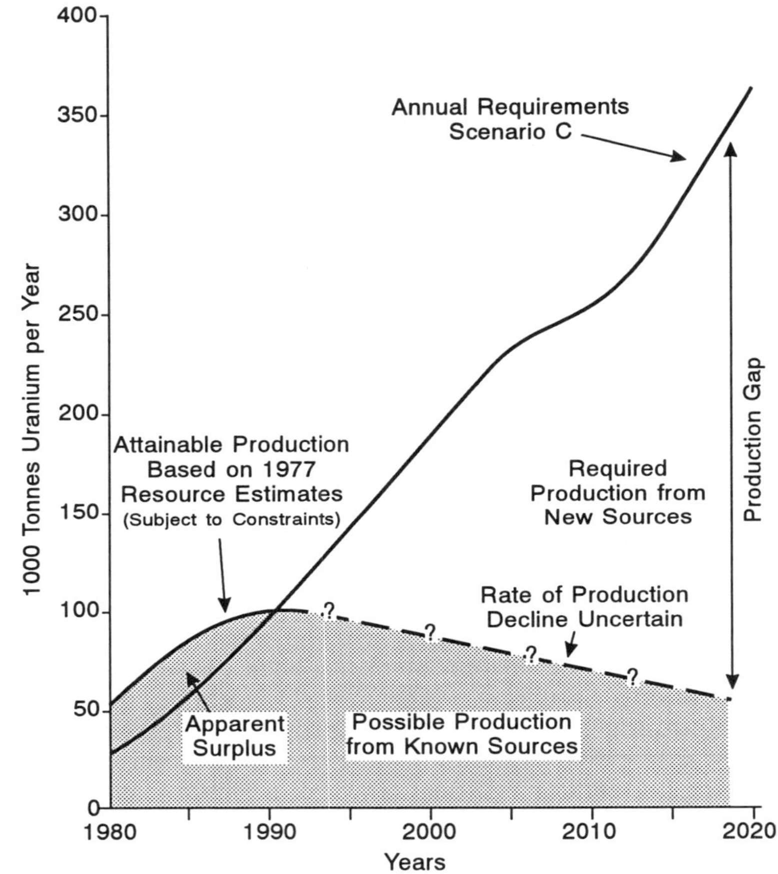

Figure 9-3: Schematic illustration of world uranium supply problem until year 2020. (Reproduced from Foster *et al.*) The quantity of uranium needed annually for initial fueling of new reactors and refueling of the existing reactors is plotted on vertical scale. Scenario C is the one which assumes relatively fast industrial growth in developing nations and the use of fast breeders as expected to be economically justified. In this scenario a production gap would appear before the end of the century and would grow worse with time. This perceived shortage of uranium, as well as the perceived difficulty in mining sufficiently large amounts of coal, are main reasons why the study by the World Energy Conference (Bloodworth *et al.*) has considered the historical growth of doubling energy consumption every 15 years as impossible to achieve.

economic potentials are naturally significantly lower, so these fuels could not possibly satisfy the global need for energy. In addition to the fact that they are on global scale in short supply, they are all, in most applications, more expensive than nuclear power or fossil fuels.

Solar Energy. The technical potential of solar energy is enormous. Our sun supplies the earth with energy which is several orders of magnitude higher than the presumed "saturation" amount indicated in Figure 9-1. In spite of this, it is difficult to see how solar energy could provide significant amounts of needed energy during the next half-century, if ever.

The basic difficulty with solar energy is the Second Law of Thermodynamics. Solar energy is diffuse, about one kilowatt per square meter, and so much only when the sun shines. (As a comparison, fission energy is "dense," about one hundred megawatts per cubic meter.) For most applications, essentially all except heating of buildings, we need "dense" energy. For instance, to use solar energy for electricity production, it must first be concentrated, and this requires work. Work costs money. Thus, solar electricity is intrinsically expensive. Various engineering designs of solar power stations clearly show this fact (Starr). As a result, solar electrical power stations have not yet become commercially feasible. No company in the world is selling them, and it is difficult to see how they could ever produce electricity cheaper than fission reactors.

Aside from being very expensive, solar electricity has two more disadvantages. One is that the sun shines only during the day, and it is bright only if there are no clouds. Therefore, a practical method of storing large amounts of electricity would have to be found before solar electricity could be used day and night, sunshine or rain. The second disadvantage is that densely populated countries would have to use a sizeable fraction of their land to collect enough sunshine. For instance, if the population of India doubled, if each person used as much electricity as we use in North America and Scandinavia today, and if all that electricity was to be produced by technology which converts most efficiently the solar energy into electricity, that of mirror collectors, then close to two percent of the Indian subcontinent would have to be covered by mirrors and access paths to these mirrors. In the case of countries with less sunshine, like England, a much greater fraction of the British Isles would have to be covered with mirrors and auxiliary power-related facilities. This is clearly impractical.

If all solar power stations were located in various deserts around the world and the electricity generated was used to produce hydrogen gas, then, in principle, the world could get all its energy from the sun. But this scheme is quite expensive and not practical to be fully deployed within the next few decades, if ever.

Optimistically, at best, the solar energy will be able to provide a certain fraction of electricity in some selected areas of the world, like southern California or Egypt. For instance, solar electricity, whether produced by mirrors or photovoltaic cells, might be practical to use for "peaking" at the time of day when the sun is strong and demand for space cooling is high. The bulk of energy used by the world will still have to be produced using other energy sources. Pessimistically, the solar-produced electricity will never become commercially practical, not even for peaking in southern California or Egypt.

Accelerator Breeder. If the use of nuclear energy expands as needed by a world that industrializes itself along historical lines, a shortage of uranium would develop early in the next century (Foster *et al.*). A solution to this difficulty is a new know-how, the technology of accelerator breeders (Davies; Grand *et al.*).

A schematic of an accelerator breeder complex is shown in Figure 9-4. In such a system, a very intense beam of protons (say, 300 milliamperes) is accelerated to high energies (about 1000 MeV) in a linear accelerator. This beam then enters a target structure made primarily of uranium and/or thorium. Each high energy proton creates up to 100 neutrons which then either fission uranium and/or thorium and create heat, as in any nuclear reactor, or become absorbed in uranium 238 creating plutonium 239, or in thorium 232, creating fissionable uranium 233. The heat developed in the target is sufficiently large to operate an electric power station, which would supply all the electricity needed by the accelerator. In this way, the accelerator, target, and power station would be a self-contained

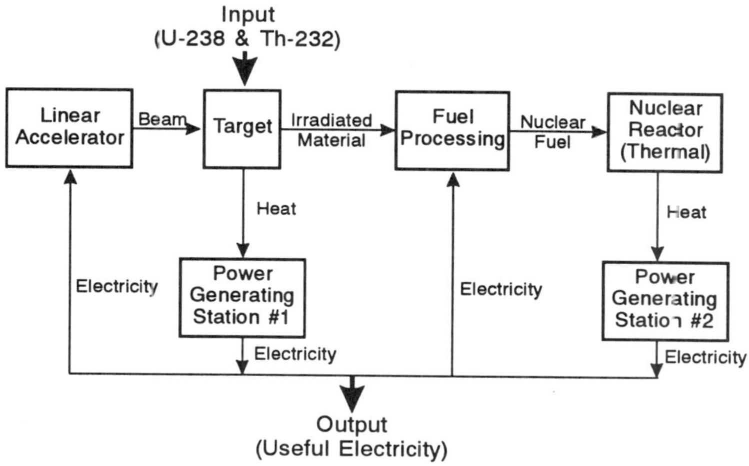

Figure 9-4: Schematic representation of an accelerator breeder complex designed for conversion of thorium 232 into uranium 233 and uranium 238 into plutonium 239.

plant producing enough energy out of non-fissionable uranium 238 and thorium 232 to drive itself. In addition, the accelerator breeder would convert sizeable amounts of these two elements into easily fissionable plutonium 239 and uranium 233. These two could then be used as fuel either in thermal reactors or in fast breeders. Because accelerator breeders could produce as much plutonium as needed, the anticipated shortage of plutonium for the first fueling of breeders would be overcome.

This technology does not exist yet. It is a scientifically feasible technology, but its engineering feasibility has not yet been demonstrated. No accelerators with 300 mA beams at 1 GeV energy have been constructed. There are good reasons to believe that they can be made, but nothing other than some early design work has been done.

The development of this technology would enable the fast, worldwide deployment of breeders, as well as the full exploitation of thorium resources. This technology should be of particular interest to India, a country which is relatively poor in any other energy resource but thorium.

Nuclear Fusion. Nuclear fusion, once harnessed, will provide mankind with almost unlimited amounts of energy. Unfortunately, this process is so complicated that even the scientific feasibility of an energy producing fusion reactor has not yet been demonstrated. Thus, it would be too optimistic to expect that this technology would provide significant amounts of energy, if any, within the next half a century.

Uranium from Oceans. Ocean water contains 3.3 parts per billion of dissolved uranium. As oceans and seas contain about 1.4 billion cubic kilometers of water, uranium resource in ocean water is estimated to be about 4.5 billion tons, two hundred times more than the estimated sources of high-grade uranium ore. If it were possible to extract economically a sizable fraction of this uranium, and it might be (Schenk *et al.*), then this amount of uranium alone would be sufficient to power our affluent world for a few hundred years using just thermal reactors. In that case, the development of accelerator breeders and even extensive use of fast breeder reactors would not be necessary for the next few centuries.

Energy and Environment

There exists no ideal technology, a technology that would not present certain dangers to humankind and the environment. Energy technologies, nuclear power included, are no exception.

All harmful effects produced by various energy technologies can be divided into three broad groups: (1) local and/or short-term effects; (2) regional and/or medium-term effects; and (3) global and/or long-term effects.

Under local, we shall understand events whose spatial extent is less than a few kilometers; under regional, those events whose extent is up to a few thousand kilometers; and under global, events which affect the whole earth.

We shall consider short-term events those whose effects last for a few days or weeks, medium-term those whose characteristic times are years, and long-term events those that persist for times comparable to, or longer than, the human lifetime.

Events of local significance and short duration are usually

called accidents, sometime disasters. They are of the most direct and immediate concern to people. They produce instant injury or death and easily visible damage to property and the environment. Victims of these accidents are either people working with particular technology or the general public. The Bhopal disaster, or the collapse of a hydroelectric dam are obvious examples of local effects with short-term duration.

Acid rain is a typical example of regional effects whose spatial extent is hundreds and thousands of kilometers and whose slow health effects impact the time-scale of a decade or more.

There exist only two long-term global effects related to energy production. These are the productions of nuclear waste and carbon dioxide. The first effect is, in fact, more regional than global, but it is certainly of long duration. The carbon dioxide pollution is truly global in its nature and of duration much longer than human lifetime.

The proliferation of nuclear arms does not fall into any of the above three categories. It is the case of a gross misuse of a technology which appears as a result of human relations, rather than the coupling between technology and the environment.

Local/Short-Term Effects

Non-nuclear Technologies. Every technology has its own type of accidents. For instance, the coal industry is well-known for its mining accidents which kill anywhere between a few and a few hundred miners. Hydroelectricity is known for its high rate of construction accidents and occasional dam failures.

Inhaber (Inhaber, Table K-31) lists the numbers of dead due to various major dam disasters, and concludes that up to 1980 (including the Morvi Dam accident), somewhere between 5,500 and 7,000 people have been killed worldwide in hydro dam accidents. These numbers, reproduced in Table 9-1, illustrate the often ignored fact that dams are very dangerous structures and that "clean" and renewable hydropower is not all that "clean."

Many more examples of energy related accidents, not associated with nuclear power, could be given. A very extensive survey of risks due to various technologies has been given by Cohen and Lee.

Table 9-1: Major Disasters Likely Associated with Hydroelectricity Generation (from Inhaber)

Place	*Date*	*Number Dead*
Vajont, Italy	1963	2,600–3,000
Morvi, India	1979	1335
Gleno, Italy	1923	600
St. Francis, U.S.A.	1926	426–450
Kiev (Babi Yar), U.S.S.R.	1961	145
Koyna, India	1967	180
Vega de Tera, Spain	1959	123–150
Sella Zerbino, Italy	1935	100–111
Oros, Brazil	1960	30–100
Coedty, Wales	1926	20–60
Teton, U.S.A.	1977	9–11
Bhakra, India	1959	10
Colorado Dam, U.S.A.	1900	8
Necaxa No. 2, Mexico	1909	4

Nuclear Accidents. A major concern about nuclear power is the possibility of catastrophic accidents. Although the Three Mile Island accident did not kill anybody (Kemeny), the Chernobyl accident did (USSR). This was not entirely unexpected. In 1957, scientists at Brookhaven National Laboratory estimated (AEC) that if one-half of the radioactive inventory of a 500-megawatt (thermal) reactor were somehow released into the atmosphere in the vicinity of a populated city, and if atmospheric conditions were unfavorable, then in one particular scenario 3,400 deaths, 43,000 acute illnesses, and tens of thousands of induced latent cancers could occur. Present day power reactors are three to eight times larger, thus, theoretically, the damage could be much worse. For several reasons this first and very crude study was too pessimistic. It did not consider what the plausible physical and chemical processes were that could spread half of the reactor core into the atmosphere outside the very strong reactor containment building. Also, this study did not attempt to estimate how probable such catastrophic accidents may be, nor did it take into account the fact that nuclear reactor accidents are slowly developing affairs; thus, the evacuation of population is possible to perform.

In the Rasmussen report (NRC) a set of complex calculations were performed in order to estimate the probability of a nuclear (power) reactor accident that could kill a certain number of people. The results of their study are reproduced in Figure 9-5a, b where they are compared with risks due to other man-made and natural accidents.

It has been hotly debated how good these estimates are (see for instance APS; Shea). In recent years, more studies about the safety of certain types of nuclear reactors were made and the conclusions are less pessimistic than the first Brookhaven estimate and the Rasmussen report (cf. Levenson and Rahn; Lederman; Wilson *et al.*).

Regional/Medium Term Effects

Acid rain. A nice introduction to this topic was presented by Likens *et al.*, and a detailed review is given in OTA.

The following news report illustrates vividly the difficulty with acid rain. In an article "Rain of Troubles," C.K. Groves writes, "During several recent storms in Wheeling, West Virginia, the rainfall measured 1.5 on the pH scale, a level far more acid than vinegar." Vinegar has a pH level of 3.0. Thus, the Wheeling rain was 30 times more acid than vinegar, and about five times more acid than lemon juice!

Acid rain kills lakes, forests and shortens the life expectancy of people. The results of some recent studies using no threshold assumption (OTA) indicate that in the eastern part of the United States alone acid rain is a contributing factor in about 51,000 deaths each year from various heart and lung diseases. In some older studies (Inhaber), this number is much smaller, but still in the range of several thousand per year.

Sulfur and nitrogen oxides which produce acid rain do not have to be dumped into the environment. The technology for the control of acid rain exists but it is quite expensive to implement. The report (OTA) states that something like four billion 1980 dollars a year would be needed to reduce the acid emissions in the eastern half of the U.S. to one-half the present value (see Figure 9-6). This estimate was echoed in a short news

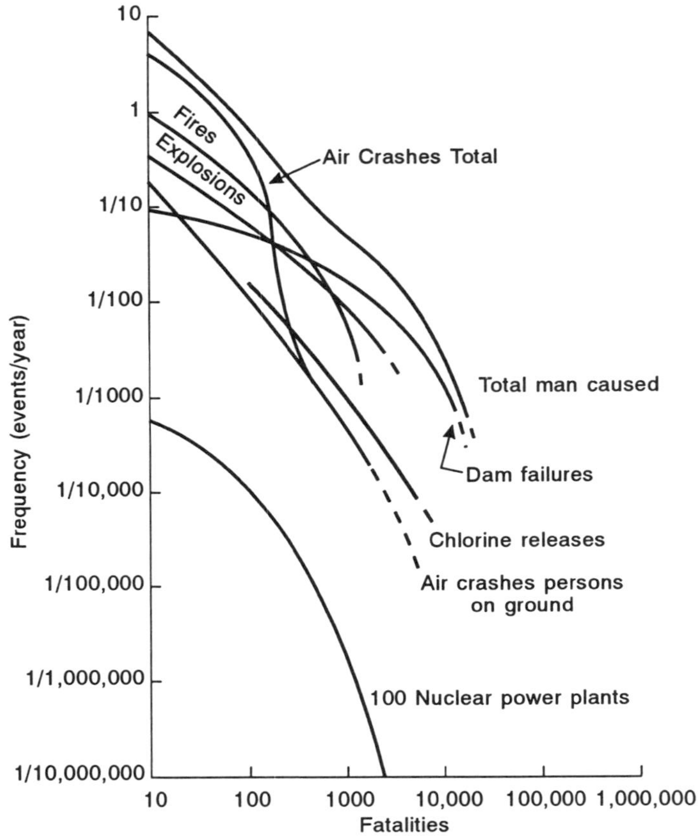

Figure 9-5a, b: Probability distribution (frequency) of events occurring in one year in which more than a certain number of people are killed, plotted as a function of the number of fatalities. On this page is plotted the frequency of man-made disasters, and on the facing page, of natural disasters. The curves for 100 nuclear power plants are plotted on basis of theoretical calculations. (Reproduced from NRC.)

report (*New Scientist*) where it was stated "... the electric industry's research institute (in the U.S.) predicted that cutting (acid producing) emissions (from power stations) by 50 percent would raise prices (of electricity) by anything up to 50 percent."

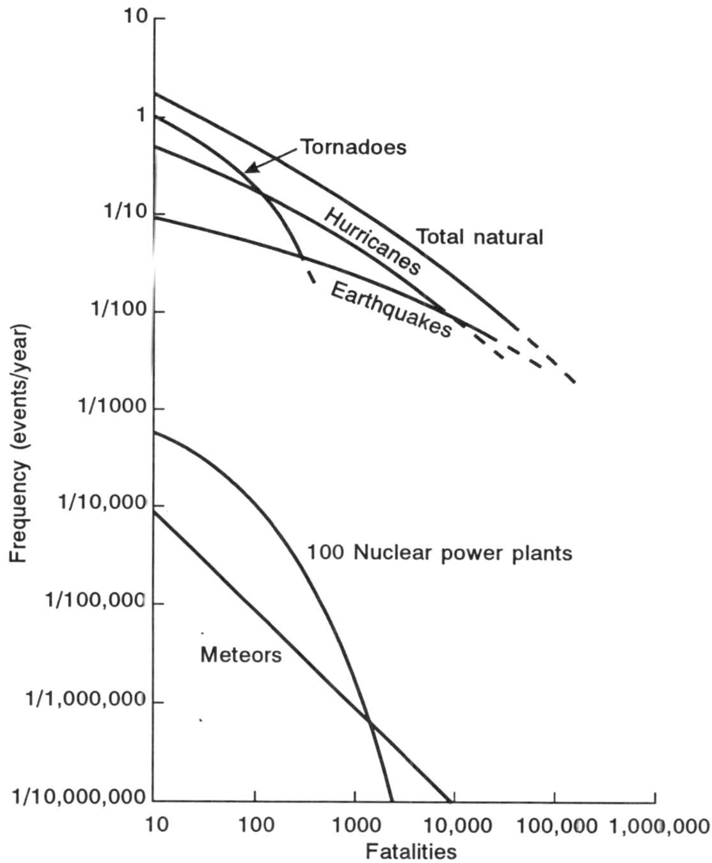

Figure 9-5b. See figure 9-5a for caption.

Radiation Effects of Nuclear Power. One of the concerns related to nuclear power is the release of radioactive substances into the environment and health effects associated with it. Long-term effects of low levels of ionizing radiation on the human body are not known (BEIR). These effects may be harmful, may be nonexistent (if there is a threshold), or may be even beneficial (Luckey, 1979: Luckey, 1988; Sagan, 1987a).

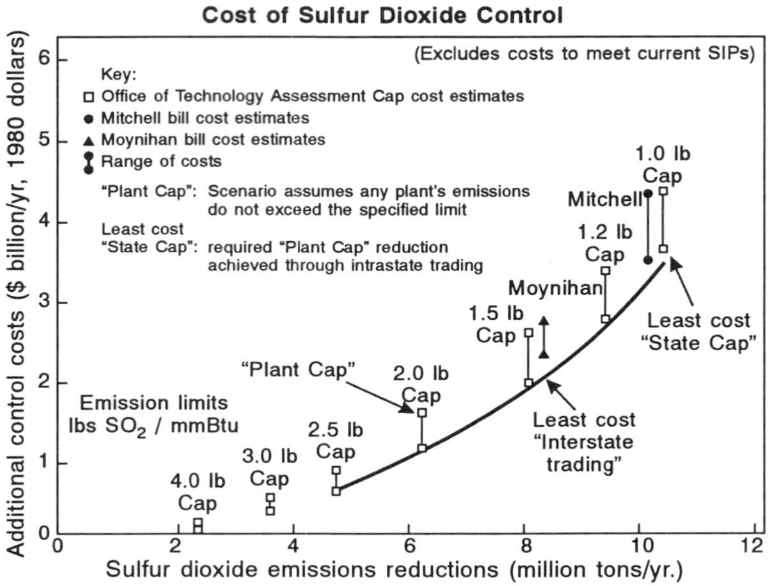

Figure 9-6: Estimated cost of sulfur dioxide controls in the U.S. coal-fired electric power stations. (Reproduced from OTA.) It has been estimated that the total amount of sulfur dioxide emissions in the 31 states of the Eastern U.S. is as much as 22.4 million tons per year, of which 16.5 million tons are produced by electric utilities (OTA).

This topic has been discussed at some length in several submissions to this conference (Totter; Luckey, 1988), as well as by Sagan (Sagan, 1987b) in a guest editorial to the special *Health Physics* issue on Radiation Hormesis (Sagan, 1987a). In this section the emphasis will be given on various interpretations of dangers produced by intentional or accidental releases of radiation by the nuclear power industry.

Most frequently the linearity hypothesis, based on certain assumptions about harmful effects of radiation (BEIR) is used to relate the collective exposure (expressed in person-rems) to cancer induction. In this way, an estimate for the number of people "killed" in a specific radiation release is arrived at. For instance, in the case of the Chernobyl accident, the expected collective dose from external radiation exposure to the population in several regions of the European part of the U.S.S.R. (U.S.S.R., Table 7.2.9) over the 50 years following the accident (including 1986) was estimated to be 38 million person-rems. (Adding internal radiation doses may double or triple this amount.) Assuming validity of the linearity hypothesis and using certain specific numerical factors, one calculates that about five thousand cancer deaths would be produced in a few decades. In a similar manner, when calculations are made for other parts of the world, several tens of thousands of "fatalities" are attributed to the accident. Such estimates create an impression that the fatalities had actually happened, that people already had been sentenced to death. They completely ignore that fact that there may not be any fatalities at all as the linearity assumption may not be correct and nature may not be obeying it. Therefore, these estimates about people "killed" by various radiation releases are factually incorrect and unnecessarily alarming . (For an update, see Jovanovich, 1991.)

Another way to talk about the effects of low levels of radiation on human health is to compare them with effects of smoking (Cohen and Lee; Marshall). The risk of death from exposure to *seven millirem* of radiation is about equal to the risk of death from smoking *one* cigarette in one's lifetime, not per day. This comparison is certainly better than the one which calculates the number of people "killed," but it also has its weaknesses. It assumes validity of the linearity hypothesis for both the radiation effects and the effects of smoking. (Smoking of a few dozen cigarettes a day for many years has been statistically related to lung cancer incidence, but not smoking of a few cigarettes in one's lifetime.)

More appropriate comparisons that involve no assumptions or extrapolations from high to low doses can be made. Physically, the effects of radiation created by nuclear power technology are qualitatively no different from effects created by natural background radiation. Thus, a comparison of effects due to these two sources is meaningful and free from any assumptions.

The cosmic component of natural background radiation is responsible for about 30 millirem of exposure per year at sea level. It doubles at about the 1,200-meter level and rises to 500 millirem at 12 kilometers (UNEP). (1 mrem of additional exposure is equivalent to moving up for about 70 meters, that is onto the 16th floor of a high-rise building.) Therefore, receiving an additional 30 millirem every year during one's lifetime is equivalent to moving about 1,200 meters up the hill or making about six transatlantic flights every year in a subsonic jet aircraft. In the Soviet report on the Chernobyl accident (USSR), it was estimated that 75 million people will be receiving, on the average, an additional eight millirem per year of external exposure during the next 50 years (excluding 1986) as a result of the Chernobyl accident. This means that each person will be receiving each year as much additional radiation as if he moved about 500 meters up the hill, or if he traveled one or two times every year from Kiev to New York and back. (If exposure due to internal radiation is added, this height should be doubled or tripled.)

Certain regions of the world have higher terrestrial background radiation than average. For instance, the state of Colorado, due to more radioactive ground and higher elevation, has about 150 millirems per year of background radiation, more than the average for the world. Thus, anyone from areas with normal background radiation, who takes one- to two-month vacations in Colorado every year, will receive as much extra radiation as those people in the U.S.S.R. who will be receiving 10–20 millirems per year of external and internal radiation due to Chernobyl fallout.

In Figure 9-7 are represented trends with time of annual effective dose equivalents from three man-made sources of radiation expressed as a percentage of average doses received from natural sources (UNEP). These three sources are diagnostic medical examinations, fallout due to past atmospheric nuclear bomb tests and exposures due to nuclear power production, which include not only public but also occupational exposures. Dash-dotted line at 250 percent represents the average doses regularly received by the residents of Colorado due to higher natural background in that region. For comparison purposes, the estimated average additional dose commitments to be received by 75 million Soviet citizens in the western part

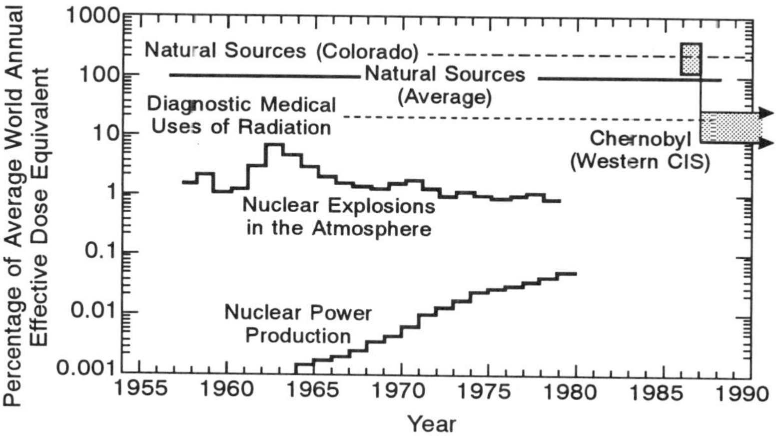

Figure 9-7: Trends with time of annual effective dose equivalents from three man-made sources of radiation expressed as a percentage of average doses received from natural sources. Doses due to nuclear power production include not only public, but also occupational exposure. (Reproduced from UNEP.) Dash-dotted line at 250 percent represents the average doses regularly received by the residents of Colorado due to higher-than-average radioactivity in the ground and high altitude of the state. For comparison purposes, the estimated additional average dose commitments to be received by the 75 million Soviet citizens in the western part of the U.S.S.R. due to fallout from Chernobyl are also shown. (For an update on the Chernobyl accident effects, see Jovanovich, 1991.)

of the U.S.S.R. due to fallout from Chernobyl are also shown. The heavy line at the bottom of the shaded band represents exposure due to external radiation (as given in U.S.S.R., Table 7.2.9). The shaded band was added to indicate that internal exposures have to be added. The band was somewhat arbitrarily widened to three times the estimated external exposure, to be on the pessimistic side. (As a rule of thumb, it is considered that exposure due to internal radiation would be about as high due to external radiation.) It should be mentioned that the U.S.S.R. report gives only two numbers for the collective exposure—one for 1986, and the other for the next 50 years.

Global/Long Term Effects

Radioactive Nuclear Waste. This topic has been discussed at length in Cohen (1977) and in Cohen (1983, Chapters 5 and 6). It is also the topic of a chapter in this volume (Chapter 8).

The critics of nuclear power state that the ultimate disposal of nuclear waste has not yet been solved, which is essentially true. No waste from nuclear reactors has yet been irretrievably disposed of and no specific process for its disposal or destruction has been selected yet. All nuclear waste produced until the present time has been temporarily stored in one form or another, usually in pools at the reactor sites. Why is this done? Briefly, the waste decays all by itself and the longer it is kept, the easier it is to handle. Temporary storage is safe and cheap. Thus, it is better to wait until safer and cheaper disposal technology is developed.

What are the possible technologies that might be used or developed in order to dispose of nuclear waste permanently? A summary of possible options is schematically presented in Figure 9-8. The full discussion of options indicated in this figure is given in Jovanovich (1986c), sections D and E.

Carbon Dioxide Pollution. When fossil fuels are burned, besides other wastes (ash, sulfur dioxide, etc.) carbon dioxide is produced. While other wastes are either contained at the source (ash), or can be contained if we wanted to pay for it (acidic oxides), carbon dioxide is released into the atmosphere and we have no feasible way of containing it.

Our atmosphere contains at present about 340 ppm (parts per million) of carbon dioxide and every year, by burning fossil fuels, we add to it one more ppm (Woodwell). If we keep increasing our fossil fuel consumption as in the past, doubling it every fifteen or so years until all fossil fuels are used up, we will increase the carbon dioxide content of the atmosphere as indicated in Figure 9-9 (Oeschger and Siegenthaler). While it is very easy to calculate the amount of released carbon dioxide from the knowledge of the amount of fossil fuels burnt, it is not possible to predict how much of it will be absorbed by the biosphere and the oceans. Two plausible biota growth factors are assumed in Figure 9-9, giving the range of possible carbon

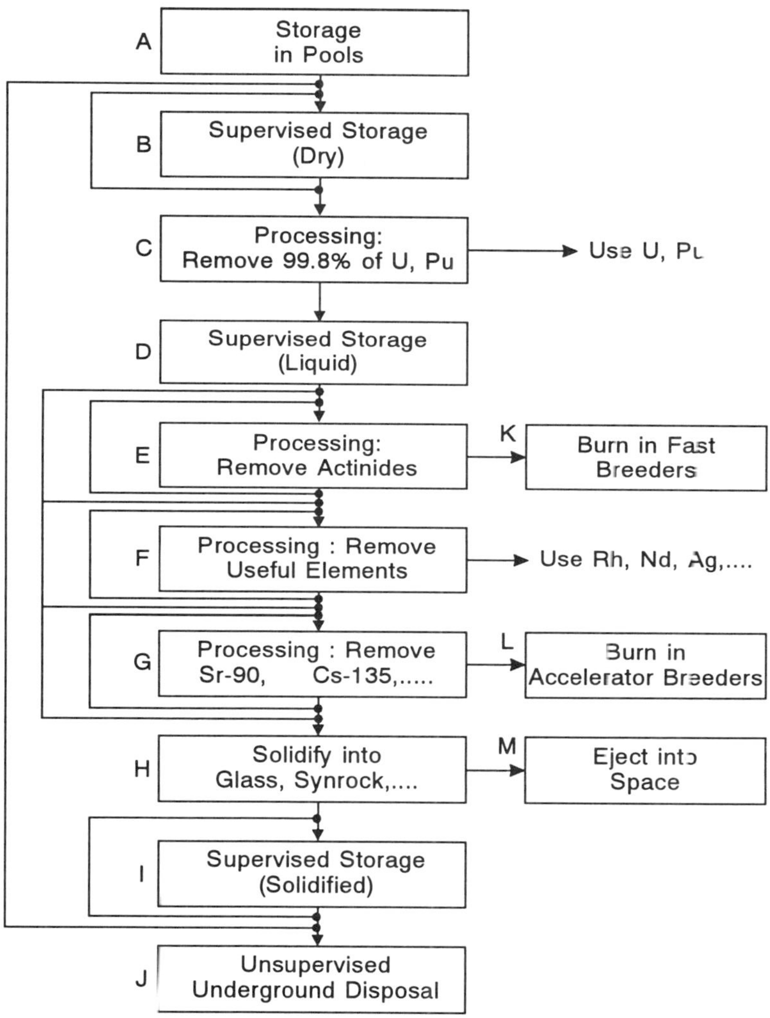

Figure 9-8: Schematic representation of possible options, present and future, that exist in dealing with nuclear waste (from Jovanovich, 1986c).

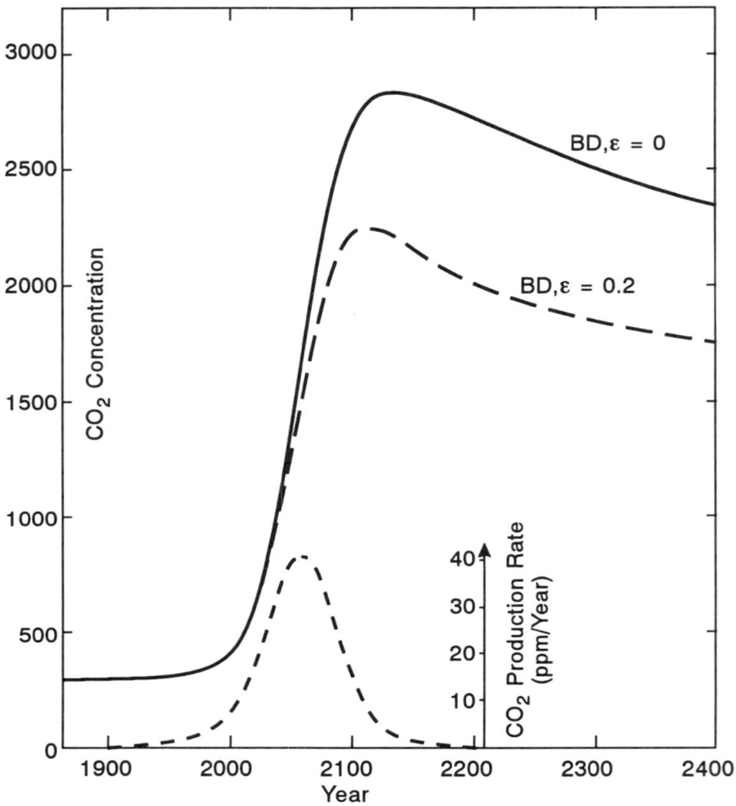

Figure 9-9: Prediction of the carbon dioxide concentrations in the atmosphere if all recoverable fossil fuels are burnt. Lowermost dotted curve assumes carbon dioxide production rate as it would be required by a world industrializing itself along historical lines and being powered by fossil fuels as much as possible until they are exhausted. The concentration curves are given for two different biota growth factors. The dashed curve with ε=0.2, predicts a large biota growth that does not seem probable. (Reproduced from Oeschger and Siegenthaler.)

dioxide concentrations. This figure shows that if the industrial revolution keeps spreading to the presently undeveloped parts of the world, as discussed earlier in this article, and if this industrial growth is fueled by coal and other fossil fuels, then by the end of the next century the world will increase carbon dioxide concentration in the atmosphere close to ten times!

What would be the effects of such a dramatic increase in the carbon dioxide content of our atmosphere? Nobody will know for sure until the "experiment" is done, and then it will be, of course, too late. But, something is known for sure: By adding carbon dioxide into the air, we are creating the so-called "greenhouse effect," that will make the earth warmer and wetter. Nobody knows by how much. At present, the best theoretical estimates indicate that by doubling the amount of carbon dioxide, the average global temperature would increase between two and four degrees centigrade (NRC). Higher amounts of carbon dioxide would increase the temperature even more. On the basis of some simple qualitative considerations, as well as some detailed theoretical calculations, it follows that the temperature increase in polar regions would be above the global average (NRC). If this warming is sufficiently large, and the existing estimates are that it will be, all ice in the polar regions would melt and the level of all seas and oceans would rise at first, within the next few decades or centuries, by five to six meters, and then, probably after several millennia by about seventy meters (Mercer; Barth and Titus).

One also could expect that higher average global temperature would produce large changes in regional climates. Deserts and fertile regions would move about. How much, where and how quickly these changes would occur nobody can predict for certain (Bernard; Hansen *et al.*; Revelle; Rotty). It is reasonable to expect that some regions will become warmer and wetter, others warmer and drier. Some changes will be beneficial, others catastrophic. There will be winners and losers.

The earth's climate is a very complicated system and relatively sudden changes are possible. We can only speculate on consequences of various climatic changes. For instance, it would be reasonable to ask what would happen to people in India if the monsoon rains shifted farther east and if the Indian Thor Desert shifted, in a period of few decades, over a sizeable portion of the Indian subcontinent? What would happen if the North American "bread basket," the Midwest, would turn into a desert, as it might (Bernard)? While, of course, there is no way of claiming that such terrible shifts in rain patterns would occur, the atmospheric sciences cannot prove that something like that would not happen.

It may not be necessary to speculate about the future in order to convince ourselves of the gravity of the problem. It may be sufficient to look at the present. The carbon dioxide component of the atmosphere, a component that has a very strong effect on the regulation of the temperature of the globe, has been changed by about 20 percent during the last 150 years, and by 8 percent during the last 30 years. During the last decade or so, sub-Saharan Africa has been hit with a series of severe droughts (Glantz). It has been reported (Cross, 1983) that the Sahara has been extending southward at the rate of 10,000 square kilometers per year and that contributing factors to this fast desertification are over-population, over-grazing and poor agricultural practices (Cross, 1983, 1984). But, as Ellsaesser has asked, could it be that the movement of the Sahara desert is the first effect of our changing the atmospheric carbon dioxide content by 20 percent? We, of course, do not know. The present atmospheric sciences cannot give us an answer to this question. But the possibility is clearly there. We have changed the environment in a significant manner without knowing what the consequences are going to be.

Drought and flood always have been among the worst natural calamities that could befall people. The large increase in the fossil fuel consumption will almost certainly induce them on an unprecedented scale. By changing the content of carbon dioxide in the atmosphere, we are gambling with the earth's climate and human lives on a grandiose scale.

Summing It All Up

Where can energy needed by an affluent future world come from? What technologies are environmentally acceptable? The short review of energy technologies and their environmental impact from the previous two sections enables us to discuss and answer these questions.

Available Technologies

On the time scale of the next half-century, fossil fuels (mainly coal), fission breeder reactors (together with accelerator breeders), and solar electricity are the only technologies that could theoretically power an increasingly affluent world. (If extraction of uranium from sea water proves to be technically and economically feasible, there would be no need to use accelerator breeders, possibly not even breeder reactors.)

Renewable energy sources, while attractive for many reasons, besides being more expensive than coal and uranium, are not sufficiently abundant to power an affluent world.

Solar energy is abundant but also intrinsically expensive. Unless science and technology make some unexpected breakthroughs, it is unlikely that solar-produced electricity will ever be economically competitive with nuclear electricity. Thus, practically, solar energy will not be able to power an increasingly affluent world during the next half-century or so.

On the time scale beyond the middle of the next century, fossil fuels would be used up, but fusion energy might become available.

Based on the present technical availability and economic competitiveness, excluding environmental effects and possible scientific breakthroughs (they are obviously unknown, unpredictable and cannot be counted upon), only coal and fission (thermal and breeder) reactors will be available in abundance until the middle of the next century, and only breeders and possibly fusion reactors beyond that.

The Least Bad of All Technologies

> None yet e'er drank a honey'd draught
> Unmixed with a cup of bitter gall,
> And cup of gall for honey equally doth call,
> That so, the mixture one may easier drink.
>
> Nyegosh (1811–1851)
> "The Mountain Wreath" (verses 563–66)

A technological utopia does not exist. Each of the technologies is "bad" in some respects. Thus, the only meaningful question to ask is, which technology is "the least bad," which technology has the least amount of "bitter gall?"

In this section we shall compare negative aspects of the most promising technologies in order to find "the least bad."

Nuclear and Hydro Accidents. When one speaks of nuclear accidents, a comparison ought to be made with accidents caused by other technologies. Comparing nuclear with hydro dam failures is inviting, as hydropower is usually considered as the best of all energy technologies—cheap, "clean," and renewable.

Until now, all nuclear power stations around the world have produced about one-sixth as much power as all hydroelectric power stations since the first one was built at Niagara Falls. Among the general public there has not been a single "prompt" fatality until now due to nuclear power reactor accidents. (At Chernobyl [USSR] there were 30 occupational but no prompt public deaths.) As discussed on page 305, dam failures occur every few years somewhere, sometimes with thousands of people being killed in a single failure (Table 9-1). Thus, until now, nuclear power has had hundreds of times fewer "prompt" (public and occupational) fatalities per unit of electricity produced than the renewable hydroelectric power. (Nuclear power has not had a single prompt public fatality thus far—a remarkable achievement.)

During the Chernobyl accident, roughly a thousand square kilometers of land were contaminated and evacuated, possibly for a few decades, maybe even longer. Radioactive contamination of land is certainly unique to nuclear energy. On the other hand, it must not be forgotten that hydro dams also render large tracts of land unusable by flooding them—essentially forever. For instance, a big hydro dam on the Dnieper River near Kiev has flooded just about the same area of land as the Chernobyl accident has contaminated. Of course, there are some economic benefits in creating a large lake, while there are no known benefits in contaminating land with radioactivity. But the flooding is for as long as the river is used for power production, and that means forever if the world is to use renewable energy as much as possible. Radioactive contamination, while very long-term, is not forever. (For an up-to-date discussion of the Chernobyl accident, see Jovanovich, 1991.)

By comparing prompt damage in accidents, nuclear power,

including Chernobyl, has been so far a lesser evil than the best renewable electricity source we have—hydro-electric power.

Radiation Releases and Acid Rain. Potential long-term damage due to accidental and intentional radiation releases should be compared with the dangers due to acid rain.

If we assume that the linearity assumption is valid for radiation effects, then the Chernobyl accident will "kill," according to some estimates, during the next 50 years, close to 100,000 people around the world. Similarly, if we assume that a no-threshold assumption is valid for the effects of acid rain, then every year, in the U.S. alone, 50,000 people are "killed" by acid rain. (See Jovanovich, 1991.)

Thus, the accidental Chernobyl spill was not as bad as routine emission of gases contributing to acid rain.

Nuclear and Fossil Fuel Wastes. In some respects nuclear waste is similar to carbon dioxide. When fossil fuels are burned, carbon dioxide is produced. When a nuclear reactor operates, nuclear waste is produced. (In the context of this paragraph we shall assume that radioactive waste means any radioactive material produced by the nuclear technology, not just the spent nuclear fuel.)

It is appropriate to compare the effects on the environment of these two wastes for two limiting cases, that of an all-fossil-fueled world and of an all-nuclear world.

If we assume an affluent world fueled by fossil fuels, then we shall have a carbon dioxide scenario as depicted on Figure 9-9. The effects would be a very large warming of the earth and unknown changes in the global climate.

If we assume an affluent world fueled by nuclear fission power, then we will not be able to avoid an increase in effective radiation doses due to radiation spills and occupational exposure to radiation. Figure 9-7 tells us how much the effective background radiation has been increased by the existing nuclear power industry. In an affluent and all-nuclear world the number of nuclear reactors would be increased by, say, three hundred times over present. If we assume no improvements in nuclear technology (a very pessimistic assumption), then the effective natural radiation doses generated by the nuclear industry would be also increased by about three hundred times. With reference to Figure 9-7, we see that in that case, the average effective natural radiation doses would be increased by 10 percent, to the level it was at the height of atmospheric nuclear testing in the early 1960s. We should also remember that this increased exposure level would be still far less than the present natural background radiation exposure in Colorado.

It is indeed hard to see why increasing background radiation by less than 10 percent, as we have done once already in the past, would produce more risk and harm to the environment and humanity than increasing carbon dioxide concentration of the atmosphere by close to ten times!

Nuclear power is again the lesser of the two evils, if it is, in this context, an evil at all. We should again remember that if radiation has a hormetic effect on human health, then increasing background radiation could be good for us.

Comparing All Technologies. Inhaber made an extensive study of risks associated with various energy technologies. In it he considered not only risks due to the operation of power plants, but also risks involved in processing of raw materials, manufacturing of components and construction of plants. His main conclusions are reproduced in Figure 9-10. The histogram relates fatality rates (public and occupational deaths were combined) in various industries per unit of energy produced. According to that study, electricity produced by natural gas is the safest, the "least bad," while nuclear power is the second best.

Only on this comparison does nuclear power not win out (i.e., as the "least bad"). On the other hand, it should be noted that this comparison does not take into account the risks due to carbon dioxide pollution produced by burning natural gas, as these risks cannot be quantified, and also that he assumed linearity for estimating risks due to low levels of radiation.

Nuclear Proliferation and Energy Shortage. The potential connection between the development of nuclear power and the proliferation of nuclear arms is generally considered to be a very serious problem and unique to nuclear power. This problem, however, is more political than technical in its nature and, thus, it should be compared with political problems that would

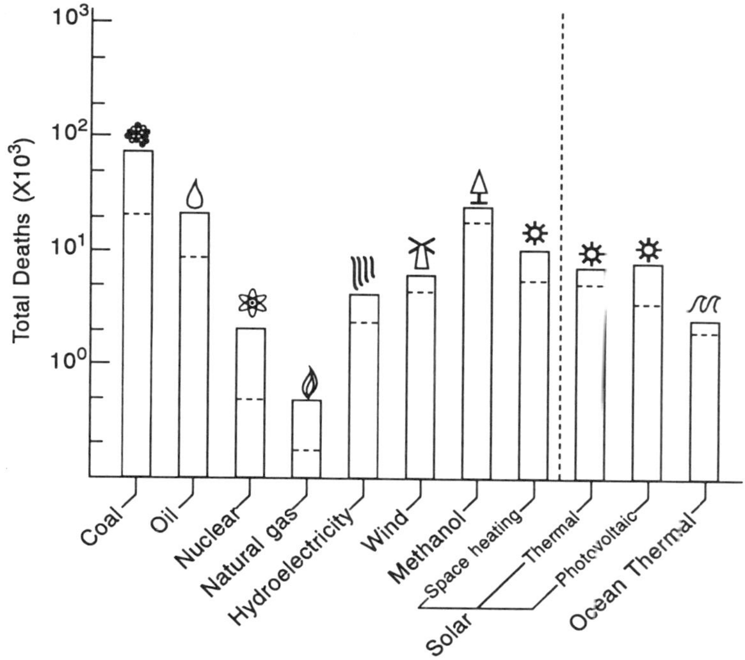

Figure 9-10: Total (public and occupational) deaths, times 1,000, per megawatt-year as a function of the energy system. The tops of the bars indicate the upper end of the range of values; the dotted lines within the bars, the lower. Those bars to the right of the vertical dotted line indicate values for technologies less applicable to Canada. (Reproduced from Inhaber.)

be created if a severe world shortage of energy developed and the various nations would have to agree on how to share the available energy.

It is possible to argue that a potential nuclear arms proliferation represents a much smaller danger to the world than an energy shortage. A world energy shortage would create reasons to fight wars, conventional or nuclear (Jovanovich, 1986a). Mere possession of nuclear arms is not dangerous if there are no reasons to fight a war. The United States, Britain, and France all have nuclear arms but, clearly, none of them, at present, feels threatened by the nuclear armies of the other two.

A Desirable Scenario

At present, close to 90 percent of all commercial energy consumed in the world is produced by burning fossil fuels. Ideally, due to the greenhouse effect, the burning of fossil fuels should be kept at the minimum, with nuclear power producing the bulk of the rest. How fast should the use of nuclear power grow in such an ideal case?

At present, a little under one quarter of primary commercial energy consumption in the world is used for electricity production. In some developed countries, more than two-fifths of their energy is used for that purpose, and the fraction is still growing. Thus, we shall assume that a half-century from now an affluent world will use, say, about two-thirds of its total energy for electricity production.

In Figure 9-11, in a schematic manner, is outlined a desirable, although probably not a very realistic, scenario for the future. Actual past and assumed future world commercial energy, and total hydro and nuclear electricity consumption are shown in a schematic manner. Electricity is expressed as the amount of energy needed to produce it. Present production of hydro and nuclear electricity accounts for about one-quarter of all electricity produced, the remaining three-quarters coming from the burning of fossil fuels.

A simple extrapolation of the total electricity consumption into the future is made on the assumption that at "saturation" our affluent world will be almost an "all-electrical" world, so that electricity consumption will increase over the present by about 25 times, although the increase in total energy consumption will be only 10 times. The curve for hydroelectricity is fairly arbitrarily extended into the future until a reasonable saturation is reached, a saturation which assumes that about three-quarters of the technically feasible hydro sites will be developed (see Figure 9-1). The curve for nuclear electricity has been extended into the future so that by the year 2020 all electricity will be produced by nuclear and hydro power. The heavy line represents the amount of fossil fuels that will still be used for purposes other than electricity production. It was obtained by approximately subtracting lines for hydro and nuclear electricity from the curve for total energy consumption.

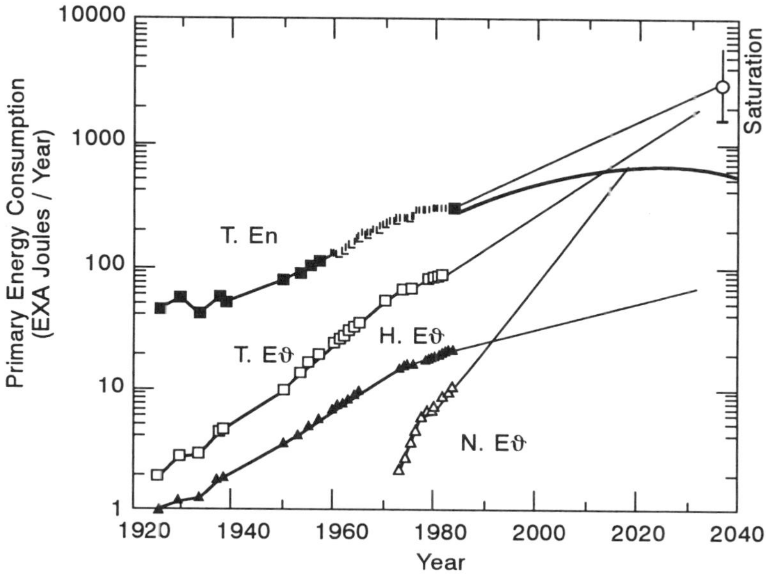

Figure 9-11: Actual past and assumed future world commercial energy (dark square) and total (square), hydro (dark triangle) and nuclear (triangle) electricity consumption are shown in a schematic manner. Electricity is expressed as the amount of fossil fuels needed to produce it, assuming thermodynamic efficiency of 36 percent. (At this efficiency 10 megajoules of fossil fuel energy are needed to produce one KWh of electricity.) The assumption is made that the future growth of total energy and total electricity consumption will continue as in the past until a saturation level of 10 times the present energy consumption (as in Figure 9-1) and 25 times present electricity consumption is reached. A postulated growth of nuclear electricity with doubling times of about six years was chosen so as to have only a modest growth in the use of fossil fuels, as shown by the heavy continuous line.

This exercise was performed to see what should be done in order to minimize further growth in fossil fuel consumption and at the same time to allow for the expansion of economies in the undeveloped and developing world along historical lines. The assumption of no fossil-fueled electrical power stations by 2020 essentially implies an immediate ban on new construction of these stations—in 35 years all stations that are now in existence will be retired from service as old and obsolete. The line on nuclear electricity then shows that nuclear power stations should be built at a rate doubling nuclear capacity every six years, which is similar to the growth of nuclear power since 1976 (cf. Figure 9-2). In this optimistic scheme, by the year 2030, there will be something between fifty thousand and one hundred thousand reactors operating in the world.

Even under these assumptions, the fossil fuel consumption, that is, carbon dioxide production, would double sometime in the first quarter of the next century and stay constant or decrease slowly afterwards. Climatic consequences of such an increase in fossil fuel consumption may not be trivial, but they would certainly be much smaller than if carbon dioxide content increased as indicated in Figure 9-9.

The above desirable scenario of fast growth is technically feasible only on the assumption that the technology of accelerator breeders can be developed on time, or that the extraction of uranium from ocean water proves technically and economically feasible. An affluent world fifty years from now will almost certainly have difficulty with liquid fuels for transportation, but liquefaction of coal, use of ethanol, non-polluting electric cars for short trips and trolley-trucks for long trips should be able to make up for the shortage of oil that would likely develop well before fifty years are up.

Whether such a desirable and optimistic scenario is feasible from the point of view of political and economic conditions in the world is, of course, another matter, which is beyond the scope of this article.

Conclusion

If existing energy resources and technological know-how to use them are used wisely, an affluent world with ten billion people and five billion cars can be fueled for centuries to come without producing unacceptable insult either to the environment or to the people themselves. Whether food resources, minerals and other raw materials exist to sustain an affluent world of ten billion people and five billion cars has not been discussed in this article, but there is reason to believe that they do exist. If so, then there are no physical limitations that would prevent the whole world from becoming at least as affluent as North America is today.

During the next few decades, the only energy sources economically available in sufficiently large quantities to power an increasingly industrialized world are coal and nuclear power. Renewable energy sources, including solar power, are either quite expensive, in short supply, or both. Nuclear fusion has not been harnessed for energy production yet, and there is simply no way it could make a significant contribution as an energy source within the next few decades.

The burning of coal threatens the global climate. Thus, it would be prudent to limit its use.

Nuclear thermal and breeder reactors are able to provide needed power, whenever and wherever needed, in an environmentally acceptable (the "least bad") manner and at a reasonable price (cheaper than coal) for thousands, even millions of years.

Unfortunately, the world is not introducing nuclear power as fast as needed in order to avoid risks due to carbon dioxide-induced climatic changes. Every country is solving its own energy problem in its own way without any regard for the effects on the globe as a whole. Futhermore, an irrational anti-nuclear movement is delaying the development of nuclear power in many countries that are very heavy carbon dioxide polluters. Such attitudes should not continue much longer. The production of energy in general, and carbon dioxide pollution in particular, have become of worldwide importance. They should be tackled at an international level.

What the world urgently needs is a moratorium on building new fossil-fueled power stations—not on building nuclear power stations.

Acknowledgments

It is my pleasure to express gratitude to Dr. H. Ellsaesser for inviting me to participate in the 15th International Conference on the Unity of Sciences and present this paper.

I also would like to thank Dr. A. Weinberg, Dr. Ralph Rotty, and others at the Institute for Energy Analysis at Oak Ridge for many enlightening discussions I have had with them while visiting the Institute during my sabbatical leave of absence.

References

AEC. 1957. Technical Possibilities and Consequences of Major Accidents in Large Nuclear Power Plants. WASH-740, U.S. Atomic Energy Commission.

APS. 1975. The Study Group on Light-Water Reactor Safety, "Nuclear Reactor Safety—The APS Submits its Report." *Physics Today* 28(7):38.

Barth, Michael C. and James G. Titus. 1984. "Greenhouse Effect and Sea Level Rise—A Challenge For This Generation." New York: Van Nostrand Reinhold Company.

BEIR. 1980. Committee on the Biological Effects of Ionizing Radiations, "The Effects on Populations of Exposure to Low Levels of Ionizing Radiation: 1980." Washington, D.C.: National Academy Press.

Bernard, Harold W. Jr. 1981. "The Greenhouse Effect." New York: Harper and Row.

Bloodworth, I.J. *et al.* 1978. "World Energy Demand to 2020—The Full Report to the Conservation Commission of the World Energy Conference." Latham, N.Y.: Science and Technology Press (Published for the World Energy Conference by IPC).

British Petroleum. 1984. "BP Statistical Review of World Energy."

Cohen, B.L. 1977. "The Disposal of Radioactive Wastes from Fission Reactors." *Scientific American* 236(6):21.

__________. 1983. "Before It Is Too Late." New York: Plenum Press.

__________. 1985. "Putting 'Risk' in Perspective." *Fusion* 7(2):37.

__________ and I-Sing Lee. 1979. "A Catalog of Risks." *Health Physics* 36:707.

Cohen, Jerry J. 1988. "Nuclear Waste Disposal: The Nature of the Problem." (Chapter 8, this volume.)

Cross, Michael. 1983. "Last Chance to Save Africa's Topsoil." *New Scientist* 99:289.

__________. 1984. "UN Admits Failure to Halt Deserts." *New Scientist* 102:3.

Darmstadter, J. 1971. "Energy in the World Economy." Baltimore: The John Hopkins Press (published for Resources for the Future Inc.).

Davies, John. 1977. "Conserving Uranium Without the Fast Breeder." *Nature* 270:376.

Deffeyes, Kenneth S. and Ian D. MacGregor. 1980. "World Uranium Resources." *Scientific American* 242(1):66.

Ellsaesser, Hugh W. 1985. "The Sahelian Drought—First Evidence of a CO_2-Induced Climatic Change?" Report UCRL-92499. Livermore, CA: Lawrence Livermore National Laboratory.

Encyclopedia International. 1974. New York: Grolier, Inc.

Foster, J.S., M.F. Duret, G.J. Phillips, J.I. Veeder, W.A. Wolfe, and R.M. Williams. 1978. "Nuclear Resources—The Contribution of Nuclear Power to World Energy Supply, 1975–2020." Latham, N.Y.: Science and Technology Press (Published for the World Energy Conference by IPC).

Glantz, Michael H. 1987. "Drought in Africa." *Scientific American* 256 (6):34.

Grand, P., K. Batchelor, J.R. Powell, and M. Steinberg. 1977. "The Accelerator Breeder, An Application of High Energy Accelerators to Solving Our Energy Problems." *IEEE Transactions on Nuclear Science*, NS-24(3):1043.

Groves, C.K. 1980. "Rain of Troubles." *Science*, July/August, 76.

Hafele, Wolf, Jeanne Anderer, Alan McDonald, and Nebojsa Nakicenovic. 1981. "Energy in a Finite World: Paths to a Sustainable Future." Cambridge, MA: Ballinger Publishing Company. (Report by the Energy Systems Group of the International Institute for Applied Systems Analysis.)

Hansen, J., D. Johnson, A. Lacir, S. Lebedeff, P. Lee, D. Rind, and G. Russell. 1981. "Climate Impact of Increasing Atmospheric Carbon Dioxide." *Science* 213:957.

Inhaber, H. 1982. "Energy Risk Assessment." New York: Gordon and Breach.

Jovanovich, J.V. 1986a. "Future Energy Consumption in Less Developed

Countries and Associated Macro-Risks." Presented at the International Symposium on Total Risk and Benefit Impact of Energy Alternatives held at the University of Waterloo, Waterloo, Canada, May 20–23. Published in "Energy Alternatives: Benefits and Risks," H.D. Sharma (ed.) University of Waterloo Press, Waterloo (1990).

_________. 1986b. "Energy Needed For Continuing Industrialization of Developing Countries." Proceedings of Energy and Development, International Scientific Conference, Zagreb, Yugoslavia, 29 September to 3 October, 1986, Beograd, Yugoslavia: Yugoslav Scientific Forum, 913.

_________. 1986c. "Spent Nuclear Fuel—A Waste or a Resource." Unpublished lecture note. Major part of this lecture note was published under the title: "How to Live with Nuclear Power: The Problem of Spent Nuclear Fuel," *Physics in Canada,* March 1989, p. 68.

_________ and R.M. Rotty. 1986. "A Plausible Future of Rapid Energy Growth in China." *Energy* 11:829.

_________. 1991. "The Chernobyl Accident: Five Years After, Part I: The Explosion," *Physics in Canada.* March 1991, p. 69; "Part II: Radioactive Releases and Consequences," *Physics in Canada,* July 1991, p. 116.

Kemeny, John G. 1979. Report of The President's Commission on the Accident at Three Mile Island, the Need for Change: The Legacy of TMI, October 1979, Washington, Washington, D.C.: U.S. Government Printing Office.

Lederman, Luis. 1985. "Probabilistic Safety Assessment: Growing Interest." *IAEA Bulletin* (Autumn):51.

Levenson, M. and F. Rahn. 1981. "Realistic Risk Estimates." *IAEA Bulletin* 23(4).

Likens, G.E., R.F. Wright, J.N. Galloway, and T.J. Butler. 1979. "Acid Rain." *Scientific American* 241(4):43.

Luckey, T. Don. 1988. "Hormesis and Nurture with Ionizing Radiation." (Chapter 7, this volume.)

_________. 1979. "Hormesis with Ionizing Radiation." Boca Raton, Florida: CRC Press, Inc.

Marshall, Walter. 1982. "Talking about Accidents." Paper presented at the IAEA Int. Conf. on Nuclear Power Experience, Atom, 312, 210, (October 1982).

Mercer, J.H. 1978. "West Antarctic Ice Sheet and CO_2 Greenhouse Effect: a Threat of Disaster." *Nature* 271:321.

NRC. 1975. "Reactor Safety Study—An Assessment of Accident Risks in U.S. Commercial Nuclear Power Plants." Report WASH-1400 Washington, D.C.: U.S. Nuclear Regulatory Commission.

__________. 1983. Report of the Carbon Assessment Committee, "Changing Climate." Washington D.C.: Board on Atmospheric Sciences and Climate, National Research Council, National Academy Press.

Nyegosh, P.P. (1811–1851) "The Mountain Wreath." Rendered into English in 1970 by James W. Wiles, Westport, Connecticut: Greenwood Press, Publishers (1970).

Oeschger, H. and U. Siegenthaler. 1978. "The Dynamics of the Carbon Cycle as Revealed by Isotope Studies." In Jill Williams, "*Carbon Dioxide, Climate and Society,*" p. 45, Proceedings of a IIASA Workshop, Feb. 21–24, 1978, Luxenburg, Austria: Elmsford, N.Y.: Pergamon Press.

OTA. 1982. Office of Technology Assessment, Congress of the United States, "The Regional Implications of Transported Air Pollutants," July 1982.

Revelle, Roger. 1982. "Carbon Dioxide and World Climate," *Scientific American* 247(2):35.

Rippon, Simon. 1984. "Accident Source Terms—Cause for Less Concern." *Atom* 327:2.

Rotty, Ralph M. 1984. "The Nature of the CO2 Problem Certainties and Uncertainties." *Environmental Progress* 3(4):253.

Sagan, Leonard A. 1987a. "Special Issue on Radiation Hormesis." *Health Physics* 52(5).

__________. 1987b. "What is Hormesis and Why Haven't We Heard About it Before?" *Health Physics* 2(5):521.

Schenk, H.J., L. Astheimer, E.G. Witte, and K. Schwochau. 1982. "Development of Sorbers for the Recovery of Uranium from Seawater." *Separation Science and Technology* 17(11):1293.

Shea, R.F. 1978. "Prose and Controversy"; Clamshell Alliance, "Nuclear Accidents: A Look at the Record." *IEEE Transactions on Nuclear Science* NS-25(5):1098. W.H. Arnold, "Perspective on Nuclear Accidents: A Look at the Record," *ibid.*, 1101.

Starr, Chauncey. 1970. "Role of Solar Energy in Electric Power Generation." In "Current Issues in Energy—A Selection of Papers," Elmsford, N.Y.: Pergamon Press.

This Week Report, "U.S. Study Says Acid Rain Can Be Beaten," *New Scientist* 99:9.

Totter, John R. 1988. "Cancer: A Result of Incidental Exposures or of Senescence, or Both?" (Chapter 11, this volume.)

UNEP. 1985. "Radiation: Doses, Effects, Risks," United Nations Environment Program, P.O. Box 30552, Nairobi, Kenya.

USSR. 1986. U.S.S.R. State Committee on the Utilization of Atomic Energy, "The Accident at the Chernobyl's Nuclear Power Plant and Its Consequences." Information compiled for the IAEA Experts' Meeting, 25–29 August 1986, Vienna. Translated by the U.S. Department of Energy as a Working Document for the Post-Accident Review Meeting.

Wilson, Richard *et al.* 1985. "Report to The American Physical Society of the study group on radionuclide release from severe accidents at nuclear power plants." *Reviews of Modern Physics* 57(93).

Woodwell, G.M. 1978. "The Carbon Dioxide Question." *Scientific American* 238:34.

TEN

THE DISTANT ENERGY FUTURE

Mark P. Mills

Throughout the past decade, a tremendous amount of effort has been devoted to identifying future energy supplies and the patterns associated with environmental impacts. My objective in this article is not to present another analysis of the technologies involved, but to look at the broad context of such analyses and explore the underlying philosophies that may pose methodological errors in long-term projections.

In broad historical terms, the developed world has only recently left an era dominated by wood and muscle power and entered one that is dominated by fossil energy and mechanical power. But we are now moving rapidly into another era, one in which electrical energy and informational power prevail.

Propelling the transition are inexorable forces imprinted on the fabric of our civilization: (1) population growth, (2) the pursuit of affluence, and (3) urbanization. The rate of increase in world population has declined, but the global population is still climbing. By 2050, it will be up to eight billion. Even with a substantial improvement in the energy efficiency of the developing world, it will not be possible to offset the simple energy demands for the mere subsistence of the 2.5 billion additional people that will exist in the developing world.

The pursuit of affluence, no matter how affluence is measured, is an integral part of our culture. Those who have it try to maintain or improve it while those without it strive for it. Most of the world does not have an affluent lifestyle, but in an information age, most will have access to images of how affluent people live.

The usual measure of affluence is GNP per capita. This measure is useful because it counts such factors as plumbing, irrigation, medical care, etc., and it also tallies luxuries such as transportation and other conveniences. Among developed countries, the differences in energy usage per GNP are relatively small. But the difference between energy usage in developed countries compared with that of developing countries is very significant, about twenty to one. Any modest increase in energy use or affluence in the developing world will be strongly reflected in overall energy use.

The third factor, urbanization, is still going on in the developed world. Despite idealized views of the rural life of the gentleman farmer, people are continuing to locate to the "fringe cities" of metropolitan areas. But urbanization is especially lively in developing countries where people are anxious to give up their hoe and hard lifestyle for a chance to reside in an urban center which they see as an opportunity to improve their lives, their children's lives, and those of their grandchildren. The obvious result of urbanization is bigger energy demands. Urban centers require not only more energy per capita, but a higher quality of energy.

Most energy studies incorporate these trends and conclude with goals or recommendations relevant to future energy needs and environmental impacts. The goals usually boil down to: reducing the amount of energy used in developed nations; keeping the increase in energy usage in developing countries to a minimum; and reducing the environmental impacts of energy usage. These are not mutually exclusive, nor are they achievable through central planning. The central planning economies fail to accomplish any significant goals in the economic sector of the development of nations. In any case, the issues are far more complex than can lend themselves to planning. More candidly, people's biases and basic needs influence plans much more than intellectual analyses of environmental impacts.

The first goal—to reduce the amount of primary energy used

per capita in developed nations—has been underway for decades. In the United States, for example, the amount of primary energy per unit of GNP has been declining. This largely unsung phenomenon has been occurring for 130 years and is obviously not a consequence of the oil embargo period. Instead, the downward curve reflects technological progress. Everything from improvements in steam cycle engines to developments in electricity drives continuing reduction in the amount of energy used per unit of GNP.

Attempting to reduce the total amount of energy used in developing nations is impractical if not impossible. Increasing amounts of energy will be needed per person in those nations in the near future and beyond. Moreover, fossil fuels—oil in particular—are the most attractive fuels for developing nations because they are simple, tremendously flexible and can be easily used to accomplish a multitude of tasks.

The goal to reduce environmental impacts is in harmony with technology induced reduction of energy use per capita. As stated earlier, improved efficiency is a natural consequence of improved technology. Where you find upward trends in technologies of energy utilization and production, you find declines in resource consumption. And, if fewer resources are used, this is the first indication that there are fewer environmental impacts.

There has been a series of interesting and complex projections made about the impact of energy utilization on the environment. As computer models become more sophisticated, each successive project group that looks at the energy question comes up with a more complicated projection. Unfortunately, the number of variables still exceeds by orders of magnitude the number that can be included in any models that are used for long-term projection. This fact underscores the reality that inherent in human nature is the inability to predict the future.

Consider transportation as an example. If, at the time of the deployment of the first DC9, an expert were asked to project the future aircraft fleet and its impact on the environment and energy, would he have anticipated a fleet of wide-body jets like the 747s? Possibly, because the technology was conceivable and was, in fact, anticipated. The time cycle was reasonable and the art predictable. However, at the time of the first propeller-driven aircraft, it would have been extremely difficult to predict the fleet of wide-body jets. To press the point, at the dawn of the automobile age, it would have been impossible to predict the energy needs and fossil fuel resources required to sustain the transportation needs of world civilization or even one country in the year 1990. We could not have anticipated the transportation systems that now exist, such as the world-wide fleet of wide-body jets, which consume tremendous amounts of energy. Nor would we have anticipated the phenomenal economic and social benefits. The bottom line is that current energy use and the associated environmental impacts could not have been predicted sixty years ago. Decisions regarding which energy resources to tap were driven by the technologies; technologies were not developed to meet the projections.

Two exhaustive long-term projections of energy needs were made by the International Institute for Applied Systems Analysis in Vienna (1981–82) and the Institute for Energy Analysis in Oak Ridge (1985). These were exceptionally thorough and objective by most standards, and they still come under fire for their biases and lack of objectivity; that is, for the assumptions researchers made to arrive at the projections.

These types of analyses are intriguing, but over the long term they tend to be meaningless. The time frames are up to sixty to seventy years into the future. In these time frames, a tremendous number of changes can happen. The variations in the prediction of the total primary energy that the world will require in 2050 run from no change to fifteen times more energy consumed. It's such an immense range that it seems these predictions mean little.

Fifteen to twenty years ago, people made energy projections for the 1980s. Those who got it right are now looked upon as sages, but too often, their conclusions were based on erroneous reasoning. Consider the soft-energy advocates, who in the early 1970s were predicting a U.S. energy use of about what we use today. They had the only low-path projections available—everyone else forecasted higher levels. If one reviews their models, however, one will see that their supporting conclusions were not borne out.

They forecasted a reduction in the use of centralized energy for the production of electricity and a penetration of so-called renewable energy source. The opposite happened. Renewable energy sources made no significant penetration in the last twenty years in the U.S., and the centralized projection of electricity increased dramatically. But, as mentioned earlier, primary energy use declined per unit of GNP. The primary influencers, unpredictable to the soft-energy advocates, were economic dislocations, technological progress, industrial stagnation, a recession, and the change in behavior patterns in response to high prices.

Let's continue briefly with the short term, the next ten to twenty years. Industrial and social inertia and political preferences will dominate events. In this period, it will be possible for people to behave counterproductively and get away with it. That is, for short periods of time, distortions in the economy will allow moves that make no energy-economic sense. Consider the U.S. federal tax subsidies associated with renewable technologies. An analysis of the nature of these subsidies found they yielded about ten to twenty times more federal dollars per unit of energy delivered than all of the conventional technologies. Naturally, the marketplace responded accordingly. Many windfarms, cynically called "tax farms" by some, were built. When the subsidies ended, so did the windfarms, by and large. Over the long haul, Natural Law catches up. Very diffuse energy sources which use the kinds of primary technology that exist today will not have a significant impact on U.S. and global energy needs.

All indicators show that in the near future, the world (the U.S. especially) will continue to depend heavily on oil. This will be driven by the dynamics of the world marketplace. Disincentives associated with capital-intensive replacements for world oil indicate that oil will be the preferred fuel. The fact is that the Middle East harbors at least two-thirds of the world's cheap oil. There are tremendous reserves there, but exploration activities are limited, so the resource is underutilized. There may be quite a few additional cheap "elephant" oil fields in the Middle East. As long as lifting costs are less than a dollar a barrel, the odds are very good that oil will be an attractive fuel, especially on the margin where all brokerage occurs, both in the developed and developing countries.

Long-term parameters are easier to describe because they are conceptually broad brush with fewer details. There are four parameters that drive possibilities in the very long term: natural limits; the nature of global resources; technological synergisms; and technological serendipity.

First, let me mention what I mean by natural limits. Up until the 1940s, we did not adequately understand the nature of the fission process and were thus unable to predict that one could use nuclear fission to make energy. Ignoring the contribution of nuclear fission was not a violation of natural law, but simply the result of unenlightenment. On the other hand, proposing energy systems that are in essence perpetual motion or require more energy to be put in than is produced violates natural laws. One natural reality is that high-density energy systems are better than low-density energy systems.

The second element, global resources, presents a much more complex problem which can be broken down into three areas: conventional, unconventional, and unanticipatable. In the conventional area, there are tremendous quantities of resources, more than are traditionally recognized. Geologists point out that advances in petroleum geology in the past ten to fifteen years are greater than they have been in over fifty years. There are over 800 petroleum regions in the world, of which only 260 have been explored and developed. Even in a "mature" oil province—the U.S.—the finding cost per barrel of oil has declined from $13 in 1981 to under $4 today.

The unconventional category holds tremendous opportunities. Resources are just waiting for technologies to tap them. Shale oil has possibilities, as does deep gas. There are opportunities inherent in the use of hydrogen should inexpensive hydrolysis techniques be developed and so on.

In the unanticipatable category of resources, one cannot pick—or anticipate—a resource for obvious reasons. But consider a wild card. The energy inherent in lightning from electrical storms is phenomenal. It may some day be possible to capture and store that electricity. It sounds like science fiction, but sixty or seventy years ago, today's technologies were science fiction.

The third element in exploring long-term parameters is technological synergisms. This is an area where energy analysts generally fail because it is impossible to predict the unanticipatable and because this area often involves another realm of expertise. Synergisms between an information society and electricity usage already herald phenomenal changes in energy use patterns in the world, certainly in the developed world. The capability to manipulate and use energy will profoundly change our ability to do things. The amount of energy we consume will be reduced as we become able to use it more precisely and with greater control. The best energy source or medium for this is electricity. Over the last decade and a half, trends and patterns of the western world all point to a big reduction in primary energy use per unit of product produced—at the same time increased uses of electric energy per unit of product produced are evident.

In material production, this trend is unbroken and is driven by the flexibility of electricity and the controls made possible by computers. The computer revolution is still in its infancy. When it is fully mature, the impact on energy use will be profound.

The fourth long-term parameter may be technological serendipity. What are the items that will occur under serendipity? Who knows? Consider the development of a room temperature superconductor. A room temperature superconductor would significantly change energy distribution and use patterns. It would not only reinforce the synergism of electrification but would make auto electricity a reality. The electric car, electric transportation, all manner of things, will become practical. There would be very big changes, changes we can't anticipate because they are beyond the realm of imagination. The advances in the last sixty years have been phenomenal—there is no reason to doubt that we will see equal or greater advances in the next sixty years.

Another "wild card"—the force of gravity. One truism of civilization and of science is that when we finally understand a force, a short time later, we learn how to use it. Once the phenomenon of electricity and magnetism was understood, it wasn't long before practical applications arose. It took us a while to figure out what the strong nuclear forces were and to start to use them, and we still haven't peaked in this realm. We don't know much about the weak nuclear forces, or about gravity. When these forces are understood, it is a natural consequence that we start exploiting them. The implications are astounding.

There are other possibilities, like organic computers, which introduce a whole new realm of informational density and information management. Again, this will have impacts on industry, society, infrastructure, and how we do things.

Finally, I offer a concluding note about long-term projections. In the next fifty years or so, our understanding of natural systems is bound to become more sophisticated and drive technology advances at a rate more rapid than that of the past half century, a period which itself hosted huge technological development. The simple extension and full utilization of the synergisms that we now understand can foreshadow dramatic changes in energy utilization, most of which will be in the developed world in the areas of electrification and informational technology.

ELEVEN

CANCER: INCIDENTAL EXPOSURES, SENESCENCE, OR BOTH?

John R. Totter

Introduction

Discussions of cancer epidemiology usually start from a foundation of assumptions that are rarely explicit enough to provide a base from which a firm argument can be built. In this discourse I will attempt to state as clearly as possible the assumptions, usually only implied, that are necessary to arrive at the conclusions one sees or hears about almost daily in the media.

The current most popular hypothesis on the origin of cancer was epitomized most succinctly by Dr. Robert Hoover in his oral presentation at the New York Academy of Science Symposium in March 1975. He stated that "... every cancer has a cause ..." Although one can hardly take exception to a simple restatement of the principle of cause and effect, in the context of his other words, it was clear he meant we could find the cause, eliminate it, and therefore be free of that type of cancer, thus reducing the total cancer burden of the population by that much. For this to be a reasonable conclusion to draw, one must make a rather complex set of assumptions and find support for them. Most simply, one might assume that humans are potentially immortal, all causes of death are "accidental" in the sense that they are not inevitable, and then it would follow that Dr. Hoover was presenting a conclusion that is a necessary consequence of these premises. Where is the scientific support for the hypothesis that our bodies are immortal?

Disease of Senescence

Some causes of death are clearly accidental; these include traumatic injury as well as most of the infections and parasitic diseases we die of, because at least the exposure is not inevitable. If we are not immortal then some of the mortal diseases must be diseases of senescence. The categorization of the causes of death we use is all-inclusive and there is no category that is labeled "old age," or "senescence" in this country. Therefore, an operational definition of the classes of diseases that are to be labeled "diseases of senescence" must be made.

One way to do this is to examine mortality tables and discover which diseases tend to be very infrequent in early life but highly prevalent in old age and associated with death. If the relationship is such that it can be expressed as a smooth mathematical function, its frequency rising steadily with age, then that disease is a candidate for designation as a disease of senescence. If all, or at least all of the quantitatively important causes of death, are treated in this fashion and the results expressed by some parameter such as median age at death, they can be compared as in Tables 11-1 and 11-2, which list causes of death of women in England and Wales for 1861 and 1964 respectively. The data are from the compilation "Causes of Death" (Preston, Keyfitz, and Schoen, 1972). The last three categories listed in both tables are the only causes of death that fulfill the two criteria (fatal and rising smoothly with age) listed above for "senescent" causes of death. Death from influenza, pneumonia, and bronchitis fit the two criteria in the 1964 table but not in the 1861 table. In most other countries examined, only the last three causes of death listed in Tables 11-1 and 11-2 appear to be suitable as candidates. Preston (1976) has earlier shown the special relationship of the "other and unknown" category to cardiovascular disease and cancer.

The acceptance of these three groups of causes of death as the main contributors to "senescent" deaths is not in qualitative conflict with current views on cancer. With the exception of a few vocal extremists, many epidemiologists accept the view that there are some "spontaneous" cancers. The problem then becomes one of differentiating the fraction of the mortality from cancer that is "spontaneous" (chiefly owing to senescence) and for the remainder determining the causes (presumably from detectable external environmental conditions or from "life-styles" that can be altered or removed entirely). If this part of the problem can be solved, there still remains a less obvious problem: Why, assuming that we find a large fraction of cancer

Table 11-1: The Median Ages at Death and the Fraction of the Population Dying from the Listed Causes of Death

	England and Wales Women 1861	
Cause of Death	*Approximate Median Age at Death From Cause Listed (years)*	*Fraction of the Newborn Population*
Certain Diseases of Infancy	0.5	0.03986
Diarrheal Diseases	3.7	0.04603
Other Infections & Parasites	4.3	0.12199
Maternal	31.0	0.01257
Respiratory Diseases & Tuberculosis	32.3	0.11618
Violence	37.0	0.01676
Certain Degenerative Diseases	50.0	0.00677
Influenza, Pneumonia, & Bronchitis	58.0	0.12495
Cancer	59.3	0.03173
Cardiovascular Disease	66.0	0.13251
Other and Unknown	71.0	0.35065
		1.00000

Table 11-2: The Median Ages at Death and the Fraction of the Population Dying from the Listed Causes of Death

England and Wales Women 1964

Cause of Death	*Approximate Median Age at Death From Cause Listed (years)*	*Fraction of the Newborn Population*
Certain Diseases of Infancy	0.5	0.00899
Diarrheal Diseases	77.0	0.00591
Other Infections & Parasites	66.5	0.00333
Maternal	31.0	0.00073
Respiratory Diseases & Tuberculosis	64.5	0.00165
Violence	75.0	0.03754
Certain Degenerative Diseases	75.0	0.02180
Influenza, Pneumonia, & Bronchitis	81.0	0.09778
Cancer	70.3	0.17068
Cardiovascular Disease	80.2	0.56632
Other and Unknown	78.7	0.08527
		1.00000

mortality to be, in fact, the result of senescence, does cancer mortality appear to be "independent" of cardiovascular mortality? In other words, if most or all of the diseases classified as belonging to the category of cardiovascular disease are the result of senescence, shouldn't all of the diseases classified as cancer behave in the same statistical manner if they are also found to be chiefly the result of senescence? The answer to this question is very complex and involves yet another subject—the cause(s) of aging itself.

Current Views on Cancer Causation

Before proceeding further with the relationship of cancer to aging, additional information about the origin of the current popular view of cancer causation will be briefly reviewed. Most problems of the type posed by cancer occurrences are attacked in the laboratory by reductionist means. The first clues pointing to possible laboratory study arose from observations of cancers associated with certain occupations that involved exposure to chemicals. The most striking of these, but by no means the only one, was the association of scrotal cancer with chimney sweeping, pointed out by Sir Percival Pott. Chemicals implicated as specific causes of human cancer were certain nickel carbonyls (Doll, Morgan, and Speizer) and 2-naphthylamine (Friedell; Hueper). These discoveries led to successful searches for other cancerogenic agents. The findings gradually gave rise to a general belief that all or almost all cancers may indeed arise from exposures to chemicals. If exposure to cancerogenic chemicals could be prevented or avoided, then perhaps, it was thought, most cancers could be eliminated.

Another and quite unrelated series of studies provide additional reasons to consider seriously that cancers are entirely "accidental" in the sense that their causes might be avoided. These studies arose from interest in geographical differences in the occurrence of different types of cancer. For some years there was a section of the National Cancer Institute, under the leadership of H.L. Stewart, that gathered data on incidence and mortality from cancer on a worldwide basis and evaluated the completeness of the data (see Dunham and Bailar for an example). The chief (but not the only) early proponents of the idea that the differences in incidence or mortality from specific cancers in different localities must be, at least in part, attributable to differences in exposure to external agents, were John Higginson, and Richard Doll (1977). Doll, in his review, lists cancers which show ratios of about five to one to more than two hundred to one between high and low incidence areas.

Both of these investigators now prefer to attribute most cancers (at least 60–70 percent) to "lifestyles," perhaps because they recognize that it is difficult to identify sufficient concentrations of known carcinogens in the environment to account for more than a few percent of the observed total cancer burden. It is clear, however, that both Higginson and Doll still support the idea that cancers are independent of each other and of cardiovascular disease. This assumption is a vital part of their arguments.

The Problem of Independence of Causes of Death

It is difficult to explain to people unfamiliar with statistical procedures or with epidemiology, the effect that the assumption of independence has on both calculations and conclusions concerning the effects of environment on lifespans. Perhaps the following story will help introduce the problems to the reader who may be unfamiliar with the importance of the assumption to epidemiology:

> A Martian biologist was given a grant to spend one year in the U.S.A. to determine the secret of human longevity. Being asexual itself (as all Martians are) and never having encountered sexual reproduction in populations it had studied, it was somewhat perplexed by the noticeable differences in the two major classes of humans it saw. It was far too busy gathering data to study the matter closely but did manage to mark down whether the subjects habitually wore dresses or trousers, and insured that the otherwise random sample had equal numbers of the two classes even though it was evident that many more wore trousers than dresses.
>
> Recognizing that the quickest way to study the problem of longevity was to prepare a life table, it proceeded to gather data from a suitable sample of the U.S. population for 1964. There wasn't time to do a complete study, but it did manage to get death rates from all causes, and separately for cancer and cardiovascular disease for all ages. The median age at death for the Martian's sample was determined to be 75.59 earth years. There was just enough time to prepare Table 11-3.
>
> By the time this Table was finished, the Martian Embassy notified the investigator that, owing to the large national budgetary deficits, its support would be terminated immediately and it must return to

Table 11-3: Summary of Statistics at the Median Age of Death

		Cause of Death		
Subject Class	*Median Age at Death*	*Cancer*	*Cardiovascular Disease*	*All Causes*
			Fraction of the Population	
Trousers	73.8	.053	.141	.284
Dresses	77.99	.050	.104	.216

Mars. The report was written in great haste with the conclusion that the wearing of trousers imposed a serious environmental hazard, the effects being shown mainly by a six percent increase in the cancer risk and a greatly elevated cardiovascular death rate. The average life shortening, owing to trousers, that was observed was estimated to be 4.19 years.

The Chief of Environmental Health at Interplanetary University, who received a copy of the report and knew all about sex, observed that the Martian's results would agree with his own records if 33 percent of the Martian's "trousers" group were women.

This story illustrates many of the problems that confront the epidemiologist, especially when he attempts to assign causes for the differences that he sees in his samples. In the story, the epidemiologist did not know that genetic differences might be the underlying cause of the difference in longevity, as well as some of the difference in wearing apparel. The Martian assumed that it was seeing differences that were independent of constitution. Even people who have access to knowledge of the differences that arise from differing genetic constitutions sometimes prefer to ignore them. For example, Miller and Gerstein prefer to attribute the male-female difference in longevity entirely to differences in smoking habits, a conclusion that demands complete independence of causes of death. Many epidemiologists take a less extreme stand and attribute one-half or somewhat less to cigarette smoking prevalence among males. However, is it possible that they (for that matter, most of us) are as blind as the Martian toward some constitutional difference in those who can and usually do become addicted to cigarettes and those who do not? Following R.A. Fisher (1959) and others, P.R.J. Burch (1975, 1978, 1984) has repeatedly tried, without much success, to convince skeptical epidemiologists that they have no sound logical foundation upon which to base their contention that smoking cigarettes shortens the lifespan by several years. In a recent letter to the editor of *Chemical and Engineering News*, Richard J. Hickey also has reiterated his dissatisfaction with the basis of the U.S. surgeon general's attack on tobacco use.

The conclusions that are usually drawn about cigarettes causing life shortening are squarely based on three assumptions:

first, that lung cancers are caused by cigarette smoking; second, that deaths from the smoking-associated lung cancers and cardiovascular diseases occur at younger ages than the smokers would otherwise have succumbed; third, and most importantly, that cigarette smokers constitute a valid random sample of the (male) population.

If the third point made above were to be resolved, the other two also would be quickly resolvable. To illustrate, let us look again at data from the 1964 life tables for U.S. males (Preston, Keyfitz, and Schoen, 1972). From these data, it can be calculated that the median age at death of the shorter-lived one-half of the male population is 59.28 years while for the longer-lived one-half it is 79.93 years (these are the life table ages at which 25 and 75 percent respectively of the population have succumbed) for a difference of 20.65 years. Combining these two linearly in different proportions allows us to estimate, rather crudely, the divergence from randomness that would give any desired shortening (spurious) of lifespan that might occur if the cigarette smoking cohort were to be unevenly divided between the two halves. If it is evenly divided, the median age at death becomes 69.61 years. (The actual median age is 70.81. This difference is small, compared with the 20.65 years and attributable to the error involved in assuming a straight line survival curve.) If the division is 65:35, that is, 15 percent more of the total smokers are in the short-lived group, the calculated median age is given by:

$$0.65 \times 59.28 \text{ yr} + 0.35 \times 79.93 \text{ yr} = 66.51 \text{ yr}$$

or 3.1 years shorter than the 69.61 years given for the 1:1 ratio. Of course, in a real, well-designed epidemiological study there would be many refinements built in to try to minimize the errors but this crude estimate shows how close the distribution of smokers into the two halves of population, long-lived and short-lived, must be in order to calculate a reasonably accurate median age at death for the smokers.

There is no direct way to resolve this difficulty. We cannot bring the dead smokers back to live their lives over again without smoking to see what their lifespans would then be. Many indirect methods have been tried but they are, without exception, plagued by the same difficulty detailed above or by an analogous one.

Is there any reason to believe that smokers are not a valid random sample of all of the genetic constitutions that occur? It may be speculated that most people do not adopt a costly habit unless they derive some benefit from it. Could it be that the benefit is the relief of some perceptible stress that is strongly associated with a genetic constitution that is, in part, the cause of a short life?

The problem of independence can be understood better if we compare the epidemiologists' needs with those of the insurance actuary. Both have available the same information and both use many of the same statistical techniques to organize the data. The question that should be uppermost in the epidemiologists' mind, if he is comparing the longevity and causes of death of two individuals living in two separate but not distant communities, is if these two people were to be exchanged would they have the same or different longevities; the same or different causes of death? If it were possible to have the two live their lives over from birth but each in the other's environment, we would be able to say something about how much the environment affected their longevity and causes of death and how much stemmed from their different genetic constitutions.

On the other hand, the actuary has much less interest in whether the genetic constitution largely determines longevity. He deals with large numbers of individuals and he knows that the mortality curve is unlikely to change rapidly. He can, therefore, deal with his insurance problem with little worry that his calculations can be rapidly outmoded. Whether the environment or heredity determines the shape of the curve is not important for his calculations. He can use the assumption of independence of causes of death to estimate the number of deaths from other causes when one or two causes are removed without making a significant error in the estimate, even if the assumption is not biologically correct. He has no responsibility for attempting to define the mechanism of aging.

It is when we wish to attack the problem of cause or to promote means to eliminate a cause of death that the assumption of independence begins to assert an overriding importance.

With the possible exception of sufficiently large studies on identical twins reared in different environments, epidemiological studies cannot offer definitive proof of independence or dependence of different causes of death because the direct experiment of bringing deceased people back after death to relive their lives in a different environment cannot be accomplished. The inability to provide a scientifically acceptable resolution to the problem of independence places most of the current hypotheses attributing causal relationship to factors that correlate well with cancer incidences and mortalities into an intellectual limbo. This opens the way for promotional fervor to overcome natural scientific skepticism and for some dedicated believers to misuse their power as reviewers of manuscripts to suppress scientific dissent and degrade the system of self-correction of science through peer review. This aspect of the problem was noted by Efron in her review *The Apocalyptics.* Many grave consequences follow from this distortion of the scientific effort. In particular, "campaigns" planned to eliminate this or that cause of death may result in much ill-aimed and costly efforts detracting attention from the routes that must eventually be followed to achieve the real and usually worthwhile goals of the campaigns. It would be ironic if someone who almost was within grasp of a solution of a key problem in aging was frustrated from doing so by failing to benefit from the soothing effect of a good smoke at a crucial moment because his workplace was off limits for smokers.

Senescence as a Cause of Cancer

The alternative to the hypothesis of "complete" independence is "complete" dependence. That is, all causes of death attributable to senescence are strictly determined genetically. Of course, genetic causes exert themselves differently in different environments. The problem is to determine to what degree and in what manner does the environment have to change in order to change, by a given amount, the age at death of a person dying from a genetically determined cause of death? It is pointless to ignore the environment with whichever hypothesis one is concerned.

It should be evident that the choice between hypotheses concerning the relative importance of heredity and environment in determining the lifespan of individuals and populations cannot be made on a basis of solid proof. That is not available and the choice we make will have to depend first on the degree to which a hypothesis can explain without stress findings that can be accepted as factual. Second, it must fit comfortably within broader aspects of biology, not conflicting with generalizations that are basic to our understanding of biology.

With these thoughts in mind, some comparisons will now be made between the currently popular hypothesis placing emphasis on "environmental" (or "lifestyle") causes of cancer and a hypothesis placing a much larger dependence upon constitutional (genetic) causes. It should be clear that what is being striven for is a reasonable compromise that accounts for the facts, with the outcome expected to be somewhere between the extremes of 100 percent "accidental" (as defined above) and 100 percent "natural" causes leading to mortality from cancer.

The first question to be considered concerns the form of the mortality curve for cancer as compared with other causes of death tentatively placed in the "senescent" group, as indicated above. It can be noted that if we accept the idea of the deaths being mostly "accidental," i.e., not due to senescence, we have to explain the mathematically regular increase in death rate with age. This requires an additional *ad hoc* hypothesis, that of a series of latent periods with just the right length to make the curve ascend smoothly.

Alternatively, if we accept the constitutional hypothesis, the cancer mortality curve should resemble that obtained from the other "senescent" causes which, in fact, it does, although it is not identical with them. In particular, the median age at death is lower and the variance differs from that for cardiovascular disease. Because individual components of the cardiovascular group also differ among themselves, this does not seem to present a difficult problem. The shape of the curves are near enough to the normal curve to qualify as showing the usual type of variability expected to result from genetic heterogeneity.

The lower median age of death from cancer allows us to examine this difference for variability inasmuch as time series data are available for several countries. Some of these available from *Causes of Death* (Preston, Keyfitz, and Schoen, 1972) are shown in Table 11-4. The high correlation of the differences in median ages at death, for both men and women between "all" senescent causes and cancer, provides good evidence that the differences are not accidental but rather must be constitutional. Also, the fairly close concordance between median ages at death for cancer in men and women suggest that only a small difference could result from the rather large differences in environment (including smoking) encountered by men and women over the course of the years represented in Table 11-4. This argument, of course, may not hold for other components of "lifestyle."

The secular changes in median age at death seen in Table 11-4 may be in small part arithmetical, but a close examination reveals that spurious effects, if any, must be small. The changes are largely owing to real changes in lifespan under the envi-ronmental conditions extant during the lifetimes of the populations. If they are largely owed to the current environment, there is little support for any argument that conditions have deteriorated during the fifty to one hundred years preceding 1964.

On the contrary, it appears much more likely that genetic changes have occurred rapidly during the last few hundred years and these have become manifested (i.e., phenotypes have more closely approximated genotypes) since the Industrial Revolution began increasing the general prosperity of the various populations.

The Relationship of Quantitative Nutrition to Senescence

Charles Dickens has written of a person's fiscal balance: "Annual income twenty pounds, annual expenditure nineteen, nineteen and six; result happiness. Annual income twenty pounds, annual expenditure twenty pounds ought six; result misery."

Nature accounts for an animal's energy balance just as rigidly. If we have a smaller energy income than outgo, she will demand her pound of flesh. If it is the other way (and what may

Table 11-4: Changes with Time in Median Age at Death from Cancer and from "Senescent" Causes

	Women			Men		
Year	*Cancer years*	*"Senescent" Causes years*	*percent**	*Cancer years*	*"Senescent" Causes years*	*percent**
			Australia			
1911	66.085	73.358	(62.8)	66.035	70.964	(57.3)
1933	68.281	75.657	(68.3)	69.512	73.021	(79.3)
1964	71.291	77.880	(71.3)	69.406	72.043	(78.5)
			England & Wales			
1861	59.342	69.350	(47.2)	62.733	67.141	(42.5)
1901	62.575	69.269	(51.3)	63.767	68.883	(42.9)
1931	67.062	73.522	(65.8)	67.006	71.757	(59.4)
1964	70.338	78.381	(81.7)	68.790	72.380	(76.8)
			Italy			
1881	60.499	63.698	(34.3)	62.818	63.492	(32.0)
1901	62.324	68.344	(46.6)	64.595	68.388	(43.8)
1921	64.003	72.593	(55.7)	65.198	72.061	(53.0)
1931	64.794	73.396	(54.7)	65.725	72.453	(56.5)
1964	70.202	78.570	(75.2)	68.060	74.289	(81.4)
			Japan			
1899	59.171	65.263	(57.8)	61.499	63.290	(55.2)
1908	59.897	66.638	(53.9)	62.398	65.202	(52.5)
1940	61.126	70.054	(54.3)	62.214	66.328	(50.1)
1951	63.693	74.180	(65.9)	64.601	71.450	(59.9)
1964	68.131	77.523	(80.8)	67.686	73.434	(75.2)
			USA			
1900	61.754	68.540	(44.9)	65.553	68.889	(39.6)
1930	67.724	71.588	(59.9)	65.657	70.449	(53.9)
1964	70.044	78.663	(85.1)	68.833	72.211	(81.4)
Mean for listed countries						
	64.917	72.273		65.630	69.906	
SD	4.023	4.643		2.585	3.305	
Correlation Coefficient*[(r)]	0.925			0.870		

* The numbers in parentheses are the approximate percentage of those surviving to age 1 that die of "senescent" causes. For this table, "senescent" causes include cancer, cardiovascular diseases, and other and unknown causes. Cancer is not excluded because its contribution to the total deaths is relatively small and other and unknown category must include some cancer as well as cardiovascular disease.

be even worse) she will add the pound of flesh. These are temporary expedients, however, and there are more subtle adjustments that go on over the long term. We can see why there must be other ways to adjust by estimating weight that would have to be added if we consumed each day a 10 percent excess in Calories. A pound of flesh represents about 600–1,500 Calories depending on how much is fat. This is one-third to about two-thirds of a day's caloric requirement for a sedentary adult male. If another 500–1,000 Calories are needed to convert the rest to flesh, we see that an extra pound of flesh represents about one day's intake of food. At 10 percent excess, ten days should add one pound, one year 36.5 pounds, and ten years 365, etc. We do not see many individuals that weigh more than 400 pounds, even though we may be reasonably sure that many people eat in excess of 10 percent more than their daily requirement for maintenance of weight. Those of us who are old enough to remember the depression of 1929–1942 know that people ate at least 10 percent less than they do now, on the average. Not many died of starvation but fat people nearly disappeared from view during the decade of the 1930s.

As we shall see, the way our bodies adjust to differences in intake of food has an important bearing on our lifespans and, consequently, on the rates at which members of a population die of senescent causes, including cancer.

What mechanisms are available to ameliorate the effects of too great a caloric intake? Several things quickly come to mind: Increased physical activity; increased conversion to heat and elimination of the excess heat; and increased reproductive activity. The mechanisms for accomplishing all of these things are reasonably well-known. An increased production (or altered composition) of the thyroid hormones results in what appears to be a reduced efficiency of conversion of energy to muscular activity, thus increasing the heat output relative to the muscular work done. Altered functioning of the pituitary, perhaps secondary to the need for more thyroid activity, may change the reproductive cycle, preparing for increased reproductive activity. In any case, the individuals in the population must respond to the signal from the environment that indicates an abundance of food. Those species which lacked an appropriate response must have long since been crowded out by those that developed the more efficient mechanisms for responding appropriately. In humans, abundant food reduces the age at which young women become able to bear children. Also, it reduces the intervals between childbirth because it reduces the length of time young must be nursed to ensure their survival, and nursing inhibits conception. These responses inevitably lead to an increase in the population of early maturing, rapidly reproducing individuals relative to the late maturing, slow reproducing ones. When food is scarce (relative to the energy required to gather it), the early maturing, rapidly reproducing contingent would be most sensitive to the lack and tend to die early, while the late maturing, slow reproducing group should be able to subsist on less, and consequently, in the long run, produce more offspring than the shorter-lived group. In this way, it can be seen that reproductive rate should, through natural selection, rapidly become negatively correlated with lifespan. The direct observation of this reciprocal relationship in experimental animals was probably first made by Hall and Marble, who worked with laying hens.

It can readily be appreciated that the scenario outlined above could account for a population "boom and crash" cycle under certain environmental conditions. Under other conditions, it might lead to a fairly steady equilibrium or to any combination between these extremes. The scheme allows for entrainment of cycles into externally driven fluctuations such as an annual cycle. It seems quite likely that the cycling between extremes, whether steady or not, is a major source of variance in lifespans within populations. It also should be a source of variance in size as explained elsewhere (Totter, 1985), and a source of intermittent or continuous pressure for an increase in size over long periods of time.

Radiation Hormesis

If we combine the foregoing suggestions with what is known of radiation hormesis in animals, we can lay a solid foundation upon which to base a hypothesis that accounts for senescence operationally. Unfortunately, it does not give a solution for this problem at the molecular level, but it has much to offer at the physiological level and points to the type of research needed for progress in understanding aging at both a physiological and a molecular level.

When x- or gamma rays are absorbed in water or a water-containing medium, electrons are freed from the water molecules following which the water decomposes into hydrogen atoms and hydroxyl radicals as well as into hydrogen$^+$ and hydroxl$^-$ ions. Unless it has been carefully excluded, the oxygen in solution scavenges the hydrogen atoms to produce superoxide radicals and hydrogen$^+$. The superoxide radicals can combine with others of the same kind and hydrogen$^+$ to give hydrogen peroxide and oxygen. The hydrogen peroxide is a powerful oxidizing agent because at tissue pH and in the presence of traces of certain metals (iron, copper, etc.) it gives rise to hydroxyl radical that may produce Fenton-type reactions. The damage to cells and cell components caused by the products of radiolysis are called "indirect effects" because they can occur to molecules that do not suffer direct ionization through absorption of radiation. Up to about 70 percent of the measurable tissue damage from low linear energy transfer absorption may result from the indirect effect of ionizing radiation (Lea).

Oxidizing Radicals and Senescence

Evidence has been accumulating for the past thirty to forty years that indicates the presence of, and the necessity for, oxygen radicals in metabolizing tissues. These radicals are similar to or identical with the ones produced by the absorptions of x- or gamma radiation by oxygen containing water. (For reviews, see Rodgers and Powers; Gottlieb, Langmuir and Totter; Ames). It has long been believed that exposure to ionizing radiation produces many of the same biological effects as are seen to result from the passage of time (Harman, D.; Lindop and Sacher). Several investigators have advanced the idea that "aging" is, in fact, largely or entirely the result of the deleterious actions of oxidizing radicals produced during metabolism. The well-known relationship between basal metabolic rate and lifespan in endotherms is in good accord with the hypothesis (Sacher).

If the conclusion is correct that oxidizing radicals are responsible for aging, then we should see some manifestation of small doses of whole-body ionizing radiation that mimic the effects of food. They are not difficult to find. One of the most universal effects of a meal is the suppression of appetite. Reduction of appetite is a well-known effect of whole-body, head or trunk irradiation. It is true that the dose-effect relationship for this manifestation is not well explored. However, the effect is no more likely to exhibit a threshold than any other known effect of radiation exposure. It has been suggested that this appetite effect is responsible for the hormetic effect of radiation (Totter, 1985). If appetite is suppressed by nearly continuous low-level exposure to gamma rays, one would expect to see, as one result, the effects of the reduced food intake. The most conspicuous of these effects is lifespan extension as seen in the experiments of Lorenz *et al.* (1952). Their results may be compared with the food restriction experiments of McCay *et al.* and many others.

An abundance of food tends to "rectangularize" mortality curves as may be seen in human populations undergoing the "demographic transition." This result is also seen in the Lorenz experiments. Of course, no calories useful to the animal are obtained from the irradiation, so weights may not increase as fast as they would with more food. However, an additional physiological change consistent with the "food substitution" idea was seen by Lorenz *et al.* This was the increase in body fat seen in irradiated animals, as compared with non-irradiated controls.

The animals receiving the larger semi-continuous radiation exposure (0.002–0.088 gray/day) had quite steep survival curves with shortened lifespans, in accord with the steepened survival curves in well-fed humans.

The chief difference in results between the humans and the irradiated mice were that the largest increase in death rates in mice occurred because of cancer rather than cardiovascular diseases. It has been suggested elsewhere (Totter, 1985) that this difference may be owing to difference in normal flux levels of oxidizing radicals between the surroundings of the nucleus and those of the cell walls in tissues.

The oxygen radical hypothesis of aging offers many attractive avenues of research. It also offers a basis for understanding how longevity (or senescence) has been molded by selection. The strategies such as repair, reduction of radical flux, etc., that are necessary to promote longevity are easy to enumerate. It must be remembered that the most important natural one was increase in size, which permitted a reduction in basal metabolism—the source of the radicals.

As indicated at the beginning of this section, the key problem to the organism is to use the energy available to it to insure its own survival and that of sufficient offspring so that the species persists. There are ways to manipulate the requirements for energy to insure species survival. Those animals that were not lucky enough to find the right combination are no longer here. The very small mammal has to gather a prodigious amount of food to keep warm and have enough excess energy to produce frequent large litters of young. Its high metabolic rate ages it rapidly. It has no time to invent ways to gather and store food for winter so that it could reduce its reproductive rate and so acquire a somewhat longer lifespan.

The large mammal or bird requires much less food for unit weight so it can put relatively more energy into insuring a longer life. It cares for its young and protects it against predation so it need not reproduce so many times. These savings permit improved protection against the smaller number of radicals formed so that it ages more slowly.

The Future

Humans are in a transition period now—evolution is proceeding at a prodigious rate. We are changing both in size and longevity. We have, owing to organization and mechanical transport, severely reduced the energy (our own) formerly required to gather food. Our young seldom die before reproductive age. We could reduce our reproduction so as to just maintain numbers. The energy savings could be used to increase longevity—but must we wait for the evolutionary change —could we cheat a little? The trouble with waiting for the evolutionary change is that it is likely to go in an undesired direction. If food remains abundant and easily obtained, the signal to our physiology (and the endocrine glands will respond) will be to reproduce rapidly and age quickly. One can already discern this tendency in the statistics for some developed countries. For example, the reduction in age at puberty among young women is well established (Wyshak and Frisch).

On the other hand, if we eat sparingly and exercise vigorously, the signal to the endocrines is that food is difficult to obtain and calories must be husbanded to insure that we can use them to obtain food. Consequently, our reproductive rate and our aging rate diminishes. The diseases of senescence are postponed and our lives lengthened. The adjustment of the endocrines to the environmental signal created by the scarcity or abundance of food and our "energetic" response is best seen in the slope of the life-table mortality curves that reflect the change in reproductive success of the short-lived compared with the long-lived descendants.

If we owe our longevity to the action of natural selection in eliminating the short-lived and favoring the long-lived, how does it come about that mortality curves do not follow a normal distribution as do many other attributes whose variance is governed by selection? The answer is that some do, although not many appear to behave in this manner. The reason most do not may be found in the fact that, as noted above, most populations are undergoing strong selection just at a time when most of our useful data are being gathered. Selection tends to distort curves which under relaxed conditions would be normal as depicted, for example, in Wilson and Bossert's "Primer of Population Biology" (Wilson and Bossert).

Favoring this explanation is the fact that mortality curves that largely consist of death from senescent causes are easily converted to normal curves by a simple systematic operation on the age axis. That is, the data normalized and converted to normal equivalent deviates may be compared with ages raised to some power less or greater than one. (If the best fit is given by the power one, then the curve is naturally a normal one.) The degree to which the power differs from one to get a good fit to the curve is a measure of the degree of selective activity of the living conditions. This systematic shrinking or stretching of the age-axis is akin to use of a logarithmic normal distribution that shrinks the age axis as the age increases. The "age-power" method (Totter, 1981), however, is much more general than the log-normal distribution. The conversion of life-table data to formulas describing the parameters of normal curves from which the life-table data may be accurately reproduced (correlation coefficients are almost always above 0.999) is a great convenience and allows many comparisons that are not easily made with other mathematical operations, such as conversion to Gompertz or Weibull curves. Figure 11-1 illustrates one use of the method to depict the frequency distribution of women that died of senescent diseases in England and Wales in 1901 or in 1964. Other comparisons may be seen in an earlier publication (Totter, Adler, and Storer). One can see that in spite of great differences in rates for the diseases, the relationship of the fraction dying at any age is the same in 1964 as it was in 1901, except for the age displacement of ten years, owing at least in part, to increase in longevity in the four or five generations between the old and the more recent data collection. Precisely the same display could be made for the total cancer or total cardiovascular deaths with only the median ages at death differing for the two.

The results shown in Figure 11-1, typical of many such comparisons that have been made, do not speak in support of a chiefly "accidental" type of environmentally caused death from either cancer or cardiovascular disease. They do strongly support the idea that whatever effect the current environment has, its expression is determined by the genetic composition of the population. The frequency distribution of deaths from both cancer and cardiovascular disease at different ages among the members of the population are largely fixed by this composition. The genetic composition was, of course, affected quite rapidly by the environment encountered by the ancestors of the current generation. The fact that the gene frequency may be rapidly changed by environmental conditions does not imply that we may be able to change the conditions at will. The quantitatively most important attribute of the environment is the accessibility of food and it is the caloric intake that is of overriding importance in affecting the lifespan and the death rates from all senescent causes of death.

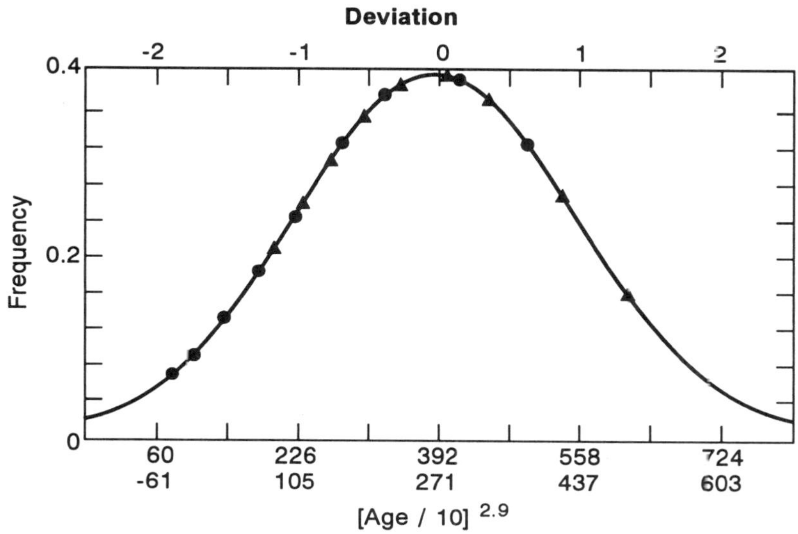

Figure 11-1: Comparison of the Relative Frequencies of Death of Women from Senescent Diseases at Each Age in 1901 (circle) with That of Those That Died of Similar Diseases in 1964 (triangle).

The upper scale at the bottom of the figure is for the 1964 life table; the lower scale is for the 1901 life table. The median age at death for 1964 is (approximately $\frac{1}{392^{2.9}}$ x10 = 78.4; for 1901 it is $\frac{1}{271^{2.9}}$ x 10 = 69.0. The scale at the top of the figure is given in normal equivalent deviates (sigmas). The fraction of the total deaths that were from senescent diseases was 0.513 in 1901 and 0.817 in 1964.

References

Ames, B.N. 1983. "Dietary Carcinogens and Anti-carcinogens." *Science* 221:1256–64.

Burch, P.R.J. 1975. "Problems in Interpretations of Cancer Statistics With Special Reference to Lung Cancer." *Journal of the Society of Occupational Medicine* 25:2–10.

__________. 1978. "Are 90 Percent of Cancers Preventable?" *IRCS Journal of Medical Sciences* 6:353–356.

__________. 1984. The Surgeon General's "Epidemiologic Criteria for Causality." Reply to Lilienfeld. *Journal of Chronic Diseases* 37.2:148–156.

Doll, R. 1977. "Strategy for Detection of Cancer Hazards to Man." *Nature* 265:589–596.

__________, L. Morgan, and F.E. Speizer. 1970. *British Journal of Cancer* 24:623.

Dunham, L.J. and J.C. Bailar. 1968. "World Maps of Cancer Mortality Rates and Frequency Ratios." *Journal of the National Cancer Institutes* 41:295–334.

Efron, E. 1984. *The Apocalyptics, Cancer and the Big Lie.* New York: Simon and Schuster, 258–69 and 474–5.

Fisher, R.A. 1959. *Smoking. The Cancer Controversy. Some Attempts to Assess the Evidence.* Edinburgh: Oliver and Boyd.

Friedell, G.H. (ed.). 1969. "The Epidemiology and Etiology of Human Bladder Cancer." *Journal of the National Cancer Institute* 43:215–307.

Gottlieb, S.F., I.S. Longmuir, and J.R. Totter (eds.). 1983. "Oxygen: An In-Depth Study of its Pathophysiology." Undersea Medical Society Publication No. 62 (W.S.), 3-1-84.

Hall, G.O. and D.R. Marble. 1931. "The Relationship Between the First Year Egg Production and the Egg Production of Later Years." *Poultry Science* 10:194–203.

Harman, D. 1956. "Aging: A Theory based on Free Radical and Radiation Chemistry," J. Geront., 71:298.

Hickey, R.J. 1986. "Smoking Hazards" (Letter). *Chemical and Engineering News*, June 16, 2–3.

Higginson, J. 1976. "Importance of Environmental Factors in Cancer."

Inserm Symposia Series Vol. 52, IARC Scientific Publications, 13:52, 15–24.

Hoover, R. 1979. "Environmental Cancer, in Public Control of Environmental Health Hazards." *Annals of the New York Academy of Sciences* 329:50–60.

Hueper, W.C. 1942. *Occupational Tumors and Allied Diseases.* Springfield, Illinois: Charles C. Thomas.

Lea, D.E. 1947. *Actions of Radiations on Living Cells.* New York: The MacMillan Company, 47–64; 108–111.

Lindop, P.J. G.A. and Sacher. 1966. *Radiation and Aging* London: Taylor and Frances Ltd.

Lorenz, E., L.O. Jacobson, W.E. Heston, M. Shimkin, A.B. Eschenbrenner, M.K. Deringer, J. Doniger, and R. Schweisthal. 1954. *Biological Effects of External X and Gamma Radiation,* Part 1, R.E. Zirkel (ed.). New York: McGraw-Hill Book Company, 24–167.

McCay, C.M., L.A. Maynard, G. Sperling, and L.L. Barnes. 1939. "Retarded Growth, Lifespan, Ultimate Body Size and Age Changes in the Albino Rate After Feeding Diets Restricted in Calories." *Journal of Nutrition* 18:1–13.

Miller, G.H. and D.R. Gerstein. 1983. "The Life Expectancy of Nonsmoking Men and Women." *Public Health Reports* 98:343–349.

Pott, P. 1775. *Cancer Scroti, In Chirugical Observations 63–68.* London: Hawes, Clarke and Collins.

Preston, S.H. 1976. *Mortality Patterns in National Populations: With Special Reference to Recorded Causes of Death.* New York: Academic Press, 24–25.

__________, N. Keyfitz, and R. Schoen. 1972. *Causes of Death.* New York: Seminar Press.

Rodger, M.A.J. and E.L. Powers (eds.). 1981. *Oxygen and Oxy-Radicals in Chemistry and Biology.* New York: Academic Press.

Sacher, G.A. 1959. "Relation of Lifespan to Brain Weight and Body Weight in Mammals." In *Ciba Foundation Symposium on the Lifespan of Animals,* 115–133.

Totter, J.R. 1981. "Relative Risk of Cancer from Externally and Internally Generated Oxy-Radicals." In *Health Risk Analysis.* Richmond, C.R., P.J. Walsh, and E.D. Copenhaver (eds). Philadelphia: The Franklin Institute Press, 393–402.

__________. 1985. "Food Restriction, Ionizing Radiation and Natural Selection." *Mechanism of Aging and Development* 30:261–271.

__________, H.I. Adler, and J.B. Storer. 1985. "The Relationship of Cancer Mortality to Lifespan and Food Supply Rate." *Nuclear Science and Engineering* 90:459–466.

Wilson, E.O. and W.H. Bossert. 1971. *A Primer of Population Biology.* Sunderland, Massachusetts: Sinauer Associates Inc.

Wyshak, G. and R.E. Frisch. 1982. "Evidence for a Secular Trend in Age of Menarche." *New England Journal of Medicine* 306:1033–1035.

TWELVE

MANAGING TECHNOLOGICAL RISKS

A BIOETHICAL PERSPECTIVE

Margaret N. Maxey

In recent years several anthologies have appeared assessing "lessons learned" from some 25 years of effort in "managing technological hazards" (National Academy of Engineering; MacLean, Landau, and Rosenberg; Repetto). The more discerning will detect a new mood and refreshing attitude toward the human future and proposals for possible goals.

Among respected leaders in the environmental crusade of the 1960s, Robert M. White invites us to recall that the "new environmental ethic," which emerged during the 1970s, contrasted sharply with the pre-World War II "ethic of utility," which pursued a goal of "the greatest good for the greatest number."

> The new ethic embraced the philosophy that we must learn to live more closely in harmony with the natural environment and that our activities must foster that harmony.

White reminds us that this new ethic has been aggressively institutionalized, driven by the conviction that government regulation and strict enforcement could compel human beings to adopt activities more closely in harmony with the natural environment. New laws simply assumed that the values and social benefits of protecting the environment and human health took priority over economic costs which industry could easily accommodate. However, two decades of vigorous economic growth in the 1960s and 1970s have given way to economic constraints of the mid-1980s. It is now clearly recognized, says White, that environmental protection is not costless to society. Moreover, years of stringent government regulation obviously have failed to protect the public from "hazard spectaculars such as Three Mile Island, Love Canal and Bhopal." Acknowledging the fact that "the world is fraught with irreducible risks," White concedes that we may have to forego reducing some risks—even setting priorities in the management of a vast array of hazards—as morally necessary if we are to protect human rights to fairness and distributive equity.

These concessions indicate a significant shift in perspective. Yet, they seem tempered by a residual reluctance to concede that learning to live "in closer harmony with the natural environment" may create more problems for practical living, not to mention ethical reflection, than it solves. Several ethicists have attempted to regard a stable "balance of nature" as a sound foundation for deriving a new ethic for protecting the environment from presumed hazards of modern technology. Consider the following exhortations. Ian McHarg writes, "We must learn that nature includes an intrinsic value system." Thomas Colwell (1969: 50) offers a more expansive moral prescription:

> The balance of Nature provides an objective normative model which can be utilized as the ground of human value. . . . Nor does the balance of Nature serve as the source of all our values. It is only the *ground* of whatever other values we may develop. But these other values must be consistent with it. The balance of Nature is, in other words, a kind of ultimate value. . . . Human values are founded in objectively determinable ecological relations with Nature. The ends which we propose must be such as to be compatible with the ecosystems of Nature.

These prescriptions interpret and amplify Aldo Leopold's seminal essay, "The Land Ethic," containing his celebrated statement, "A thing is right when it tends to preserve the integrity, stability and beauty of the biotic community. It is wrong when it tends otherwise."

Environmental ethics have not developed into a monolithic mindset. One school of interpreters maintains that "rights" should be extended to include their attribution not only to sentient fauna and flora, but also to rocks and rivers, as a means of entailing human obligations not to "harm" them (Stone; Regan and Singer; Nash). Others insist upon a holistic interpretation of the total ecosystem as possessor of inherent moral worth. The Ecosystemic School derives the "good" of nature and its moral oughts for human beings from a normative "balance of nature" to be preserved by a correlative self-sustainable society, by species diversity and other imperatives (Colwell, 1972; Callicott; Rolston). Both the Extended-Rights School and the holistic Ecosystemic School are predicated on the normative principle that "Nature knows best" (Commoner). Protecting a pristine environmental quality is the paramount moral mandate. Both schools tacitly assume that the state of nature is fundamentally benign and benevolent and should be protected from detrimental encroachments inflicted by technological man. They foster the widespread popular belief that human transactions with nature, by definition, cause resource dissipation and degradation of a pre-existing pristine condition. By implication, biohazards threatening public health in the modern world are caused primarily, if not exclusively, by post-World War II technologies producing industrial sources of pollution and toxic wastes. The risks selected for paramount concern are technological risks.

In contrast to environmental ethics, bioethics is concerned with problems which result from human efforts to protect and enhance the life-sustaining qualities of the biosphere. It attempts to transcend the misleading dichotomies which have tended to pit "technological man" against a "natural environment"—e.g., rights vs. utility, risk vs. benefit, growth vs. development, hard vs. soft technologies, large centralization vs. small decentralization, among others. These dichotomies not only result in needless polarizations; they also distract the public's attention from the vital interdependence between a life-sustaining biosphere and a life-preserving material well-being of individuals who presently comprise our common humanity. Bioethics seek the means to provide basic material well-being for living persons as the sine qua non of protecting a preferred, agreed-upon environmental quality. Bioethics seek principles as a guide to setting priorities which can be implemented by public policies aimed at realistic, achievable goals. To this end, bioethics must develop a far more complex and comprehensive conceptual framework than is envisioned by current versions of "environmental ethics."

In search of this conceptual framework, this brief paper proposes to examine a puzzling anomaly as a key to understanding the cultural climate of opinion which currently sustains a cluster of beliefs about unprecedented threats of risk from "technological hazards" with their attendant polarizations. Three questions will be explored. (1) For what reasons have technological risks been selected as requiring paramount concern and management? (2) Does scientific evidence and reasoning validate or invalidate this particular process of risk selection? (3) How might a conceptual framework for bioethical inquiry be developed to foster a more comprehensive public awareness and consensus on this risk selection process?

Technological Risk Selection: Surrogate for Social Criticism?

Citizens in modern technological societies enjoy a standard of living, political freedoms, and social amenities unparalleled in history. Yet, they harbor strongly negative attitudes toward the quality of their lives and their environment. What can account for this puzzling anomaly? In less than a generation, a cultural sea-change has transformed relatively affluent citizens into fear-ridden critics who protest against technological hazards, environmental pollution, and personal contamination caused by "corporate cancer" (Nader). Science and technology, as recently as two decades ago, had been considered preeminent sources of safety and reliable instruments of human control over hostile nature. Why have they become the object of skepticism and doubt? Douglas and Wildavsky (1982a,b) ask, "Of what are Americans fearful?" The answer is ironic: "Nothing much, really, except the food they eat, the water they drink, the air they breathe, the land they live on, and the energy they use."

Sociologists and psychologists customarily attribute differing perceptions of risk to different personality traits. However, Douglas and Wildavsky maintain that dangers or risks are selected for attention dependent upon the strength and direction of social criticism and cultural bias. People pay attention to certain risks, not arbitrarily but purposefully, that is, to reenforce and conform to a specific way of life already selected on other grounds. When people disagree about "perceived" vs. "actual" risks they may be disposed to take, or to avoid, the disagreement signals divergent agendas for changing or preserving preferred forms of social organization.

We cannot account for current disputes over the discrepancy between actual vs. perceived risks if we cling to the view that real knowledge of an external world should be allocated to experts in the physical sciences, while mistaken perceptions pertain to the realm of personal psychology. According to this faulty division, dangers are assumed to be inherent in a physical state of affairs, and the risks they pose are objectively ascertainable by experts. Subjective personality traits—i.e., an individual is either a risk-taker or a risk-avoider—cannot account for a bias toward the selection of certain types of risk and not others. A subjectivist view can be nullified by two questions: Why is it that experts disagree? Why does one and the same individual fear environmental risks to the exclusion of those dangers more immediately life-threatening?

Only a cultural theory of risk selection as a product of cultural bias and social criticism can account for the anomalies we encounter in affluent technological societies. As Douglas and Wildavsky observe, "Only a cultural approach can integrate moral judgments about how to live with empirical judgments about what the world is like." The risks we select to control or mitigate, individually and collectively, are integral to the choices we make with respect to the best way to organize social relations, to protect shared values, and to devise institutional mechanisms for providing informed consent in the formulation of public policy.

In the current dispute about technological hazards and perceived threats to environmental protection, partisans accuse each other of serving the vested interests of preferred social institutions. Whether one represents the "danger establishment" or the "industrial establishment," each side accuses the other of irrational bias, of misperceptions of *real* risks, of subversion of the public interest. For a cultural theory of risk selection, the primary task is to discover the fountainhead of cultural biases which place these mutual critics in irreconcilable opposition to one another.

Origins of Social Criticism

If public attention is a reliable indicator, the most dominant, dreaded, all-pervasive threat to public health and safety over the past 25 years appears to be cancer. The ostensible cause of this ubiquitous threat was singled out in the early 1960s when, coincidentally, the environmental movement took cultural root. Carcinogens, it is widely believed, do not exist as "natural" phenomena in our world. Instead, they are considered to be artificial inventions of technological man, produced in the form of synthetic chemicals during and after World War II, and directly responsible for no less than 90 percent of all cancer deaths.

The emergence of this widely-shared belief has been chronicled with careful documentation by Edith Efron. She traces its evolution through the pages of influential books by Rachel Carson, Barry Commoner, and Samuel Epstein. Less well-known as mentor and primary instructor of popular beliefs about carcinogens is Umberto Saffiotti of the National Cancer Institute. In a 1976 paper delivered at a cancer conference, now regarded as a classic, Saffiotti set forth ideas based on his fundamental assumption, namely, that industrial chemicals are the primary source of cancer in modern technological societies:

> I consider cancer as a social disease, largely caused by external agents which are derived from our technology, conditioned by our societal lifestyle and whose control is dependent on societal actions and policies.

Since Saffiotti knew at the time he delivered his paper that most industrial substances had never been tested for carcinogenicity, his use of the personal pronoun "I consider" signaled a personal belief and not a scientific conclusion derived from a disciplined study of the data. Indeed, at that time, he was engaged in a vigorous campaign for passage of the Toxic Substances Control Act which had as its objective a program and budget making it possible to acquire such data.

Whereas less audacious colleagues were content to express similar ideas in private, Efron observes that Saffiotti "proposed, in effect, that his personal belief be endowed with the status of a toxicological axiom." Classifying data as positive, negative, and inconclusive, Saffiotti boldly defined the toxicological policy he wished to see applied in the case of inconclusive data falling into a "gray area":

> The most "prudent" policy is to consider all agents, for which the evidence is not clearly negative under accepted minimum conditions of observation, as if they were positive. . . . In other words, *for a prudent toxicological policy, a chemical should be considered guilty until proven innocent.* [Emphasis added.]

Implicit in this policy is its supportive assumption: It must be assumed that no amount of any carcinogen is safe, since no one knows the amount of it required to trigger the growth of a malignancy. In short, there is *no threshold dose* below which exposure to a carcinogen is "safe." Even one molecule of any carcinogen is presumed to be a potential biohazard.

Although Saffiotti's policy and its underlying assumptions could not claim validity as the result of scientific conclusions from available evidence, they are clearly a logical culmination of seminal ideas planted more than a decade before—ideas forming the intellectual inheritance of "official regulatory science." Among these seminal ideas, by far the most pervasive and significant is the presupposition that *nature is noncarcinogenic,* and that carcinogenic chemicals are synthetic compounds invented by the chemical industry and proliferated by all basic industries comprising our modern societies.

Carson's Vision

The non-natural causation of cancer was given widespread popular currency by Rachel Carson. Her influential vision of man's role in creating a cancerous universe is vividly described by Efron:

> The major damage, [Carson] said, had been done since the early 1940s, the period of World War II, when the world, she said, had become contaminated by an "unnatural" production of radiation, and by great numbers of "sinister" chemicals, which, in partnership, were altering the very nature of the biosphere.

Carson's choice of the term "sinister" was not arbitrary. As Efron explains, ". . . they were 'synthetic' artifacts of human intelligence; they had been manufactured in test tubes and were substances which had no natural equivalents and to which no life could adjust." The reason Carson believed this to be the case is explained further by Efron:

> By producing "unnatural" substances in huge quantities, she said, man had turned the world in which he lived into "a sea of carcinogens" that was now taking almost one of every four lives. Man was absorbing them into his body through the air and the water and, particularly, in the form of pesticides, in his food . . .

Carson's belief that human life could no longer adjust to these unnatural substances was derived from a hidden assumption, namely, that evolution has already directed and shaped the adaptation of life through eons of changing environments, to the point at which our natural environment has now reached a balanced equilibrium devoid of carcinogens—except where contaminated by man's pollution.

Carson's seminal ideas are clearly echoed by John Quarles in his congressional testimony while serving as Deputy Administrator of the EPA:

> Since Darwin, man has recognized the ability of living things to adapt to their environments. The great diversity of life in our biosphere reflects the successful resistance of man and other species to the myriad of chemicals found in nature. However, the advent of chemical technology in the past decades has introduced billions of

pounds of new chemicals that are often alien to the environment, persistent, and unknown in their interactions with living things. (cf. Efron.)

Following the passage of the Toxic Substances Control Act, an anonymous commentator on the law (*Environmental Law Reporter*), shows how entrenched the idea of unnatural chemical carcinogenicity had become:

> Any chemical molecule not found in nature may be said to have the potential to harm biological organisms from bacteria to man. This is because living organisms' internal defense and waste removal systems are not likely to be prepared to cope with substances of a kind that they and their evolutionary precursors never had to contend with in the natural universe.

From this brief review of influential thinkers who have shaped a cultural climate of opinion during the past 25 years, the cultural anomaly we are experiencing becomes more understandable. With Nature exonerated of harmful health effects, and technological man held culpable, the magnitude of risks entailed by technological hazards appears overwhelming and yet, paradoxically, reassuring. What man has wrought, man can undo or eliminate. Cancer prevention, in concert with cancer elimination by purging our natural environment, has provided powerful symbols for rallying political action. When partisans of the "danger establishment" hold the "industrial establishment" guilty-until-proven-innocent of having caused an unnatural "social disease," rebaptized as "corporate cancer" (Nader), it should be clear that an entire socio-economic system has been virtually arraigned and convicted. The demands made for legislating and enforcing an "environmental ethic" as an antidote to an ethic of utility have entailed a repudiation of the cultural values of an industrialized society.

The selection of risks from technological hazards as objects for paramount ethical concern and stringent regulatory control is the direct result of a social process shaped by several deeply entrenched beliefs:

- Nature is noncarcinogenic, benign and benevolent;
- Industrial society is malignant and malicious.

Mankind At the Turning Point, by Mesarovic and Pestel, was issued as the second report of the Club of Rome. It is instructive that its prologue was introduced with an epigraph borrowed from Alan Gregg: "The world has cancer and the cancer is Man." This provocative claim invites us to explore the status of scientific evidence which might serve to validate or invalidate the selection of risks from technological hazards as worthy of paramount concern.

Technological Risk Selection: Scientific Evidence and Irony

For reasons which defy a defensible explanation, overwhelming evidence of the existence, magnitude, and ubiquity of *natural carcinogens* has only recently been given the scientific attention it warrants. As Efron wryly remarks:

> The strange truth is that within three years of the passage of the Toxic Substances Control Act, conceived to protect the earth from Faustian man, the earth itself had been reported to be carcinogenic beyond anyone's wildest imaginings.

In 1979, a cancer researcher at the University of Wisconsin, James A. Miller, compared the types of carcinogens produced by nature with those produced by men. He wrote:

> A wider variety of carcinogens occurs in our natural environment; the term "naturally occurring" is used here in its broadest sense. The great majority of these agents have undoubtedly been present throughout evolution, some may even have facilitated the speciation of living systems (cf. Efron).

In 1983, a storm of protest greeted Bruce Ames' (1983, 1984) revealing and provocative article "Dietary Carcinogens and Anticarcinogens" and response to Letters in *Science*. If any doubt remained, he made it clear that we are virtually immersed in a sea of toxic substances inherent in nature; the human diet contains a plethora of natural mutagens and carcinogens. His findings suggest that the preoccupation with relatively small amounts of man-made effluents from various industrial processes has diverted research attention away from the most health-effective and cost-effective ways to reduce the burden of cancer through dietary and lifestyle changes. Ames' recommendations stand in sharp contrast against the Saffiotti-Epstein school of thought which holds that cancer is basically a political problem, induced by the irresponsible politics and practices of corporate capitalism.

In light of popular beliefs about unprecedented threats from exposure to radiation derived from nuclear technologies, recent publications about the ubiquity and magnitude of terrestrial radiation are a revelation. D.J. Benninson *et al.* writes:

> Terrestrial radiation is emitted from radioactive nuclides present in varying amounts in all soils and rocks, the atmosphere and the hydrosphere, and from those radionuclides that, transferred to man through food chains or by inhalation, are deposited in his tissues. . . . Even now, despite the widening use of radiation-producing devices, the widespread radioactive contamination from nuclear weapon tests and the increasing applications of nuclear energy and radioisotopes, natural sources are the main contributors to the radiation exposure of most of the human population and are likely to remain so in the foreseeable future.

Despite overwhelming scientific evidence, radiation exposures from man-made technologies—whether for medical diagnosis or electricity generation—have been singled out as a unique cause of the dread disease of cancer as well as genetic mutations affecting distant future generations. Its effects are called "insidious" because they are believed to be latent and delayed in their manifestation. An entire ethical framework for social criticism has been erected on the scaffolding of beliefs which are at odds with the actual status of scientific evidence.

Hypothetical Harm

Over 25 years have been spent in developing radiation protection philosophy and standards, years dominated by a conservative assumption: namely, that every radiation dose greater than zero entails some possibility of somatic and/or genetic harm. A linear, zero-threshold hypothesis has guided the development of radiation standards by regulatory agencies and their expert advisors. Consequently, the public has been led to believe that "there is no safe dose of radiation" and "every radiation dose is an overdose." The origin and validity of ethical objections to man-made applications of radioactive substances appears to stand or fall on the claim that any radiation exposure, no matter how trivial, has some health effect which it seems wise to assume does some harm.

It is ethically imperative to recognize that, despite a vast array of radiobiological data, there is no conclusive evidence to prove the existence or absence of a threshold. Ethical condemnations, based on the linear hypothesis, take it to be an unassailable scientific conclusion; but in fact, it is only an inconclusive theory, an extrapolated hypothesis, an ultra-conservative and protective rule of prudence.

We could not exist if the linear hypothesis were applied to and enforced upon personal lifestyle exposures to natural radiation. The absence of scientific evidence of harm from exposure to low levels of naturally occurring and man-made radiation is not due to incompetence or oversight or lack of attempts to ferret them out. Lauriston Taylor has described the situation in unambiguous terms:

> No one has been identifiably injured by radiation while working within the first numerical standards set by the NCRP and then the ICRP in 1934.
>
> Let us stop arguing about the people who are being injured by exposure to radiation at the levels far below those where any effects can be found. The fact is, the effects are not found despite over forty years of trying to find them. The theories about people being injured have still not led to the demonstration of injury and, though considered as facts by some, must only be looked upon as figments of the imagination.

Hormesis

It has been scientifically established that there are net beneficial effects from exposure to low levels of several toxic substances, e.g., copper, selenium, and fluoride. Professional ethics and logic compel one to ask if competent members of the scientific community should not examine radiobiological data through the lens of a hypothesis counter to, but equally worthy of, the attention given the linear, zero-threshold hypothesis. T.D. Luckey (1980, 1982, and this volume) has presented a strong case for evidence supporting the possibility that exposure to low-level radiation might have net beneficial effects. Indeed, it may be essential for the continued well-being of living organisms which have evolved in relation to wide variations in exposure to natural radiation.

Effects Not Unique

Additional empirical data call in question ethical arguments objecting to technological applications of radioactive substances. No one disputes the fact that there is clear scientific evidence to support this statement: "Ionizing radiation, delivered in sufficiently large amounts, can cause determinable effects or injuries to any biological system" (Taylor). Nonetheless, as Lauriston Taylor and others have repeatedly stated, any somatic or genetic effect caused by radiation also can be caused by a minimum of fifteen hundred other agents, with no possibility of positively identifying which one might be the culprit. Besides not being unique in their biological effects, radiation dose effects are not cumulative, since a process of repair or replacement of cells, both somatic and genetic in nature, has been demonstrated to occur in radiation therapy techniques (Taylor).

A Practical Threshold

Granted that there is no conclusive scientific evidence to prove the existence or absence of a threshold; nevertheless, there is an *ethically* sufficient basis for recognizing a "practical threshold" for low exposures to radiation—namely, a dose level below which the elapse of time before any likely effect could be expected to appear is in excess of one's remaining life expectancy. Raabe *et al.* have shown that, for bone tumors caused by radium in canines, there is clearly a practical threshold: If the dose is lower than 39 millirem per day, the elapse of time before any tumor could be expected exceeds the lifespan.

Etiology of Cancer

The scientific status of hypotheses about the environmental origin of cancer has proved to be as problematic as the linear, zero-threshold hypothesis. Thirty years ago, John Higginson formulated a complex theory which, through misinterpretations and distortions, has led people to believe that some cancer-causing agent lurks in everything we eat, drink, and breathe. He has attempted to correct distortions of his theory by pointing out that, when he ascribed the incidence of cancer in industrialized societies to "environmental causes," he meant the total environment—namely, cultural components of lifestyles, such as diet and behavior, agricultural practices, hygiene, and social mores—and not merely or exclusively physical chemicals.

In his major statement on cancer policy, the late Philip Handler, while President of the National Academy of Sciences, stated in a public address:

> . . . We should lay to rest the idea that it is these man-made compounds, abroad in the land, that are responsible for the fact that 25 percent of Americans die of cancer. They are not. The possible effects of all known man-made chemicals, when totaled, could contribute only a minuscule fraction of the total of all carcinogens in our population.

John Totter has reflected on the origin of spontaneous cancer. He shows that, when corrected for competing risks, mortality from cancer appears independent of the level of industrialization in a country and, thus, of its man-made pollution. He maintains that it is not among man-made agents that one should look for primary carcinogens, but instead, among all-pervasive "normal" environmental components. Totter suggests that the culprit is *oxygen*: it is a recognized mutagen; experiments have shown that it causes tumors in fruit flies; in the Ames assay test for screening carcinogens, it shows up positive.

To summarize: Popular beliefs contrary to scientific evidence have generated ethical arguments that risks from the technological application of toxic substances are a violation of human rights, fairness, and equal protection, unjustified by any foreseen and intended benefits to the many. These claims are not supported by the actual status of scientific evidence. First, the assertion that any and every exposure to toxic materials not only *can* but inevitably *does* cause somatic and genetic harm is based on a profound misunderstanding of the scientific status of the linear hypothesis. Second, no scientific—much less ethical—justification exists for singling out radiation as a unique cause of cancer or genetic mutations when over a thousand other toxic agents in common commercial use are known to be capable of producing the same health effects if exposed to them in sufficient doses. There is both an ethical and scientific basis for recognizing a practical threshold or *de minimis* dose below which risks of exposure to toxic materials are trivial and ought (legally and morally) to be ignored. Finally, since the etiology of cancer remains highly theoretical, uncertain and hypothetical, any moral condemnation of "technological hazards" is ethically unjustifiable, since it presumes as a scientific conclusion that any exposure to a suspected carcinogen actually causes harm which can and should be prevented. To the contrary, a *hypothetical harm* can entail only a *hypothetical violation* of rights, fairness, equal protection and intergenerational equity.

A tragic irony stalks public preoccupation with risks entailed by "technological hazards" and public willingness to believe that nature is benevolent and benign. We have grown accustomed to a ritual incantation of symbols of "hazard spectaculars," as White labels them: Love Canal, Three Mile Island, Bhopal, Chernobyl. Each is allegedly a catastrophic reminder that human life is beset by ominous unknowns in technologies encroaching on our otherwise benign, balanced natural world. The public is rarely confronted with several rude reminders:

In 1978, Love Canal was the subject of a frightening media exposé. Hooker Chemical was summarily tried and convicted of harming the public because of presumed mismanagement of chemical wastes. But *no immediate deaths* resulted from exposure to these wastes. Biostatisticians and epidemiologists have made frightening predictions about eventual cancers and genetic defects, but the fact remains that they are only statistical and hypothetical.

In 1979, Three Mile Island traumatized the public. Threats to public safety led to mass evacuations. But *no immediate deaths* resulted from exposures to radiation. Once again, biostatisticians and epidemiologists are quoted by prophets making terrifying predictions about hypothetical cancers and genetic defects.

In 1984, Bhopal was the object of global concern and the mecca for trial lawyers. Some *1,200 immediate deaths* were caused by chemical toxins afflicting people who, we should recall, had been allowed to move into close proximity to the plant after its construction in a sparsely populated area.

In 1986, the reactor at Chernobyl caused *31 immediate deaths of workers* who succumbed to smoke and heat while fighting the fire, with some receiving large doses of radiation. We now hear mind-boggling statistical predictions of cancer deaths among exposed populations over the next seventy years—based entirely on the linear, zero-threshold hypothesis, with little recognition of its hypothetical status.

The tragic irony surfaces when we recall that, within the same time frame as these so-called "technological catastrophes," Nature conspired to perpetrate the following events:

Item: May 1985, *10,000 immediate deaths* in Bangladesh were caused by a typhoon.

Item: September 1985, *10,000 immediate deaths* in Mexico City were caused by an earthquake.

Item: November 1985, *20,000 immediate deaths* in Armero, Colombia were caused by a volcanic eruption triggering an avalanche of mud.

Item: August 1986, three days after Soviet scientists convened an international body of experts to discuss the causes of the Chernobyl accident, *2,000 immediate deaths* at Lake Nios, Cameroon, were caused by a release of toxic gases of undetermined origin.

History is replete with stark lessons about the harm inflicted on humans and their life-sustaining biosphere by the natural environment and its heedless malevolence. The Great Flood of 1887 in China's Honan Province caused 900,000 deaths. Winds from a great hurricane in 1900 caused a storm surge in Galveston, Texas, which claimed 6,000 lives. The volcanic eruption at

Mont Pelee, Martinique, in 1902 caused 30,000 immediate deaths. These events are, indeed, "hazard spectaculars" against which human beings have tried to defend themselves by devising technologies for prevention and protection of basic material well-being. Such tragic reminders of the destructive power of nature serve not to exonerate human fragility or to excuse technological failures; but rather, they heighten a sense of our precarious achievements in technological control over natural forces. To be sure, there are worse things in life than death. An "immediate death" in prospect appears less dreaded than a lingering death from cancer which, therefore, appears more worthy of ethical argumentation concerning rights, fairness and equal protection. However, appearances are deceptive. Prolongation of life expectancy is not an absolute good; nor is exposure to technological risk an unjustifiable harm. Bioethics must negotiate the inevitable truth that values conflict, priorities must be set, risks must be taken, and that some policies will cancel out others. Before celebrating the birth of a new "environmental ethic" derived from values inherent in nature, inviting us to conform human activities to a presumed "ecosystemic balance," we should be wise enough to recognize that such an ethic dies the death of a thousand qualifications.

Toward a Conceptual Framework for Risk Selection

"Lessons learned" seems to be an increasingly popular invocation at a time when, paradoxically, we seem captivated by a new imperative: "No trials without prior guarantees against error" (Wildavsky, 1985: 204). The foregoing reflections suggest that, among major lessons as yet unlearned, the presumed dangers from "technological hazards" qualify as more shadow than substance. A more serious threat may prove to be the ethically incomplete and myopic framework in which mere hypotheses, supporting *ad hoc* and piecemeal risk selections, have been used to legitimize costly and ineffective safety regulations which Nature violates with impunity. What ethical justification exists for excessively costly expenditures of public money spent to reduce trivial risks? A social policy which denies benefits to the many, regionally and globally, based on the mere assumption that some hypothetical and statistical individual might be placed at risk, violates fairness, equity, and equal protection in ways that are rarely considered. The deficiencies in conceptual tools for assessing risks to health and safety must become part of the "lessons learned" over the past quarter-century of regulatory legislation. Let us examine some possible deficiencies in the currently accepted conceptual framework.

"Hazards"

Contrary to a common misconception, "hazards" are not baldly "there" in a physical state-of-nature nor in human transactions with it. What people regard as hazardous in any given era, reflects what they have come to know about their environment and what they value as essential, desirable, or preferable on a scale of real possibilities. In short, human beings structure hazards. It is simply not accurate to define a hazard as "toxicity of substance" or "violence of event" or "magnitude of consequences" that can be classified and predicted. A hazard exists only when, and to the degree that, harmful exposure of and assimilation by the human body or other valued living systems becomes a genuine and not merely an imaginable possibility. A possibility exists when there is an inability or failure to devise and maintain controlling actions or safeguards which govern environmental pathways of exposure.

At Risk: Benefit and Harm

Another popular misconception is that risk is a normative concept for certifying consequences to human beings which are harmful, dangerous or "bad." These contrast with consequences which are beneficial, pleasurable or "good." By implication, benefits are antithetical to risks or "risk-free." This false antithesis between risks and benefits conveys the mistaken notion that there is a way to have one without the other. We do not intend harm to ourselves. What we intend—what is "at risk"—is the possibility that an intended benefit from risk-taking may not materialize and some harm may occur. Any such harm is clearly an unwanted and unintended side effect. The proper symmetry is between harms and benefits. Risks are conceptually and morally neutral.

Degrees of Benefit and Harm

To develop appropriate bioethical principles, we must distinguish among degrees of benefit and corresponding harm. Fairness and rights pertain differently to (1) goods essential to members of society, e.g., food, clothing, shelter, water, energy, or *basic goods*; (2) goods advantageous to people, e.g., most manufacturing; (3) goods which are of peripheral value to people, e.g., aerosol deodorants for which there are substitutes at lower cost and less potential for adverse effects.

Basic harms may result if people are deprived of goods essential to subsistence and material well-being. In contrast to these goods, there are harms so trivial or negligible as to be undetectable, or in some cases unavoidable, if another *greater harm* of injury or deprivation is to be prevented. This distinction provides an ethical basis for distinguishing a category of trivial or "justifiable harm" which is in contrast to the deprivation of basic goods which, if preventable, is an "unjustifiable harm."

To illustrate, let us assume that one accepts biostatistical estimates based on a linear hypothesis. It is possible that diagnostic x-rays may cause some potentially detectable harm by "killing" or disrupting (reparable) cells. But preventing the greater harm of serious health impairment to a patient renders a comparatively negligible cellular harm ethically justifiable.

The most pernicious result of using undifferentiated "harm" language—especially when making statistical estimates to describe biological effects in an entire population—is that an unsuspecting citizen is led to believe that actual body-counts of certified injury and death to humans have been tabulated to provide biostatisticians and epidemiologists with scientific data for stating that exposure to low doses of a toxic substance not only *can* but *does* and inevitably *will* cause unjustifiable harm. To the contrary, the basis for scientific estimates is only hypothetical and speculative.

Incremental vs. Systemic Risk Analysis

It has become commonplace to consider a toxic substance or a single "hazardous technology" as if it represents only an *incremental* risk, or a simple addition to a current risk background.

To the contrary, any "new" risk reorders an entire system by displacing, offsetting or otherwise restructuring a prior pattern of benefits and harms. Only *systemic* risk analysis enables us to recognize this modification in a meaningful context.

To illustrate, several recent studies demonstrate that a positive correlation exists between income and an improved health and life expectancy (Siddall; Wildavsky, 1980; Singer; Kitagewa and Hauser). Steady economic growth in industrialized, developed societies has resulted in improved housing, better nutrition, environmental sanitation, and more plentiful food from mechanized agriculture; machines have reduced accidents in the workplace. Unless we make comparisons with pre-existing conditions, or with naturally occurring risks which would exist in the absence of actions taken by humans, we would totally falsify the fact that "new" risks cause displacements of "old" risks. What potential hazards did people live with before freons made possible refrigeration for food preservation? The incidence of ptomaine poisoning and stomach cancer has dropped dramatically. What potential hazards could we experience if drinking water were not chlorinated?—if DDT were not used against malarial mosquitoes?—if pesticides, preservatives, and now, irradiation did not protect a food supply from insects, mold, botulism, and a host of pathogens?

In a bioethical framework, systemic risk analysis is a necessary precondition for formulating social policy. A preoccupation with biological health effects biostatistically predicted, without meaningful comparisons, cannot do justice to the full spectrum of potential benefits and harms. In a bioethical perspective, we do ourselves and our posterity a grave injustice by allowing our moral concern for basic rights to health protection, and our concern for protecting the life-sustaining qualities of the biosphere, to be narrowed down to—and in too many cases trivialized by—an obsession with hypothetical health effects from a few toxic substances. This myopic preoccupation siphons public attention away from preventable causes of malnutrition, disease, and premature deaths from natural catastrophes which claim thousands of lives daily in our world.

Bioethical Principles and Priorities

Correcting deficiencies in conceptual tools might eventually mitigate public misconceptions. But a more fundamental need remains: namely, a more comprehensive bioethical framework through which to filter scientific evidence and conflicting moral claims. Whether putative risks or certifiable hazards, chronic or acute, synthetic or natural, created by nature or God or man, the operative skill is always *proper management*. Scientific understanding of physical laws must be coupled with an understanding of cultural biases and their role in political stratagems for shaping social policy.

Medical professionals have long had guidance in their conduct toward individual patients from the classic admonition, "Do no harm." It is an inescapable social reality that a public policy cannot possibly "do no harm." Even the wisest, most clairvoyant of policies will entail risks of some unintended harm to some hypothetical individual at some future circumstance. As a moving target, public policy should reflect a fundamental bioethical principle:

> Fairness and distributive equity require a policy for health and environmental protection to be based on a method of assessing risks considered in a total system, so as to prevent individuals from experiencing basic harm deemed to be unjustifiable.

With the premise that an entire spectrum of potential biohazards is taken into account, a second bioethical principle might be formulated:

> Fairness and equal protection require a policy which has as its goal an equitable management of potential sources of basic harm, that is, management which is proportioned to actual, identifiable basic harms which human effort, time, and money can reduce in the most cost-effective manner available.

This principle requires policy-making to evaluate comparisons among naturally occurring and technology-induced biohazards. It requires cost-comparisons of available methods for per capita reduction of each potential biohazard. This principle seeks to derive the most public health protection for the most people from a finite amount of money.

A number of academic ethicists object to this principle, arguing that it commits the "naturalistic fallacy" by using a utilitarian cost/benefit calculus which derives moral "ought" statements from factual or scientific ("is") reasons alone. For example, risks from low-level radiation exposure *ought* to be morally acceptable because they *are* in fact not only within the range of socially accepted risks, but are outweighed by technological benefits and economic advantages of a less costly technology. A utilitarian ethic based upon a naturalistic fallacy stands accused of being the root cause of two dangerous consequences: (1) it results in a violation of moral and civil rights to equal protection and due process; (2) it leads to a socially unjust sacrifice of public health and safety to unprincipled economic growth and technological development (Shrader-Frechette). Upon examination of this line of argument, we find a striking paradox: these ethicists reach their ethical conclusions by committing the very fallacy they repudiate. Not only do they misinterpret the linear hypothesis as an established scientific *fact*, but they base their moral claims and ethical arguments on the validity of what they presume to be an established scientific conclusion—namely, that all radiation exposure is unequivocally the cause of harm. This is a worst-case example of the naturalistic fallacy.

Even if it is valid in ethical theory, the naturalistic fallacy has nothing to do with how, in practice, ethical consideration stands in relation to the hypothetical status of scientific premises outlined above. When biohazard assessment and management make use of limited conceptual tools, such as risk analysis or harm/benefit ratios or economic cost-effective projections for optimizing the allocation of a finite amount of money, it does not follow that this exercise amounts to drawing ethical conclusions based on a fallacious conversion of a scientific "is" into a moral "ought."

To illustrate: When a risk analysis concludes that the *hypothetical* risks from exposure to low-level diagnostic x-rays are far below the *actual* risks of exposure to many other potential biohazards encountered in the environment, one is not converting an "is" to an "ought" by reaching the conclusion that one set of risks *ought* to be morally acceptable because the other *is.* This interpretation profoundly distorts the ethical problem posed by comparative risk analysis. The ethical reasoning required is that the acceptance of a greater risk should make the lesser risk equally if not more acceptable. Assessment of hypothetical vs. actual risks should induce the ethical conclusion that it is not morally justifiable to spend more public money to reduce already trivial, hypothetical risks, rather than spend it to reduce identifiably greater risks to life and health that can actually be diminished by that expenditure.

Contrary to superficial criticisms, when we compare more or less cost-effective methods of reducing risks to people and the biosphere, and compare net benefits (whether tangible and measurable, or abstract and arbitrary), we are not wielding utilitarian tools for placing a callous dollar-value on human life or impairment as a *moral judgment* of individual worth. Much less are we estimating economic losses to society as a measure of personal expendability sacrificed to achieve technological advancement or "benefits" to an abstract society. In the real world of practical decisions and setting priorities in the allocation of public money, we are maximizing the value we place on human life when we endeavor to allocate limited amounts of money in such a way that we reduce widespread harm, thereby preventing premature loss of life expectancy and diminished quality of life. It is a covenantal expression of our common humanity.

Conclusion

The past quarter-century has presented us with a puzzling anomaly, as well as an abiding lesson in the power of paradigmatic ideas. Some thinkers would have us interpret our human predicament as one of suspended animation between the figures of Faust and Frankenstein. Other thinkers interpret the human predicament as an inheritance of a timeless tension between conflicting images of evil, one symbolized by Promethean suffering and the other symbolized by Christian suffering. George Pickering describes this tension in terms of our shared covenant with fellow humans:

> On the one side are those who fear that human action has grown too large, that too much power must inevitably violate the covenant

> and invade the province of the divine. On the other side are those who fear that we will shrink from our responsibilities, intimidated by their scale, and thus fail to extend the covenant to its widest possible reach. Each side, it seems, locates responsibility where the other perceives idolatry. . . . Each side recoils from the prospect of being responsible for avoidable suffering.

As we grow in knowledge and awareness of a vast array of potential threats to our personal well-being, as well as to the life-sustaining biosphere, Aaron Wildavsky (1980) reminds us that our quest for safety and security is really not a new phenomenon. What is new, he observes, is that "the *collective* urge to risk reduction is so much greater than the sum of the individual urges it claims to represent. Each of us would do less for ourselves than we would insist that the government do for us." Rather than some laissez-faire abandonment of regulations, he recommends a more demanding set of criteria for resorting to them:

> The dice should be loaded against any "right" to protection against risk. Only if risks are palpable and remedies ascertainable and populations limitable and dangers unavoidable, should government regulate risk.
>
> "Killing people with kindness" is only a slightly subtle form of verbal aggression. Killing people with safety would be one of the supreme ironies of our time.

References

Ames, Bruce. 1983. "Dietary Carcinogens and Anticarcinogens." *Science* 221:1256–1264.

__________. 1984. Reply to "Letters." *Science* 224:760ff.

Benninson, D.J., A. Bouville, G.J. O'Brien, and J.O. Snihs. 1977. "Dosimetric Implications of the Exposure to the Natural Sources of Irradiation." *International Symposium on Areas of High Natural Radioactivity.* Ed. by Academia Brasilera de Ciencias, Rio de Janeiro: R.J.

Callicott, J. Baird. 1983. "Animal Liberation: A Triangular Affair." *Ethics and the Environment.* Ed. Donald Scherer and Thomas Attig. Englewood Cliffs, N.J.: Prentice-Hall.

Carson, Rachel. 1962. *Silent Spring.* Greenwich, Conn.: Fawcett.

Colwell, Thomas B., Jr. 1969. "The Balance of Nature: A Ground for Human Values." *Main Currents in Modern Thought* 26:50.

__________. 1972. "Ecology and Philosophy." *Philosophical Issues.* Ed. J. Rachels and F. Tillman. New York: Harper and Row.

Commoner, Barry. 1972. *The Closing Circle: Nature, Man and Technology.* New York: Bantam.

Douglas, Mary and Aaron Wildavsky. 1982a. "How Can We Know the Risks We Face? Why Risk Selection Is a Social Process." *Journal of Risk Analysis* 2/2:2–5.

__________. 1982b. *Risk and Culture: An Essay on the Selection of Technological and Environmental Dangers.* Berkeley, CA: University of California Press.

Efron, Edith. 1984. *The Apocalyptics: Cancer and the Big Lie.* New York: Simon and Schuster.

Environmental Law Reporter. 1976. "From Microbes to Men: The New Toxic Substances Control Act and Bacterial Mutagenicity/Carcinogenicity Tests." 6:10251–10252.

Epstein, Samuel S. 1972/1979. *The Politics of Cancer.* San Francisco: Sierra Club Books. Rev. ed. Garden City, New York: Anchor/Doubleday.

Gregg, Alan. 1950. "A Medical Aspect of the Population Problem." *Science* 121:681.

Handler, Philip. 1979. "Dedication Address." Evanston, Ill.: Northwestern University Cancer Center.

Higginson, John. 1979/1980. "Cancer and Environment: Higginson Speaks Out," reported by T. H. Maugh, Research News. *Science* 205:1363–66. Reprinted in *American Industrial Hygiene Association Journal* 29:33.

Kitagewa, A. and P. Hauser. 1973. *Trends and Differential in Mortality.* New Haven: Harvard University Press.

Landau, R. and N. Rosenberg, eds. 1986. *The Positive Sum Strategy.* Washington, DC: National Academy Press.

Leopold, Aldo. 1949. "The Land Ethic." *A Sand County Almanac.* New York: Oxford University Press.

Luckey, T.D. 1980. *Hormesis With Ionizing Radiation.* Boca Raton, Fl.: CRC Press.

__________. 1982. "Physiological Benefits from Low Levels of Ionizing Radiation." *Health Physics* 43:771–789.

McHarg, Ian L. 1970. "Values, Process and Form." In Robert Disch, ed., *The Ecological Conscience: Values for Survival.* Englewood Cliffs, NJ: Prentice-Hall.

MacLean, Douglas, ed. 1986. *Values At Risk.* New Jersey: Rowman & Allanheld.

Mesarovic, M. and E. Pestel. 1974. *Mankind at the Turning Point: The Second Report to the Club of Rome.* London: Hutchinson.

Miller, James A. 1979. "Concluding Remarks on Chemicals and Chemical Carcinogenesis." In *Carcinogens: Identification and Mechanisms of Action.* Eds. A. Clark Griffin and Charles R. Shaw. New York: Raven.

Nader, Ralph. 1970. "Introduction" in *Ecotactics: The Sierra Club Handbook for Environmental Activists,* ed. John G. Mitchell and Constance L. Stallings. New York: Simon and Schuster.

Nash, Roderick. 1977. "Do Rocks Have Rights?" *The Center Magazine* 32:2–12.

National Academy of Engineering. 1986. *Hazards: Technology and Fairness.* Washington, D.C.: National Academy Press.

Pickering, George W. 1987. *The Polarizing Atom: An Abuse of Science.* (In press).

Quarles, John, Jr. 1975. Testimony, "Toxic Substances Control Act," Hearings of the Subcommittee on Consumer Protection and Finance Committee on Interstate and Foreign Commerce. House of Representatives, 94th Congress. June & July. Serial no. 94-41.

Raabe, O.G., S.A. Book, and H.J. Harris. 1980. "Bone Cancer from Radium: Canine Dose Response Explains Data for Mice and Humans." *Science* 208/4439:61–64.

Regan, T. and P. Singer, eds. 1976. *Animal Rights and Human Obligations.* Englewood Cliffs, N. J.: Prentice-Hall.

Repetto, Robert, ed. 1985. *The Global Possible: Resources, Development, and the New Century.* New Haven: Yale University Press.

Rolston, Home, III. 1975. "Is There an Ecological Ethic?" *Ethics* 85/2: 93–109.

Saffiotti, Umberto. 1976. "Risk-Benefit Considerations in Public Policy on Environmental Carcinogenesis." *Proceedings of the Eleventh Canadian Cancer Research Conference.* National Cancer Institute, Toronto, Ontario.

Shrader-Frechette, K.S. 1984. "Ethics and Energy." *Earthbound: New Introductory Essays in Environmental Ethics.* Ed. Tom Regan. New York: Random House.

Siddall, Ernest. 1981. "Risk, Fear and Public Safety." Atomic Energy of Canada. ms.

Singer, Max. 1978. "How to Reduce Risks Rationally." *The Public Interest* 51:93–112.

Stone, C.D. 1974. *Should Trees Have Standing?* Los Altos, CA William Kaufmann.

Taylor, Lauriston. 1980. "Some Nonscientific Influences on Radiation Protection Standards and Practice." *Health Physics* 39:841–874.

Totter, John R. 1980. "Spontaneous Cancer and Its Possible Relationship to Oxygen Metabolism." *Proceedings of the National Academy of Sciences* 77/4:1763.37.

White, R.M. 1986. "Emerging Issues in Hazard Management." *Hazards: Technology and Fairness.* Washington, DC: National Academy Press.

Wildavsky, Aaron. 1980. "Richer Is Safer." *The Public Interest* 60:23–39.

__________. 1985. "Trial Without Error: Anticipation Versus Resilience as Strategies for Risk Reduction." Ed. M.N. Maxey, *Regulatory Reform: New Vision or Old Curse?* New York: Praeger.

EPILOGUE

Hugh W. Ellsaesser

After having planned, chaired, and edited the chapters in this volume, certain conclusions appear clear. The perception of a malaise in our society is widely shared. It centers around our methods of mass communication or information exchange and basically reveals a dissatisfaction, if not anger and fear, at the selectivity with which these have been used to bias the adoption of laws and governmental programs over the past quarter-century. What has been perceived as bias has been a consistent ignoring or downplaying of the available scientific evidence and accumulated cultural wisdom based on past history, mainly on the excuse that the alternatives are so horrible and so imminent that we cannot afford to take the chance of waiting for better information. When all else failed, "social relevance" was held up as a superior guide to policy decisions over scientific logic and accumulated cultural wisdom. In the words of Edith Efron (1984), the intellectual trash invaded the press and the law and has *become* the culture. This change in values has been accompanied by a broad-scale attack on those individuals and institutions resisting change and particularly on commercial and industrial establishments from which most practical, as opposed to theoretical, scientists used to draw their livelihood. One notable effect has been that these viewpoints rarely if ever reach the general public. Dr. Fred Singer, a scientist of widely ranging interests, stated that he had never heard of radiation hormesis before the conference. I first heard of it when I first met Professor Luckey, shortly before his book on the subject was published in 1980. Another subtle, but very important feedback from this process, was revealed in the unedited comments of Dr. Weinberg [which I hope he will forgive me for resurrecting]. He essentially said that the reason he doesn't stand up in public and say that the disposal of nuclear wastes is not a technical problem is not because he doesn't believe this, but because the *audience would not believe it.* [I.e., the cover-up stratagem of the weavers of the emperor's new clothes is operating effectively.]

As a result of this bias (or one-way filter–Ellsaesser, 1974), we now have, written in legislative concrete, inefficient and stifling regulations of arguable necessity on air and water pollution, waste disposal, use of pesticides and herbicides, control of food content, product safety, approval of drugs and development of natural resources. And in each case we have established a ponderous administrative bureaucracy now capable of perpetuating itself almost indefinitely. A particularly notable example of this rejection of science is provided by the structure created to regulate carcinogens. In the case of nuclear energy, in particular, it is easier to argue that the structure was created for the express purpose of bringing the industry to a halt (which it clearly has) than to argue that it was designed to protect the public and the environment.

The witnessing of the Iran-Contra hearings compels me to the conclusion that scientists have been caught in a political struggle (war really—in which all is fair) for control over the social agenda of our society. While the battle rages worldwide, in the United States, at least, the Congress, a myriad of pressure groups and the press have each contended for the empirical authority over the social agenda which they found it politically expedient to claim had been usurped by the executive branch of our government. The legal and bureaucratic edifice erected in the two decades following the appearance of *Silent Spring* (Carson, 1962) attests to the craftiness of the wagers of this political war and their canniness in outmaneuvering our one constitutional and sworn upholder of the overall welfare of all the public—the President. All parts of the system or establishment were penetrated and used against the system. Science was but one of many branches of society co-opted and arrogated in this political battle—which still rages.

From this can be drawn the one great and obvious law of politics: when only one side fights a political battle, it wins. This is but a restatement of the old adage: all that is required for the triumph of evil is that men of good will do nothing. It's time we recognize these as political, rather than scientific battles, and adopt similar tactics or be prepared to accept continued defeat.

The confirmation hearings for Judge Robert H. Bork were even more revealing. Clearly, the U.S. Congress is not ruled by logic and accumulated wisdom. Most of the members are but puppets for a relatively small number of pressure groups and when the strings are pulled in unison, Congress acts accordingly, regardless of the merits involved. After such a clear revelation of realities, how could anyone have faith that President Reagan, or anyone else, will ever be able to balance the U.S. budget or stop the current rush of anti-trade and anti-wealth legislation. It is no wonder that stock markets around the world tumbled like dominoes shortly after it became apparent that the full Senate was not going to confirm Bork's nomination.

Some will argue this could not happen with a free and unbiased press. This is probably true—but our press is not unbiased. When one considers that the launching and maintenance of a "big lie" requires a veritable flood of communications (which in a free country means paid commercial communications), perhaps it is no wonder that the press chooses not to dig too deeply into or to reveal gratis the factors motivating its major customers and news generators, particularly when it largely supports the objectives of the latter.

As is evident from the chapter by Professor Jovanovich, we have reached the stage at which we are being asked to choose between alternate hazards. If the individual hazards could be reliably quantified, this would be a logical method of making decisions. However, as most of the papers in this volume make clear, there still remain serious doubts as to the validity of the hazards which have been associated with past and planned human activities even though estimating formulas developed by technical advisory committees abound. Professor Jovanovich and others, using these committee-developed standards or guidelines, conclude that nuclear energy is now near the least hazardous choice per unit of energy consumed.

Insofar as this is based on actual casualties from dam failures and mine construction, transportation, and operating accidents, it is a justifiable method of decision-making. However, when it comes to weighing hypothetical consequences from the partial core melt down at Three Mile Island and those from plumes from fossil fuel power plants, such trade-offs represent an abandonment of logic and rationality. The health effects of fossil power plant effluents are extrapolated from the ratio of "excess" mortality in the so-called air pollution episodes to the estimated contemporaneous airborne concentrations of particles and gases. In actuality, these mortality peaks were, in all probability, caused by normal winter epidemics of respiratory diseases triggered by an early season extended period of cold, damp weather (Ellsaesser, 1979). On the other hand, the extensive studies collected by Professor Luckey strongly suggest that increased exposure to radioactivity up to about 10 times that now normally received would actually be beneficial rather than harmful.

Even stronger questions have been raised concerning many other hazards for which our society has adopted quantifying formulas. Many, besides Edith Efron (1984), have admitted an abandonment of logic in the effort to attribute most, if not all, cancers to man-made chemicals. Elizabeth Whelan (1984) documents a similar lack of rationality and misuse of data in the hazard evaluations of DDT and other pesticides, PCBs, dioxin, Agent Orange, Love Canal, formaldehyde, air and water pollution, radioactivity, and acid rain.

Those concerned with the depletion of stratospheric ozone are quick to convert their estimates of ozone depletion into percentage or actual increases in skin cancer incidence. To my knowledge, they have never mentioned that cutaneous exposure to ultra-violet radiation from the sun is the only natural source of vitamin D for most vertebrates, including man, and that vitamin D is essential for proper skeletal development—i.e., rickets might as well be still unknown. They also studiously avoid equating a 1 percent decrease in ozone with its equivalents of a 12-mile displacement toward the equator, or 150-foot displacement toward higher elevation, in terms of exposure to ultra-violet radiation and incidence of skin cancer (Ellsaesser, 1978). As Professor Jovanovich pointed out, increases in exposure to radioactivity can similarly be equated to a movement "up the hill," i.e., to higher elevations since exposure to cosmic rays increases with decrease in atmospheric shielding. Such an equivalence reveals the absurdity of linearly equating cancer incidence to radioactivity, since the available statistics show that cancer incidence generally decreases with altitude.

Professor Rosenqvist and Dr. Khemani (this volume) make it clear that acid rain and acid lakes were around long before we began consuming fossil fuels in large quantities and that man's impact on the acidity of rain comes largely from separating out the alkaline particles from combustion exhaust before allowing the acid-forming gases to escape into the atmosphere.

There are even logical reasons to question the degree of climatic warming computed for increased carbon dioxide from the burning of fossil fuels, i.e., in addition to the fact the observed warming to date is substantially less than that computed. Climate modelers are relying on positive feedbacks associated with the increase in water vapor expected in a warmer atmosphere for at least a two-fold magnification of the effect of carbon dioxide alone. Over half of the earth's surface, in the tropics, essentially all precipitation is from deep convection. Any amplification of precipitation in this part of the globe also will enhance the large areas of deep subsidence (compensating downdrafts of air) surrounding the small areas of deep convection. Subsiding air is dry air, and thus, over approximately half the earth's surface the positive feedback from increased water vapor content will not occur; rather a negative feedback may occur (Ellsaesser, 1984, 1987).

Of the many impacts on the planet attributed to man and his works which I have had a chance to examine, the major two that appear to be valid reasons for concern are the crowding out of other species and tropical deforestation. The former raises ethical and moral questions, as well as questions of self-interest for man through shrinking of the gene pool just as we are looking seriously at genetic engineering. The second problem exaggerates the first, as well as raising questions concerning undesirable climate changes as a result of what may prove to be drastic and irreversible declines in biospheric mass and productivity since much of the biomass of the tropical rain forest is recycled without ever entering the typically impoverished soils.

As for the major current threat to mankind, it is man himself. And I am not thinking of the potentials he has generated for nuclear or biological warfare, mass pollution of the environment, modification of the climate or the ozone layer, or terrorism by small groups of disgruntled individuals. The truly massive early and non-natural human deaths of this century have come as a result of deliberate policies of those claiming to be the legitimate governments of the people being annihilated. In both the Soviet Union and China, large numbers of people were "liquidated" following their respective social revolutions. The U.S. Senate Internal Security Subcommittee reported 35 to 45 million in the Soviet Union and 34 to 62.5 million in China. Emigres, such as Aleksander Solzhenitsyn and Antonov Ovseyenko, calculated the toll in the U.S.S.R. at 100 million. The estimates for Hitler's concentration camps are 4.5 to 6 million. In Cambodia, the Pol Pot regime is credited with liquidating 2 to 3 million out of a total population of 7 million. In the words of Les Csorba (1987): "It was a daunting lesson—more people had been killed in three years of a communist peace than in thirteen years of American war." These dwarf all other compiled causes of early death, natural and man-induced. However, if identified and compiled over comparable time periods, early deaths from parasitic diseases, such as schistosomiasis, malaria, sleeping sickness, and plague, probably exceed these. Recently, Wood (1987) gave a global estimate of 200 million afflictions annually—either recurrences or new infections—and 2 million deaths for 1987 from malaria alone. He estimated the number of current afflictions with filariasis and schistosomiasis at 300 and 200 million respectively. And *Nature* (328:660) cited 12 million cases of leprosy worldwide. If nothing else, these statistics clearly show we are not lacking in *real* problems—there is no need to invent imaginary ones.

The greatest and most imminent hazard now threatening mankind appears to be AIDS (acquired immune deficiency syndrome), not because of the numbers already affected, but because in this case, the time bomb can be clearly heard ticking. Again, the threat is not merely of an impending epidemic in the total population, but even more so in the threatened institutionalization of the concepts that neither individuals nor groups are any longer to be held responsible for anything and that the current and future productive forces of our society must shoulder the open-ended burden of providing heroic treatment and optimum comfort for those persons who are doomed—and in most cases because of their own wilful persistence on a path toward self-destruction.

However, in keeping with the illogic of the age, AIDS has so far generated more concern for individual rights to privacy and choice of lifestyle than it has concern for the welfare or the survival of the species. Another difference in the attitude towards AIDS is revealed by the arguments being made that insects cannot transmit AIDS because of the small volumes of virulent virus transmitted (Booth, 1987). This stands in sharp contrast to the "linear extrapolation to a zero threshold" which has been applied to essentially every other hazardous agent. This logical inconsistency can hardly go undetected and, if vindicated in the case of AIDS and insect bites, should surely lead to demands for reevaluations at low exposures for other hazards as well.

References

Booth, William. 1987. "AIDS and Insects." *Science* 237:355.

Carson, Rachel. 1962. *Silent Spring.* New York: Fawcett Crest.

Club of Rome. 1972. *The Limits to Growth: A Report for the Club of Rome's Project on the Predicament of Mankind.* New York: Universe Books.

Council on Environmental Quality and Department of State. 1980. *Global 2000 Report to the President.* Washington D.C.: Government Printing Office.

Csorba, Les. 1987. "Resurrection of Campus Battlefields." *World Media Report* 2(1):29.

Efron, Edith. 1984. *The Apocalyptics—Cancer and the Big Lie.* New York: Simon and Schuster.

Ellsaesser, Hugh W. 1974. "The Dangers of One-Way Filters." *Bulletin of the American Meteorological Society* 55(11):1362.

__________. 1978. "A Reassessment of Stratospheric Ozone: Credibility of the Threat." *Climatic Change* 1:257.

__________. 1979. "Air Pollution: A Different View." *Water, Air and Soil Pollution* 11:115.

__________. 1984. "The Climatic Effect of CO_2: A Different View." *Atmospheric Environment* 18(2):431.

__________. 1987. "The Sahelian Drought—First Evidence of a CO_2-Induced Climatic Change?" *Proceedings of the Eleventh Annual Climate Diagnostics Workshop,* pp. 201–204, U.S. Department of Commerce, Climate Analysis Center.

Kahn, Herman 1980. *World Economic Development, 1979 and Beyond.* Boulder: Westview Press.

__________., W. Brown, and L. Martel. 1976. *The Next 200 Years.* New York: Morrow.

Mesarovic, M. and E. Pestel. 1974. *Mankind at the Turning Point: The Second Report of The Club of Rome.* London: Hutchinson.

Simon, Julian L. 1981. *The Ultimate Resource.* Princeton: University Press.

__________ and Herman Kahn. 1984. *The Resourceful Earth.* Oxford and New York: Basil Blackwell.

Totter, John R. 1985. "Food Restriction, Ionizing Radiation and Natural Selection." *Mechanisms of Aging and Development.*

Whelan, Elizabeth M. 1984. *Toxic Terror.* Ottawa, Ill.: Jameson Books.

Wood, Michael. 1987. "Combating Malaria." *The World and I* 2(10):176.

SOURCES

The content of this volume was derived from papers presented in Committee VII ("Global 2000 Revisited: Mankind's Impact on Spaceship Earth") at ICUS XV ("Absolute Values and the New Cultural Revolution"). The symposium, one of seven sponsored by ICUS, was held at the J.W. Marriott Hotel in Washington, D.C., 27–30 November 1986 and was chaired by Honorary Chairman S. Fred Singer and Organizing Chairman Hugh W. Ellsaesser.

CONTRIBUTORS

Warren Y. Brockelman: Faculty of Social Sciences, Mohidol University, Bangkok, Thailand.

Jerry J. Cohen: Lawrence Livermore National Laboratory (retired), Livermore, California.

Hugh W. Ellsaesser: Atmospheric and Geophysical Sciences Division, Lawrence Livermore National Laboratory, Livermore, California.

Jovan V. Jovanovich: Physics Department, University of Manitoba, Winnipeg, Manitoba, Canada.

L. T. Khemani: Senior Scientific Officer, Indian Institute of Tropical Meteorology, Pune, India.

Shu Zheng Liu: Norman Bethune University of Medical Sciences, Changchun, People's Republic of China.

T.D. Luckey: Consultant in Life Sciences and Retired Professor of Biochemistry, University of Missouri, Columbia, Missouri.

Harvey F. Ludwig: Chairman, SEATEC International and Advisor on Environment Board of Thailand, Bangkok, Thailand.

Margaret N. Maxey: Biomedical Engineering Program, University of Texas at Austin, Austin, Texas.

Mark Mills: Science Concept, Inc., Washington, D.C.

Enders Robinson: Distinguished Professor of Geophysics, Department of Chemistry, University of Tulsa, Tulsa, Oklahoma.

Ivan Th. Rosenqvist: Institute of Geology, University of Oslo, Blindern, Norway.

Max Singer: The Potomac Organization, Inc., Washington, D.C.

Fred Henglun Sun: Sea-Level Research Center, Research Institute of Geography, Fujian Teachers University, Fuz-hou, People's Republic of China.

John R. Totter: Institute for Energy Analysis, Oak Ridge Associated Universities, Oak Ridge, Tennessee.

Alvin M. Weinberg: Distinguished Fellow, Institute for Energy Analysis, Oak Ridge Associated Universities, Oak Ridge, Tennessee.

LIST OF FIGURES

LIST OF TABLES

Index

D

E

F

G

S

U

V

W

X

Y

Z